John Daniel Runkle

Elements of Plane analytical Geometry

John Daniel Runkle

Elements of Plane analytical Geometry

ISBN/EAN: 9783743345928

Manufactured in Europe, USA, Canada, Australia, Japa

Cover: Foto ©ninafisch / pixelio.de

Manufactured and distributed by brebook publishing software (www.brebook.com)

John Daniel Runkle

Elements of Plane analytical Geometry

PREFACE.

IN the preparation of this book the needs of the students of the Institute of Technology have been kept in mind. The time which they can devote to the subject is limited, and yet it is necessary that they should become reasonably familiar with its more elementary and fundamental parts. For this reason, the earlier chapters are treated with somewhat more fulness than is usual. For some propositions more than one proof is given, and particular care has been taken to illustrate and enforce all parts of the subject by a large number of numerical applications. In the matter of problems only the simpler ones have been selected, and the number in every case has been proportioned to the time which students will have to devote to them. In general, propositions have been proved first with reference to rectangular axes, because, as a rule, the proofs are easier, and when necessary, the proofs with reference to oblique axes may be omitted.

The first eight chapters of this book have been used for two years by the students of the Institute. During the first year, my associate, Mr. H. W. Tyler, checked all the numerical results and made many valuable suggestions, of which I have gladly availed myself. During the past year, I have had the able assistance of Mr. D. P. Bartlett, who has also computed the answers to the greater portion of the numerical exercises. With all the care which has been taken, errors undoubtedly still exist, and I shall be truly thankful to any one who will point them out to me. I am indebted to my friend Mr. H. K. Burrison, Instructor in Mechanical Drawing, for the drawing of the cuts. The determinant notation has not been used, because students are not prepared for it; nor has it been thought advisable to prefix an elementary chapter on the subject, since the excellent book on Determinants by Professor Hanus, published by Ginn & Co., is readily accessible. I shall be very glad if any teachers and students besides those in the Institute of Technology shall find this book adapted to their needs.

My greatest indebtedness is due to the excellent work on "Conic Sections" by Professor Salmon, and to "An Elementary Treatise on Conic Sections" by Charles Smith, M.A., Fellow and Tutor of Sydney Sussex College, Cambridge, England.

In the later chapters it will be seen that I have followed Mr. Smith's book quite closely, and have taken from it many examples, with the solutions of several of them.

I hope in due time to follow this book with a brief elementary treatise on Solid Analytic Geometry.

J. D. R.

Aug. 15, 1888.

CONTENTS.

CHAPTER I. The Point: Its Position.

	PAGE
The position of a point	1
The coordinate axes	2
Exercises on the point	5
Length of lines between given points	7
Exercises on lengths of lines	10
Related points on a given line	12
Exercises on related points	15
Areas. The triangle and quadrilateral	17
Exercises on areas	20
Polar coordinates of points	21
Exercises in polar coordinates	22
Relation between rectilinear and polar coordinates	24
Exercises on rectilinear and polar coordinates	25
Rectangular projections	26

CHAPTER II. Loci and Transformation of Coordinates.

Loci	28
Classification of loci	31
Examples of loci	32
Construction of loci	33
Exercises in construction of loci	34
The intersections of loci	41
Properties of the quadratic roots	42
Exercises on the intersection of loci	43
Transformation of coordinates	46
Exercises on transformation of coordinates	49
The degree of an equation is not altered by transformation of coordinates	53

CHAPTER III. THE STRAIGHT LINE.

	PAGE
The position of a straight line	54
To determine the position of a straight line in terms of its parameters	54
Construction, the parameters being given	55
The equation of a straight line	57
The general equation of the first degree	60
Relations of parameters	62
Variable parameters	63
Exercises on the straight line	65
Relations of the parameters when the axes are oblique	73
Equations of the straight line when the axes are oblique	74
Exercises on the straight line, axes oblique	78
The polar equation of the straight line	80
Exercises on polar equations of the straight line	81
Lines subject to given conditions	82
Exercises on lines passing through given points	88
The length of the perpendicular dropped from a given point to a given line	90
Exercises on the relative positions of points and lines	92
Angles between given lines	94
Intersections of lines	101
Exercises on angles between given lines, and on their intersections	102
Lines passing through the intersections of given lines	106
Propositions relating to transversals	115
Elementary propositions relating to the triangle	116
Exercises on lines through the intersections of lines	124
Equations above the first degree which represent straight lines	127
Exercises on equations above the first degree which represent straight lines	132

CHAPTER IV. THE CIRCLE.

The equation of the circle, axes rectangular	137
Exercises on the circle	140
Equation of circle through three given points	142
Equation of circle, axes oblique	142
Exercises on the circle	143
The polar equation of the circle	145
Exercises on polar equation of circle	146
The straight line and the circle	147
Exercises on the chords of circles	150

CONTENTS.

	PAGE
Tangent to the circle	152
Normal to the circle	154
Exercises on tangents, normals, and chords	159
Poles and polars with respect to a circle	161
Exercises on poles and polars	164
Elementary propositions on systems of circles	166
Exercises on systems of circles	174
Problems on the circle	176

CHAPTER V. THE CONIC SECTIONS.

The equations of the conic sections	178
Equation of the conic, axes rectangular	179
Properties of the conic	180
Equation of conic, origin at vertex	183
Equation of conic, origin at focus	184
Equation of conic, origin at centre	184
Equation of conic, focus on directrix	186
To trace the general form of the conic from its equation	187
To construct a conic, having given its equation	188
Equation of conic when the coordinates of the origin are (h, k), and the new axes are parallel to the old ones	190
Polar equation of conic when the pole is at the focus, and the axis of X is the polar axis	193
Exercises on equations of the conic	194

CHAPTER VI. THE PARABOLA $y^2 = 4px$.

The parabola is concave towards the axis	201
The parabola and the secant line	203
The tangent and normal	204
Exercises and problems on chords, tangents, and normals	210
Poles and polars with respect to parabola	213
Diameters of the parabola	213
The parabola referred to oblique axes	215
Exercises and problems on the parabola	216

CHAPTER VII. THE ELLIPSE $\frac{x^2}{a^2} + \frac{y^2}{b^2} = 1$.

The ellipse and the secant line	220
The tangent and normal	221
The equation of chord of contact	225

CONTENTS.

	PAGE
Poles and polars with respect to the ellipse	226
Exercises on tangents, normals, and polars	226
Elementary propositions relating to tangents and normals	229
Diameters of the ellipse	233
Auxiliary circles and the eccentric angle	236
Exercises and problems on the eccentric angle of the ellipse	242
The ellipse referred to conjugate diameters as axes	244
Exercises and problems on the ellipse	246

CHAPTER VIII. THE HYPERBOLA $\frac{x^2}{a^2} - \frac{y^2}{b^2} = 1$.

Secants, tangents, normals, and polars	250
Exercises on tangents and normals	253
Diameters of the hyperbola	255
The hyperbola and the eccentric angle	262
Exercises on the eccentric angle	269
The hyperbola referred to its asymptotes	271
Equilateral or rectangular hyperbolas	273
Problems on the hyperbola	274

CHAPTER IX. THE GENERAL EQUATION OF THE SECOND DEGREE, $Ax^2 + 2Hxy + By^2 + 2Gx + 2Fy + C = 0$.

The equation $S = 0$ always represents a conic section	277
Cause the term containing xy in $S = 0$ to disappear	279
The points in which $S = 0$ is cut by a straight line	280
The condition that $S = 0$ shall be cut in one finite point and in one point at infinity	280
The characteristics of $S = 0$	281
The discriminant of $S = 0$	282
Construct the curves denoted by $S = 0$	284
Exercises	286
The form of $S = 0$ when the centre of the conic is the origin	287
The coordinates of the centre of $S = 0$	288
Transform $S = 0$ to the axes of the conic	289
Reduce $S = 0$ to its simplest form when the curve is a parabola	294
Exercises on the conic $S = 0$	298

CHAPTER X. GENERAL PROPERTIES OF THE CONIC $S = 0$. 303

Equations of the asymptotes of $S = 0$	303
The hyperbola conjugate to $S = 0$	303
Condition that $S = 0$ shall be a rectangular hyperbola	304

Locus of middle points of a system of parallel chords of $S = 0$. 304
Condition that any two diameters of $S = 0$ shall be conjugates . 305
Axes of $S = 0$ the only pair of perpendicular conjugate diameters . 306
Condition that two lines shall be parallel to a pair of conjugate diameters of $S = 0$ 306
The ratio of the products of the segments into which two chords drawn in given directions through the origin are cut by $S = 0$ is constant 306
Tangents through any point to $S = 0$ have the same ratio as the diameters to which they are parallel 307
Equation of the tangent to $S = 0$ at the point $(x'y')$ . . . 308
Equations of two tangents through any point to $S = 0$ . . . 308
Chord of contact of two tangents to $S = 0$ which meet in the point (h, k) 308
Polar of any point $(x'y')$ with respect to $S = 0$ 309
Form of $S = 0$ when coordinate axes are a tangent and a normal at any given point on the curve 309
The conic which passes through any five given points . . . 310
The foci, the directrices, and the eccentricity of $S = 0$ . . . 311
Exercises and problems 313

CHAPTER XI. SYSTEMS OF CONICS.

Similar conics 320
Confocal conics 321
Two confocals pass through any given point $(x'y')$ . . . 322
Two confocal conics cut each other at right angles . . . 322
Only one confocal of a system will touch a given straight line . 323
The difference of the squares of the perpendiculars from the centre to any two parallel tangents to two given confocals, is constant 323
The locus of the intersection of perpendicular tangents to any two confocals is a circle 323
The locus of the pole of a given straight line, with respect to a system of confocals, is a straight line 324
The coordinates of the points in which $S = 0$ and $S' = 0$ intersect 324
The system of conics which pass through the four points of intersection of $S = 0$ and $S' = 0$ 325
Only one conic of the system $S - kS' = 0$ will pass through any fifth point $(x'y')$ 325
Only two conics of the system $S - kS' = 0$ are parabolas . . 325
Only three conics of the system $S - kS' = 0$ are pairs of straight lines 326

CONTENTS.

	PAGE
If the discriminant Δ' of $S = 0$ is zero, then the system of conics $S - kS' = 0$ passes through four given points	326
The locus of the centres of the system of conics which passes through four given points	328
The asymptotes of the system of conics through four given points are parallel to a pair of conjugate diameters of the centre-locus of this system	331
Equation of the system of conics which touch the axes of coordinates	331
The equation of the system of conics which touch four given straight lines	332
Locus of the centres of the conics which touch four given straight lines	333
The opposite sides and diagonals of a quadrilateral meet in the points O, O', O'', respectively; either point O is the pole of the line passing through the other two, with respect to the system of conics which circumscribes the quadrilateral	334
The intersection of three diagonals of a quadrilateral are the vertices of a self-conjugate triangle with respect to the conic inscribed in the quadrilateral	335
Exercises and problems on systems of conics	336

PLANE ANALYTIC GEOMETRY.

CHAPTER I.

THE POINT: ITS POSITION.

1. The point is the simplest element of any geometrical figure; and the first problem is *to determine its position* on some assumed plane, such as the plane of the paper or of the blackboard on which the figure is to be drawn.

Two conditions are necessary and sufficient to determine the position of a point on a plane, and the quantities which express these conditions are called *the coordinates of the point*.

2. *To determine the position of a point when the two given conditions or coordinates are a* **distance** *and a* **direction.**
These are called the polar coordinates of the point.

Assume a point O from which to measure distances, and a straight line OX from which to measure directions.

Denote the given distance by r; then with O as a centre, and radius r, describe a circle; the required point will fall somewhere on the circumference. Next, the point must lie on a line ON, whose direction is determined by the angle $XON = \theta$. It is plain that the point of intersection P is the only point which satisfies both conditions, and is the required point, having r and θ for its polar coordinates. This method will be applied hereafter.

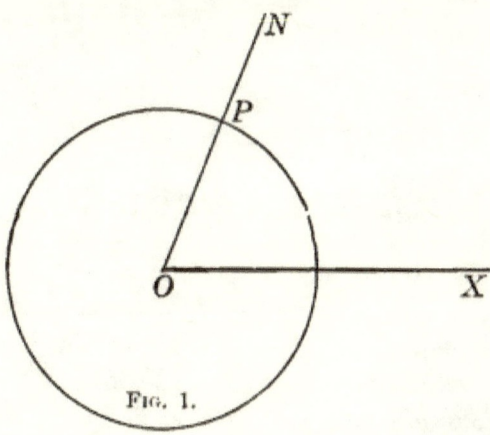

Fig. 1.

3. *To determine the position of a point when the two given conditions or coordinates are* **two given** *distances.*

These distances are called the Cartesian or rectilinear coordinates of the point.

The Coordinate Axes. — It is plain that if a point must lie at the same time on two given straight lines, it must be at their intersection, and the problem is solved by determining the positions of the two given lines upon which the required point must lie. For this purpose let us assume the positions of two straight lines XX' and YY', intersecting in the point O, as *lines or axes of reference*. The line XX' is called the *axis of X*, or *axis of abscissæ;* the line YY' is called the *axis of Y,* or *axis of ordinates;* the point O is called *the origin of coordinates,* or simply the *origin,* as it is the point from which distances are measured on both axes and in both directions. XOY is called the *first* angle; YOX', the *second;* $X'OY'$, the *third;* and $Y'OX$, the *fourth.*

The Coordinates of a Point. — Suppose, now, that the required point P must lie on a line parallel to the axis of X, and at a given *distance* b above it, measured on the axis of Y. Lay off $OM = b$, and through the point M draw a parallel PP' to the axis of X; the required point is somewhere on this line.

Again, suppose that the required point lies on a line parallel to the axis of Y, and at a given *distance* a to the right, measured on the axis of X. Lay off $ON = a$, and through the point N draw a parallel PP''' to the axis of Y. The point P, which satisfies both conditions, is at the intersection of these two given parallels, and is the required point. The position of the point P, referred to the coordinate axes XX' and YY', is therefore determined by the *two given distances* ON and OM, which are called the *coordinates* of the point; ON, its *abscissa*; and OM or PN, its *ordinate*.

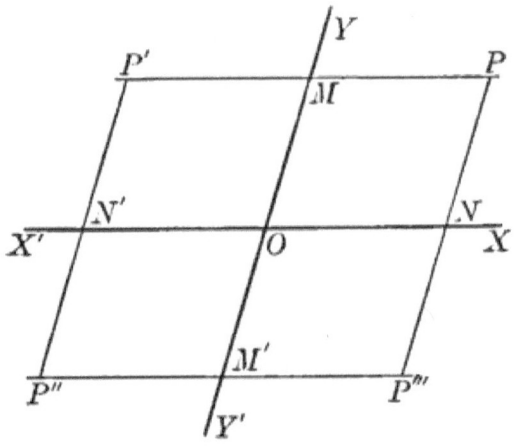

Fig. 2.

If we take $ON' = ON = a$, and $OM' = OM = b$, and through the points N' and M' draw two additional parallels, we shall have three additional points, P', P'', P''', determined by the same coordinates, a and b, which determined P.

Algebraic Signs of the Coordinates. — How shall we distinguish these four points from each other? By simply introducing the idea of *opposite directions*. Distances may be laid off on both axes in *opposite* directions from O; and if we use, as we may, the signs *plus* and *minus* to denote opposite directions, then if $ON = +a$ denotes a *positive* distance laid off on the axis of X to the *right* from O, $ON' = -a$ will denote a *negative* distance laid off to the *left*; if $OM = +b$ denotes a *positive* distance laid off on the axis of Y *above* the axis of X, then $OM' = -b$ will denote a *negative* distance laid off *below* the axis of X. In general, distances laid off on the axis of X are denoted by x, and called *abscissæ*; and distances laid off on the axis of Y are denoted by y, and called *ordinates*; and the equations of condition for the four determined points P, P', P'', P''' are

$$P\begin{pmatrix} x = +a \\ y = +b \end{pmatrix}, \quad \begin{pmatrix} x = -a \\ y = +b \end{pmatrix}, \quad P''\begin{pmatrix} x = -a \\ y = -b \end{pmatrix}, \quad P'''\begin{pmatrix} x = +a \\ y = -b \end{pmatrix},$$

or, for brevity,

$$P(a, b), \quad P'(-a, b), \quad P''(-a, -b), \quad P'''(a, -b),$$

the value of the abscissa x being written first, and the value of the ordinate y second, with a comma between them.

When the position of a point P is entirely unknown, it is written $P(x, y)$; but, if known, it may be written $P(2, 3)$, $P(a, b)$, $P(x', y')$, $P(x'', y'')$, or by any other convenient notation.

The best way to plot the point $P(a, b)$ is to lay off $ON = a$, draw a parallel to the axis of Y through N, and on this parallel lay off $NP = b$.

When the angle XOY is a right angle, the coordinate axes are called *rectangular*; and when this angle is not right, they are called *oblique*. The angle XOY is usually denoted by ω.

In the following exercises the student should carefully construct the figure for each, using any convenient unit of length. Paper ruled for the purpose is a great convenience.

In all cases rectangular axes are used, if the contrary is not stated.

EXERCISES ON THE POINT.

1. Let $P(x, y)$ be any variable point in the first angle. If x is constant and y varies, what line will the point P describe? What will be its position? As P approaches the axis of X, what change will y undergo? What will the value of y be when the point P reaches the axis of X? What further change will y undergo when P passes below the axis of X? What will the value of y be for all points on the axis of X? What equation will be true for all points on the axis of X?

2. If y is constant and x varies, what line will the point P describe? What change will x undergo as P approaches the axis of Y, and what further change when it crosses the axis of Y? What will the value of x be when P is on the axis of Y? What will the value of x be for all points on the axis of Y? What equation expresses the condition which is true of all points on the axis of Y? What is this equation called?

3. For what point will both x and y be zero? If both x and y so vary that the point P always remains at the same distance from the origin, what path will it describe?

4. Plot the points $(2, 3)$, $(-3, 5)$, $(-2, -3)$, $(4, -2)$.

5. Plot the same points, using the same axis of X and origin O, when $XOY = 60°$.

6. Connect the points $(2, 3)$, $(-3, 1)$, $(-1, -2)$ by straight lines. Does the origin lie within this triangle or without?

7. Given the points $(3, 5)$ and $(-1, -4)$. Connect them by a straight line, and show that the differences of the abscissæ and of the ordinates of the two points form the remaining sides of a triangle, which is right or oblique according as the axes are right or oblique.

8. Show from similar triangles that the points $(2, 3)$, $(1, -3)$, $(3, 9)$ lie on the same straight line.

9. Connect the points $(2, 3)$, $(-1, 2)$, $(-2, -3)$, $(1, -2)$ in order by straight lines. What is the resulting figure? Draw the

diagonals and show that they pass through the origin and are bisected by it.

10. Plot the points $(a, 0)$, $(0, a)$, $(-a, 0)$, $(0, -a)$, and connect them in order by straight lines. What will the figure be when the axes are rectangular? What when they are oblique? What is the length of the diagonals?

Ans. Square; Rectangle; Diagonals $= 2a$.

11. Plot the points $(a, 0)$, $(0, b)$, $(-a, 0)$, $(0, -b)$, and connect them in order by straight lines. What will the figure be for rectangular and for oblique axes? What the length of the diagonals?

Ans. Rhombus; Parallelogram; Diagonals $2a$, $2b$.

12. Show that the line joining the points (a, b) and $(-a, -b)$ passes through the origin and is bisected by it.

13. On what line do the points (a, b) and $(-c, b)$ lie?

14. Show that the lines joining $(a, 0)$, $(0, b)$ and $(0, a)$, $(-b, 0)$ are perpendicular to each other; so also are the lines joining $(a, 0)$, $(0, b)$ and $(b, 0)$, $(0, -a)$.

15. Show that the distance of the point (a, b) from the origin is $\sqrt{a^2 + b^2}$. Show that the line joining the points $(a, 0)$ and $(0, b)$ also equals $\sqrt{a^2 + b^2}$.

16. Show that the points $(3, 4)$, $(4, 3)$, $(-3, 4)$, $(-4, 3)$, $(-3, -4)$, $(-4, -3)$, $(3, -4)$, $(4, -3)$ are on the circumference of a circle having its centre at the origin. What is its radius?

17. On what line do all the points lie whose abscissæ equal their ordinates; that is, points for which $x = y$? Through what special point does this line pass? How does this line divide the angle XOY?

18. On what line do all the points lie for which $y = -x$? Does this line pass through the origin? How does it divide the angle of the coordinate axes?

19. Connect the points $(a, 0)$, $(0, b)$ and $(na, 0)$, $(0, nb)$ by straight lines, and show that they are parallel.

DISTANCE BETWEEN POINTS.

20. Connect the points $(a, 0)$, $(0, b)$ by a straight line. Show that the points $(\tfrac{1}{2}a, \tfrac{1}{2}b)$, $(\tfrac{1}{3}a, \tfrac{2}{3}b)$, $(\tfrac{2}{3}a, \tfrac{1}{3}b)$ are on this line.

21. Show that the lines connecting the origin with the points (a, b) and $(-b, a)$ are perpendicular to each other.

Length of Lines between Given Points.

4. *To find the distance between two given points; also the angle which this distance makes with the axis of X.*

We already know that two given points determine the position of the straight line which passes through them, and we are now to find the length of that portion of the line which lies between the two given points, and also the angle which the line makes with the axis of X.

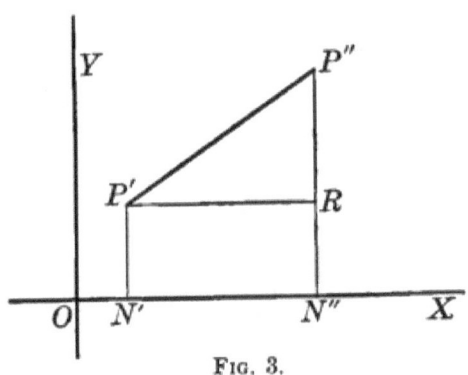

FIG. 3.

First, *when the axes are rectangular.*

Let $P'(x'y')$ and $P''(x''y'')$ be the two given points, and $P'P'' = d$. Draw $P'R$ parallel to the axis of X; then $P''P'R = \theta$ is the angle which the distance $P'P'' = d$ makes with the axis of X, because $P'R$ is parallel to this axis. *To find d and θ.*

We have

$$P'R = ON'' - ON' = x'' - x',$$
$$P''R = P''N'' - P'N' = y'' - y';$$

and by the Pythagorean proposition

$$d = \sqrt{(x''-x')^2+(y''-y')^2} \qquad (a)$$

is the required distance.

In the right triangle $P''P'R$,

$$\tan\theta = \frac{P''R}{P'R} = \frac{y''-y'}{x''-x'} = m, \text{ for brevity.}$$

$$\therefore \theta = \tan^{-1}\frac{y''-y'}{x''-x'} = \tan^{-1}m \qquad (b)$$

is the required angle. Trigonometry,* Art. 10.

If one of the points, as P', is at the origin, then $x'=0$, $y'=0$, and equations (a) and (b) become

$$d = \sqrt{x''^2 + y''^2}, \qquad (a')$$

$$\tan\theta = \frac{y''}{x''} = m.$$

$$\therefore \theta = \tan^{-1}\frac{y''}{x''} = \tan^{-1}m. \qquad (b')$$

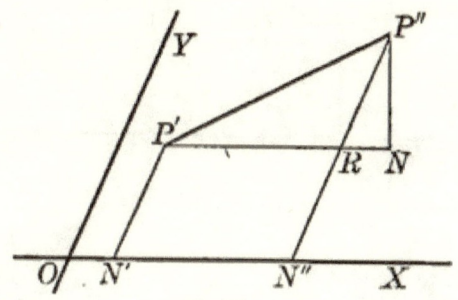

Fig. 4.

Second, *when the axes are oblique.*

In this case let the angle $XOY = \omega$; then $P''RN = \omega$, and $P''RP' = \pi - \omega$.

---

* References are to Professor Wells' Trigonometry.

DISTANCE BETWEEN POINTS.

In the oblique triangle $P''P'R$ we have the two sides $P'R$ and $P''R$, and the included angle $P''RP'$ to find $P''P' = d$. Since, as before,

$$P'R = x'' - x', \quad P''R = y'' - y', \quad P''RP' = \pi - \omega,$$

we have by Trigonometry, Art. 146.

$$d^2 = (x'' - x')^2 + (y'' - y')^2 - 2(x'' - x')(y'' - y')\cos(\pi - \omega);$$
$$\therefore d = [(x'' - x')^2 + (y'' - y')^2 + 2(x'' - x')(y'' - y')\cos\omega]^{\frac{1}{2}}, \quad (c)$$

remembering that $\cos(\pi - \omega) = -\cos\omega$. Trig., Art. 43.

Next find the value of the angle $P''P'R = \theta$, the angle which the line $P'P''$ makes with the axis of X.

In addition to the data already given we have the angle $P'P''R = \omega - \theta$, and by Trigonometry, Art. 144,

$$\frac{P''R}{\sin\theta} = \frac{P'R}{\sin(\omega - \theta)} \quad \text{or} \quad \frac{y'' - y'}{\sin\theta} = \frac{x'' - x'}{\sin(\omega - \theta)}.$$

From this equation we can find $\tan\theta$. We have

$$\frac{\sin\theta}{\sin(\omega - \theta)} = \frac{y'' - y'}{x'' - x'} = m, \text{ for brevity;}$$
$$\therefore \sin\theta = m\sin(\omega - \theta) = m\sin\omega\cos\theta - m\cos\omega\sin\theta.$$

Next divide by $\cos\theta$; then

$$\tan\theta = m\sin\omega - m\cos\omega\tan\theta;$$
$$\therefore \tan\theta = \frac{m\sin\omega}{1 + m\cos\omega}, \quad (d)$$

which gives the required angle θ through its tangent.

For $\omega = 90°$, formulæ (c) and (d) reduce to (a) and (b), as they should, since $\cos 90° = 0$.

If one of the points, as P', is at the origin, then $x' = 0$, $y' = 0$, and (c) and (d) become

$$d = \sqrt{x'^2 + y''^2 - 2x'y'' \cos \omega}. \quad (c'')$$

Second Solution for Oblique Axes.—... lar $P''N$; then in the right triangle $P''R...$

$$P''N = P''R \sin \omega = (y'' - y') \sin \omega,$$
$$RN = P''R \cos \omega = (y'' - y') \cos \omega,$$
$$P'N = P'R + RN = (x'' - x') + (y'' - y')...$$
$$\therefore \overline{P'P''}^2 = \overline{P'N}^2 + \overline{P''N}^2,$$
$$\therefore d^2 = [(x'' - x') + (y'' - y') \cos \omega]^2 + ...$$
$$\therefore d = [(x'' - x')^2 + (y'' - y')^2 + 2(x'' - x')...$$

since $\sin^2 \omega + \cos^2 \omega = 1$

Also from the right triangle $P'P''N$,

$$\tan \theta = \frac{P''N}{P'N} = \frac{(y'' - y') \sin \omega}{(x'' - x') + (y'' - y') \cos \omega}$$

by putting $\dfrac{y'' - y'}{x'' - x'} = m$.

EXERCISES ON ART. 4.

1. Find d and θ for the points $(3, 4)$ and $(-7, -8)$. If $(3, 4)$ is the point $P''(x''y'')$ and $(-7, -8)$ the point $P'(x'y')$, then by (a) and (b),

$$x'' - x' = 3 - (-7) = 10, \quad y'' - y' = 4 - (-8) = 12;$$
$$\therefore d = \sqrt{100 + 144} = \sqrt{244} = 2\sqrt{61}, \quad \theta = \tan^{-1} \tfrac{6}{5}.$$

2. Find d and θ for the same points when $\omega = 60°$. In addition to
$$x'' - x' = 10, \qquad y'' - y' = 12,$$
we have $\sin 60° = \tfrac{1}{2}\sqrt{3}$, $\cos 60° = \tfrac{1}{2}$, and formulæ (c) and (d) become
$$d = \sqrt{10^2 + 12^2 + 2\cdot 10 \cdot 12 \cdot \tfrac{1}{2}} = 2\sqrt{91}, \quad m = \tfrac{6}{5}, \quad \tan\theta = \tfrac{3}{8}\sqrt{3}.$$

3. Show that each of the following sets of points lie on a straight line:

1. $(3, 1)$, $(-2, 3)$, $(-7, 5)$; 2. $(-1, 2)$, $(2, -3)$, $(5, -8)$;
3. $(2, -1)$, $(-1, 2)$, $(0, 1)$; 4. $(3, -2)$, $(1, -1)$, $(-3, 1)$, $(5, -3)$; 5. $(0, -3)$, $(1, -2)$, $(2, -1)$, $(4, 1)$; 6. (x, y), $(x + x', y + y')$, $(x + nx', y + ny')$.

Apply formula (b). For 1, $\tan\theta = \dfrac{1-3}{3+2} = -\dfrac{3-5}{2+7} = \dfrac{1-5}{3+7} = -\dfrac{2}{5}$; therefore lines passing through any two of these points make the same angle with the axis of X. If the axes are oblique, these ratios are still equal, since the corresponding triangles are similar.

4. Connect the points $(3, 2)$, $(-2, 1)$, $(-3, -2)$, $(2, -1)$ in order by straight lines. Show that the opposite sides are parallel. Find lengths of sides and diagonals. What angles do the diagonals make with the axis of X?

5. Given the coordinates of the vertices of a triangle $(2, 3)$, $(4, -5)$, and $(-3, -6)$, to find the lengths of the sides and the angles which they make with the axis of X.

$\qquad$ *Ans.* Sides: $2\sqrt{17}$, $5\sqrt{2}$, $\sqrt{106}$.

$\qquad\qquad$ Angles: $\tan^{-1}(-4)$, $\tan^{-1}\tfrac{9}{5}$, $\tan^{-1}\tfrac{1}{7}$.

6. If the axes are oblique and $\omega = 60°$, what are the sides and angles of the last question?

$\qquad$ *Ans.* Sides: $\sqrt{52}$, $\sqrt{57}$, $\sqrt{151}$.

$\qquad\qquad$ Angles: $\tan^{-1} 2\sqrt{3}$, $\tan^{-1}\dfrac{9\sqrt{3}}{19}$, $\tan^{-1}\dfrac{\sqrt{3}}{15}$.

7. Art. 3, Ex. 10. What is the side of the square? What the sides of the rectangle?

$\qquad$ *Ans.* Square, $a\sqrt{2}$; rectangle, $2a\sin\tfrac{1}{2}\omega$, $2a\cos\tfrac{1}{2}\omega$.

8. Art. 3, Ex. 11. What is the side of the rhombus? What are the sides of the parallelogram?

Ans. Rhombus, $\sqrt{a^2 + b^2}$.

Parallelogram, $\sqrt{a^2 + b^2 - 2ab\cos\omega}$ and $\sqrt{a^2 + b^2 + 2ab\cos\omega}$.

9. What is the distance between the points (a, b) and $(-a, -b)$ when the axes are oblique? What angle does this distance make with the axis of X?

Ans. $d = 2\sqrt{(a-b)^2 + 4ab\cos^2\tfrac{1}{2}\omega}$, $\tan\theta = \dfrac{b\sin\omega}{a + b\cos\omega}$.

10. Express the condition, that the distance of the point (xy) from $(2, 3)$ is equal to 4. *Ans.* $(x-2)^2 + (y-3)^2 = 16$.

11. Express the condition, that the point (xy) is equidistant from the points $(2, 3)$ and $(4, 5)$.

Ans. $(x-2)^2 + (y-3)^2 = (x-4)^2 + (y-5)^2$; or, $x + y = 7$.

12. Show that the point $(2, 1)$ is the centre of the circle circumscribing the triangle whose vertices are the points $(3, 4)$, $(1, -2)$, $(-1, 2)$.

Related Points on a Given Line.

5. *To find the coordinates of the point which divides in a given ratio the line joining two given points.*

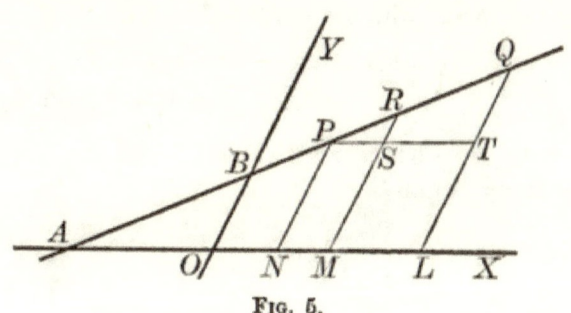

Fig. 5.

Let $P(x'y')$ and $Q(x''y'')$ be the two given points, and $R(xy)$ the point which divides the line PQ in the given ratio

$l:k$. Draw PST parallel to the axis of X. The two triangles PRS and PQT are similar, and therefore,

$$\frac{PR}{PQ} = \frac{PS}{PT} = \frac{RS}{QT};$$

or,

$$\frac{l}{l+k} = \frac{x-x'}{x''-x'} = \frac{y-y'}{y''-y'};$$

$$\therefore (l+k)(x-x') = l(x''-x'), \text{ and } x = \frac{lx''+kx'}{l+k}. \qquad (a)$$

$$\therefore (l+k)(y-y') = l(y''-y'), \text{ and } y = \frac{ly''+ky'}{l+k}. \qquad (b)$$

The coordinates (xy) are therefore found in terms of the coordinates of the two given points and the given ratio for both rectangular and oblique axes.

The two cases of most frequent use are when the distance between the two given points is bisected, and when it is trisected. For bisection $l=k$, and (a) and (b) become

$$x = \frac{x''+x'}{2}, \quad y = \frac{y''+y'}{2}. \qquad (c)$$

For trisection when R is one-third of the distance from P to Q, $2l=k$, and (a) and (b) become

$$x = \frac{x''+2x'}{3}, \quad y = \frac{y''+2y'}{3}. \qquad (d)$$

For trisection when R is two-thirds the distance from P to Q, $l=2k$, and (a) and (b) become

$$x = \frac{2x''+x'}{3}, \quad y = \frac{2y''+y'}{3}. \qquad (e)$$

Thus far the point $R(xy)$ has been supposed to fall between P and Q, giving a point of *internal* section. When the point $R(xy)$ falls to the right of both P and Q, or to the left of both, on the line through P and Q, we have a point of *external* section. In the case of internal section the directions of P to R, and R to Q, are the *same*, l and k have the same sign, and their ratio is plus; but in the case of external section, these directions are *opposite*, and the sign of the ratio $l : k$ is *minus*. If then we make either l or k minus in formulæ (*a*) and (*b*), they become

$$x = \frac{lx'' - kx'}{l - k}, \quad y = \frac{ly'' - ky'}{l - k}, \qquad (f)$$

for a point of external section.

For remembering these formulæ, note that the coordinates of either given point are multiplied by the proportional factor belonging to the segment which ends in the other.

6. *To find the points in which a line passing through two given points intersects the coordinate axes.*

Suppose that the line passing through the two given points $P(x'y')$ and $Q(x''y'')$ intersect the coordinate axes in the points A and B (Fig. 5). These are points of *external* section; the coordinates of A are $(x, 0)$, and of B $(0, y)$.

To find the x of A, make $y = 0$ in (f); then

$$ly'' - ky' = 0, \quad \therefore \frac{l}{k} = \frac{y'}{y''}.$$

Substitute this value of the ratio $l : k$ in the value of x, and get

$$x = \frac{x'y'' - x''y'}{y'' - y'} = OA. \qquad (a')$$

To find the y of B, make $x = 0$ in (f); then

$$lx'' - kx' = 0, \quad \therefore \frac{l}{k} = \frac{x'}{x''}.$$

Substitute this value of the ratio $l : k$ in the value of y, and get

$$y = \frac{y'x'' - y''x'}{x'' - x'} = OB. \qquad (b')$$

If A and B had been supposed to be points of *internal* section, the formulæ (*a*) and (*b*) would give the values of OA and OB as just found.

EXERCISES ON RELATED POINTS.

1. Given the points $(3, 4)$ and $(-4, 7)$; find the coordinates of the point $P(xy)$ which divides the line joining these points in the ratio $5 : 6$; that is, $\frac{5}{11}$ of the distance from $(3, 4)$ to $(-4, 7)$.

Apply (*a*) and (*b*), Art. 5, arranging the work as follows:

$x'y'$ $3, 4 \mid 5$ $x = \dfrac{5(-4) + 6 \cdot 3}{5 + 6} = -\frac{2}{11}$.

$x''y''$ $-4, 7 \mid 6$ $y = \dfrac{5 \cdot 7 + 6 \cdot 4}{5 + 6} = 5\frac{4}{11}$.

To find the distances which the line passing through $(3, 4)$ and $(-4, 7)$ cuts from the coordinate axes. Apply (a'), (b'), Art. 6.

$x'y'$ $3, 4$ $x = \dfrac{3 \cdot 7 - (-4)4}{7 - 4} = 12\frac{1}{3} = OA$.

$x''y''$ $-4, 7$ $y = \dfrac{4(-4) - 7 \cdot 3}{-4 - 3} = 5\frac{2}{7} = OB$.

If the required point is $\frac{5}{11}$ of the distance from $(-4, 7)$ to $(3, 4)$, or, which is the same, $\frac{6}{11}$ of the distance from $(3, 4)$ to $(-4, 7)$, then

$x'y'$ $3, 4 \mid 6$ $x = \dfrac{6(-4) + 5 \cdot 3}{6 + 5} = -\frac{9}{11}$.

$x''y''$ $-4, 7 \mid 5$ $y = \dfrac{6 \cdot 7 + 5 \cdot 4}{6 + 5} = 5\frac{7}{11}$.

2. The vertices of a triangle are $(2, 3)$, $(-6, -7)$, and $(8, -9)$.
(1) Find the coordinates of the middle points of the sides.
(2) Find the coordinates of a point on each medial line two-thirds of the distance from the vertex.

Ans. (1) $(-2, -2)$, $(1, -8)$, $(5, -3)$; (2) $(\frac{4}{3}, -\frac{13}{3})$.

3. If the line PQ (see figure, Art. 5) passes through the origin O, show that $\dfrac{x'}{y'} = \dfrac{x''}{y''}$.

4. Given $(x'y')$, $(x''y'')$, $(x'''y''')$, the coordinates of the vertices of a triangle. Show that the medial lines meet in the point

$$x = \frac{x' + x'' + x'''}{3}, \quad y = \frac{y' + y'' + y'''}{3}.$$

5. Given the vertices $(3, 6)$, $(-5, 7)$, $(6, -4)$ of a triangle. In what point do the medial lines meet? Apply the answer of the last example. *Ans.* $(\frac{4}{3}, 3)$.

6. One end of a line is at the point (a, b), and the coordinates of a point one mth of the distance from (a, b) to the other end of the line are (p, q). Find the coordinates of the other end.

Ans. $x = m(p - a) + a, \quad y = m(q - b) + b$.

7. Any side of a triangle is cut in the ratio $m : n$, and the line joining this point to the opposite vertex is cut in the ratio $l : m + n$. Find the (xy) of this last point of section.

Ans. $x = \dfrac{lx' + mx'' + nx'''}{l + m + n}, \quad y = \dfrac{ly' + my'' + ny'''}{l + m + n}$.

8. Given $x'y'$, $x''y''$, $x'''y'''$, $x^{iv}y^{iv}$, the coordinates of the vertices of a quadrilateral.

(1) Find the middle points of the opposite sides.
(2) Find the middle points of the diagonals.
(3) Find the middle points of the lines joining the centres of the opposite sides.
(4) Find the middle of the line joining the centres of the diagonals.
(5) Find (xy) for $(x'y')$, $(x''y'')$, $(x'''y''')$, as in Ex. 4; join this point to $x^{iv}y^{iv}$; find coordinates of the one-fourth point on this line nearest to (xy).
(6) The points found in (3), (4), and (5) are the same:

$$x = \frac{x' + x'' + x''' + x^{iv}}{4}, \quad y = \frac{y' + y'' + y''' + y^{iv}}{4}.$$

(7) Continue the method of (5) to n points, and find

$$x = \frac{x_1 + x_2 + \cdots + x_{n-1} + x_n}{n}, \quad y = \frac{y_1 + y_2 + \cdots + y_{n-1} + y_n}{n}.$$

This point (xy) is the arithmetical mean of the n points; it is also the centre of gravity of n equal masses situated at these points.

AREA OF TRIANGLES.

9. In what ratio will the line joining the origin and the point $(x'y')$ divide the line joining the points $(x''y'')$ and $(x'''y''')$? Suggestion: Let $k:l$ be the required ratio. The point in which the two lines meet is
$$\frac{lx'' + kx'''}{l+k}, \quad \frac{ly'' + ky'''}{l+k}.$$

But these coordinates are proportional to x' and y'.

Ans. $\dfrac{k}{l} = \dfrac{y'x'' - x'y''}{x'y''' - y'x'''}$.

Areas.

7. *To find the area of a triangle in terms of the coordinates of its vertices.*

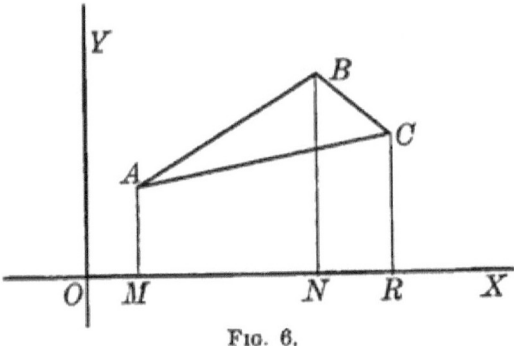

Fig. 6.

First, *when the axes are rectangular.*

Let $A(x'y')$, $B(x''y'')$, $C(x'''y''')$ be the vertices. Then the area
$$ABC = MABN + NBCR - MACR.$$

But the area of the trapezoid
$$MABN = \tfrac{1}{2}(AM + BN)(ON - OM),$$
or,
$$= \tfrac{1}{2}(y' + y'')(x'' - x').$$
$$\therefore NBCR = \tfrac{1}{2}(y'' + y''')(x''' - x'');$$
$$\therefore MACR = \tfrac{1}{2}(y''' + y')(x''' - x');$$

18   PLANE ANALYTIC GEOMETRY.

$$\therefore \text{Area } ABC = \tfrac{1}{2}[(y'+y'')(x''-x') + (y''+y''')(x'''-x'')$$
$$- (y'''+y')(x'''-x')],$$
$$= \tfrac{1}{2}[y'(x''-x''') + y''(x'''-x') + y'''(x'-x'')]$$
$$= \tfrac{1}{2}[y'x''-y''x'+y''x'''-y'''x''+y'''x'-y'x''']$$ \quad (a)

If the three points are on the same straight line, the area of the triangle is zero. Therefore

$$y'x'' - y''x' + y''x''' - y'''x'' + y'''x' - y'x''' = 0$$

is the condition that the points $x'y'$, $x''y''$, $x'''y'''$ are on the same straight line.

Second, *when the axes are oblique.*

In this case, $MN \sin\omega$ is the perpendicular distance between the parallel sides AM and BN of the trapezoid $MABN$, instead of MN. The area of each trapezoid just found must then be multiplied by $\sin\omega$; therefore the area of the triangle, when the axes are rectangular, must be multiplied by $\sin\omega$ to get its area when the axes are oblique.

If either point, as $A(x'y')$, is at the origin, then $x'=0$, $y'=0$, and the area of the triangle

$$OBC = \tfrac{1}{2}[y''x''' - y'''x'']$$

for rectangular axes, and

$$= \tfrac{1}{2}[y''x''' - y'''x''] \sin\omega$$

for oblique axes.

8. *To find the area of a quadrilateral in terms of the coordinates of its vertices.*

First, *when the axes are rectangular.*

Let $A(x'y')$, $B(x''y'')$, $C(x'''y''')$, $D(x^{iv}y^{iv})$ be the coordinates of the vertices, and let $P(xy)$ be any point within the quadrilateral. Then it is obvious that

AREAS OF POLYGONS.

$ABCD = PAB + PBC + PCD + PDA.$

But, by Art. 7,

$PAB = \tfrac{1}{2}[yx' - y'x + y'x'' - y''x' + y''x - yx''],$
$PBC = \tfrac{1}{2}[yx'' - y''x + y''x''' - y'''x'' + y'''x - yx'''],$
$PCD = \tfrac{1}{2}[yx''' - y'''x + y'''x^{iv} - y^{iv}x''' + y^{iv}x - yx^{iv}],$
$PDA = \tfrac{1}{2}[yx^{iv} - y^{iv}x + y^{iv}x' - y'x^{iv} + y'x - yx'].$

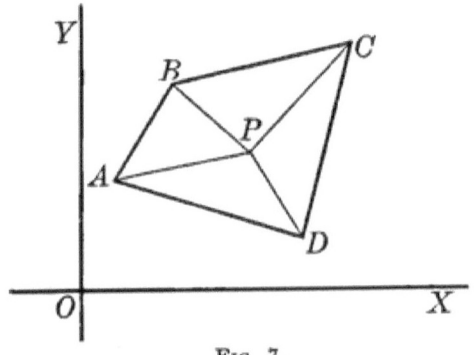

FIG. 7.

By adding and cancelling like plus and minus terms,

$ABCD = \tfrac{1}{2}[y'x'' - y''x' + y''x''' - y'''x'' + y'''x^{iv} - y^{iv}x''' + y^{iv}x' - y'x^{iv}],$

the required area. The coordinates (xy) disappear, as they should, since the area does not depend upon the position of P.

When the axes are oblique, this area must be multiplied by $\sin \omega$.

If one of the points, as $D(x^{iv} y^{iv})$, is at the origin, then $x^{iv} = 0$, $y^{iv} = 0$; and the area of the quadrilateral

$OABC = \tfrac{1}{2}[y'x'' - y''x' + y''x''' - y'''x'']$

for rectangular axes, and must be multiplied by $\sin \omega$ for oblique axes.

In the same way we can find the area of any polygon.

EXERCISES ON AREAS.

1. Find the area of the triangle (2, 1), (4, 3), (2, 5). Apply last formula of (a), Art. 7.

Write the points and products in vertical rows as follows:

Points.	1.	2.
$x'y'$	$y'x''$	$-y''x'$
$x''y''$	$y''x'''$	$-y'''x''$
$x'''y'''$	$y'''x'$	$-y'x'''$

The first product in row 1 is the y of the first point by the x of the second point, and the remaining products in this row are obtained by increasing the primes in order till x' of the first point appears. The first product in row 2 is obtained by interchanging the primes and sign of the first product in row 1, and then the row is completed by advancing the primes in order and making all the products in row 2 minus. In this way we can write down the area of any polygon. We have then for the given example:

Points.	1.	2.		
2, 1	1, 4 − 3, 2	4 −	6	
4, 3	3, 2 − 5, 4	6 −	20	
2, 5	5, 2 − 1, 2	10 −	2	
		20 −	28	

We get −8 for the double area of the triangle, or +8 by interchanging the signs of the rows. But the sign is immaterial, as we only require the absolute area.

2. Compute the same example by the first formula of (a), Art. 7.

3. Find the areas of the triangles whose vertices are (3, 8), (−5, 7), (−3, −2), and (−5, 4), (−3, −6), (5, −4).

Ans. 37 and 42.

4. If in any triangle the point of intersection of the medial lines be joined to the vertices, the three triangles thus formed are equal.

5. Find the area of the pentagon (3, 4), (−4, 1), (−3, −3), (1, −5), (5, −2). *Ans.* $50\frac{1}{2}$.

6. Find the areas of the quadrilaterals:
(2, 2), (−2, 3), (−3, −3), (1, −2); (1, 2), (3, 4), (5, 3), (6, 2); (3, 4), (−7, 1), (2, −5), (4, 0); (5, 6), (−3, 2), (4, −5), (6, 1).

Ans. 20; $\frac{11}{2}$; 50; 50.

Polar Coordinates of Points.

9. *Polar coordinates of a point.*

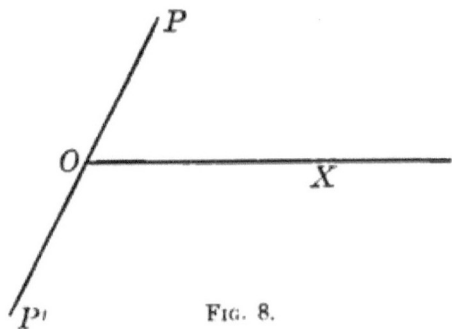

Fig. 8.

Instead of two linear coordinates we may use a line and an angle, that is, a *distance* and a *direction*, to determine the position of a point.

If any point O be taken as an origin, and OX as the line from which directions are measured, then the angle of direction XOP and the distance OP, taken together, determine the position of the point P. The angle XOP is called the *vectorial angle*, and is usually denoted by θ; the distance OP is called the *radius vector*, and is usually denoted by r; the polar coordinates of the point P are (r, θ). The point O is called the *pole*, OX the *initial* side, and OP the *terminal* side of the vectorial angle.

When the radius vector is laid off on the terminal side of the vectorial angle, it is *positive*, but if laid off in the opposite direction, it is *negative;* that is, if $OP' = OP$, the polar coordinates of P' are $-r$ and θ. But OP' is the terminal side of the angle $\pi + \theta$; therefore r and $\pi + \theta$ are also the coordinates of P'. The vectorial angle is usually counted *plus* from OX to the left, as in Trigonometry; but if it is counted in the opposite direction, then $XOP' = -\theta'$, and OP' is the terminal side of the angle; therefore r and $-\theta'$ are also the coordinates of P'.

10. *To find the distance between two points given by polar coordinates.*

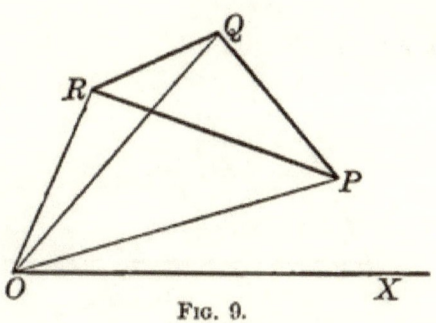

Fig. 9.

Let $P(r, \theta)$ and $Q(r', \theta')$ be the two points; then $OP = r$, $XOP = \theta$, $OQ = r'$, $XOQ = \theta'$, and the angle $POQ = \theta' - \theta$; therefore, by Trig., Art. 146,

$$\overline{PQ}^2 = r^2 + r'^2 - 2rr'\cos(\theta' - \theta). \tag{a}$$

11. *To find the area of a triangle in terms of the polar coordinates of its vertices.*

Let $P(r, \theta)$, $Q(r', \theta')$, $R(r'', \theta'')$ be the three vertices; then

$$\text{Area } PQR = POQ + QOR - POR.$$

But $\quad$ Area $POQ = \tfrac{1}{2} OP \cdot OQ \sin POQ$, by Trig., Art. 150,
$$= \tfrac{1}{2} rr' \sin(\theta' - \theta).$$

∴ Area $QOR = \tfrac{1}{2} r'r'' \sin(\theta'' - \theta')$;

∴ Area $POR = \tfrac{1}{2} rr'' \sin(\theta'' - \theta)$;

∴ Area $PQR = \tfrac{1}{2}[rr' \sin(\theta' - \theta) + r'r'' \sin(\theta'' - \theta')$
$\qquad - rr'' \sin(\theta'' - \theta)].$ $\tag{a}$

EXERCISES IN POLAR COORDINATES.

1. If P is on the initial line, what is the length of PQ?

$$\text{Ans. } (r^2 + r'^2 - 2rr' \cos\theta')^{\frac{1}{2}}.$$

2. If P is on the initial line, and Q on the perpendicular through the pole, what is the length of PQ?   *Ans.* $(r^2 + r'^2)^{\frac{1}{2}}$.

3. If Q is on OX, and P below it, what does the area of the triangle become?   *Ans.* $\frac{1}{2}[rr' \sin\theta + r'r'' \sin\theta'' - rr'' \sin(\theta'' + \theta)]$.

4. For what values of θ will the distance PQ be least and greatest? What are the corresponding values of PQ?
    Ans. $\theta = \theta'$ and $\theta = \theta' - 180°$; $r - r'$ and $r + r'$.

5. Find the distance between two points whose polar coordinates are $(2, 40°)$ and $(4, 100°)$.   *Ans.* $\sqrt{12}$.

6. What is the area of the triangle, the polar coordinates of whose vertices are $\left(1, \frac{\pi}{6}\right)$, $\left(3, \frac{\pi}{3}\right)$, $\left(4, \frac{2\pi}{3}\right)$?   *Ans.* $3\sqrt{3} - \frac{5}{4}$.

7. If $P'(r', \theta')$ is a fixed point, and the point $P(r, \theta)$ must always be at the same distance a from $P'(r', \theta')$, what equation expresses the condition?   *Ans.* $r^2 + r'^2 - 2rr' \cos(\theta - \theta') = a^2$.

This is the equation of a circle having $P'(r', \theta')$ for its centre; $P(r, \theta)$, for any point on its circumference; and a, equal its radius.

8. If the pole is at the centre, what does the equation of the circle become?   *Ans.* $r = a$.

9. If the initial line passes through the centre of the circle, what does its equation become?   *Ans.* $r^2 + r'^2 - 2rr' \cos\theta = a^2$.

10. If the pole is on the circumference, and the initial line is a secant, what does the equation of the circle become?
    Ans. $r = 2a \cos(\theta - \theta')$.

11. If the pole is on the circumference, and the initial line is a diameter, what does the equation of the circle become?
    Ans. $r = 2a \cos\theta$.

12. If the pole is on the circumference, and the initial line is a tangent, what does the equation of the circle become?
    Ans. $r = 2a \sin\theta$.

Relation between Rectilinear and Polar Coordinates.

12. *To find the relations between the (xy) and (r, θ) coordinates of the same point.*

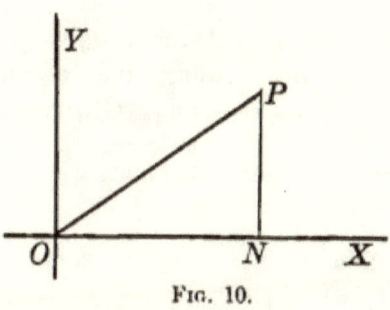

Fig. 10.

First, *when the axes are rectangular.*

Let $P(xy)$, (r, θ) be the point; let the pole be at the origin O; and let the axis of x be the initial line. Then

$$OP = r, \quad XOP = \theta, \quad PN = y, \quad ON = x;$$

and, by Trig., Art. 12,

$$x = r \cos \theta, \qquad y = r \sin \theta, \qquad (a)$$

which give x and y in terms of r and θ.

Next find r and θ in terms of x and y.

Squaring equations (a), and adding, we get

$$x^2 + y^2 = r^2 (\sin^2 \theta + \cos^2 \theta) = r^2.$$
$$\therefore r = \sqrt{x^2 + y^2}. \qquad (b)$$

By dividing equations (a) we get

$$\tan \theta = \frac{y}{x}, \text{ or } \theta = \tan^{-1} \frac{y}{x}. \qquad (c)$$

Second, *when the axes are oblique.*

In this case, $XOY = \omega$, and $OPN = \omega - \theta$.

In the oblique triangle OPN we have, by Trig., Art. 144,

$$\frac{x}{\sin(\omega-\theta)} = \frac{y}{\sin\theta} = \frac{r}{\sin\omega}.$$

$$\therefore x = \frac{r\sin(\omega-\theta)}{\sin\omega}, \quad y = \frac{r\sin\theta}{\sin\omega}. \qquad (d)$$

Next find r and θ in terms of x and y.

By (c'), Art. 4, $OP = r = (x^2 + y^2 + 2xy\cos\omega)^{\frac{1}{2}}.$ $\qquad (e)$

By division, equations (d) give

$$\frac{\sin\theta}{\sin(\omega-\theta)} = \frac{y}{x} = m.$$

By (d'), Art. 4, $\tan\theta = \dfrac{m\sin\omega}{1+m\cos\omega} = \dfrac{y\sin\omega}{x+y\cos\omega},$ $\qquad (f)$

and r and θ are found in terms of x, y, and ω.

EXERCISES ON RECTILINEAR AND POLAR COORDINATES.

1. What are the rectangular coordinates corresponding to the polar coordinates $\left(2, \dfrac{\pi}{2}\right)$, $\left(3, \dfrac{\pi}{3}\right)$, and $\left(-5, \dfrac{\pi}{4}\right)$.

Ans. $(0, 2)$, $\left(\dfrac{3}{2}, \dfrac{3}{2}\sqrt{3}\right)$, $\left(-\dfrac{5}{\sqrt{2}}, -\dfrac{5}{\sqrt{2}}\right).$

2. What are the polar coordinates of the points $(3, 5)$, $(-2, 7)$, $(6, -2)$ when $\omega = 60°$? When $\omega = 90$?

Ans. $\left(7, \tan^{-1}\dfrac{5\sqrt{3}}{11}\right)$, $\left(\sqrt{39}, \tan^{-1}\dfrac{7}{\sqrt{3}}\right)$, $\left(\sqrt{28}, \tan^{-1}\left[-\dfrac{\sqrt{3}}{5}\right]\right);$

$\left(\sqrt{34}, \tan^{-1}\dfrac{5}{3}\right)$, $\left(\sqrt{53}, \tan^{-1}\left[-\dfrac{7}{2}\right]\right)$, $\left(\sqrt{40}, \tan^{-1}\left[-\dfrac{1}{3}\right]\right).$

3. If the pole is at the origin O, and the axis of X makes an angle a with the initial line, find the rectangular coordinates of a point P whose polar coordinates (r, θ) are given.

Ans. $x = r\cos(\theta - a)$, $y = r\sin(\theta - a)$.

If the axes are oblique,
$$x = \frac{r\sin(\omega + a - \theta)}{\sin \omega}, \qquad y = \frac{r\sin(\theta - a)}{\sin \omega}.$$

4. If the rectangular coordinates of the pole are (a, b), and the initial line is parallel to the axis of X, find (xy) in terms of (r, θ), and (r, θ) in terms of (xy).

Ans. $x = a + r\cos\theta,$ $\qquad y = b + r\sin\theta,$
$$r = \sqrt{(x-a)^2 + (y-b)^2}, \qquad \theta = \tan^{-1}\frac{y-b}{x-a}.$$

5. Equations (a), Arts. 4 and 10, give
$$(x-x')^2 + (y-y')^2 = r^2 + r'^2 - 2rr'\cos(\theta' - \theta).$$
$$\therefore \cos(\theta' - \theta) = \cos\theta'\cos\theta + \sin\theta'\sin\theta. \quad \text{Art. 12.}$$

Arts. 11 and 7 give, as double areas of the triangle OPQ,
$$rr'\sin(\theta' - \theta) = y'x - x'y.$$
$$\therefore \sin(\theta' - \theta) = \sin\theta'\cos\theta - \cos\theta'\sin\theta.$$

6. Change to polar coordinates the following equations in rectangular coordinates:

$x^2 + y^2 = 5mx.$ *Ans.* $r = 5m\cos\theta.$
$x^2 - y^2 = a^2.$ *Ans.* $r^2 \cos 2\theta = a^2.$

7. Change to rectangular coordinates the following equations in polar coordinates:

$r^2 \sin 2\theta = 2a^2.$ *Ans.* $xy = a^2.$
$r^2 = a^2 \cos 2\theta.$ *Ans.* $(x^2 + y^2)^2 = a^2(x^2 - y^2).$
$r\cos^2 \tfrac{1}{2}\theta = a.$ *Ans.* $x^2 + y^2 = (2a - x)^2.$
$r = a\cos^2 \tfrac{1}{2}\theta.$ *Ans.* $(2x^2 + 2y^2 - ax)^2 = a^2(x^2 + y^2).$

13. *Rectangular projections.*

In the right triangle OPN, Art. 12,
$$x = r\cos\theta, \qquad y = r\sin\theta;$$

x and y are the rectangular projections of r upon the axes; and, in general, when θ is the angle which any straight line r makes with any assumed axis, then $r\cos\theta$ is the projection of r upon this axis; and $r\sin\theta$ is the projection of r upon an axis perpendicular to the assumed one. As r is taken plus, the signs of these projections will depend upon the factors $\cos\theta$ and $\sin\theta$. But we know from Trigonometry that when θ, counted in either direction, terminates in the first or fourth quadrants, $\cos\theta$ is plus, and minus when θ terminates in the second or third quadrants; while $\sin\theta$ is plus in the first and second quadrants, and minus in the third and fourth.

14. *In any triangle the sum of the projections of any two of the sides upon the third side as an axis equals this side.*

Let a, b, c be the sides, and A, B, C the opposite angles, then by Trig., Art. 155, Ex. 2,

$$a = b\cos C + c\cos B,$$
$$b = c\cos A + a\cos C,$$
$$c = a\cos B + b\cos A,$$

as was to be proved. A simple construction will also show the truth of this proposition.

15. *The projection of either side of a triangle upon any assumed axis equals the sum of the projections of the other two sides upon the same axis.*

Let θ be the acute angle which any side, as a, makes with the assumed axis. From Art. 14,

$$a\cos\theta = b\cos C\cos\theta + c\cos B\cos\theta.$$

But
$$0 = \pm b\sin C\sin\theta \mp c\sin B\sin\theta;$$

adding
$$a\cos\theta = b\cos(C \mp \theta) + c\cos(B \pm \theta),$$

as was to be proved, since $C \mp \theta$ and $B \pm \theta$ are the angles which the sides b and c make with the assumed axis.

These propositions may readily be extended to any polygon.

CHAPTER II.

LOCI: TRANSFORMATION OF COORDINATES.

16. *Loci.*

A *locus* is the path of a moving point. If the point always moves in the same plane, in obedience to some constant law, it will describe either a straight line or some plane curve, the character of which will be determined by the law which governs the motion. It is obvious that the locus will be a straight line if the point must always move in the same direction; it will be a circle if the point during all its motion must remain at a constant distance from a fixed point. In these cases the point in all its different positions is subject to the same condition; that is, the relation between its coordinates in any one position will be the same for all positions. If, then, we can algebraically express the relation which the coordinates of the moving point bear to each other in any one of its positions, the equation thus obtained will be the equation of the locus. This equation will be true for all the points in the path or locus, but for no others. As we have already seen, if the point is subject at the same time to two conditions, it is fixed in position and becomes an isolated point.

Let us now consider a few very simple cases. Suppose the point must so move that its ordinate y is always zero; this condition is expressed by the equation $y = 0$, and the locus is obviously the axis of X. If its abscissa is always zero, then $x = 0$ expresses the condition, and the locus is the axis of Y. Again, if the point so moves that its ordinate must always equal a constant quantity b, then $y = b$ is the equation of the locus, which is a straight line parallel to the axis of X. Also, $x = a$ expresses the condition that the moving point has a constant abscissa; the locus is a straight line parallel to the axis of Y. Take another simple case. Suppose the condition is that the ordinate y of the moving point must always equal its abscissa, then $y = x$ is the algebraic expression of this condition, and is called the equation of the locus.

Next, let us see what kind of a locus this equation represents. First, we see that when $x = 0$ we also have $y = 0$; but the point

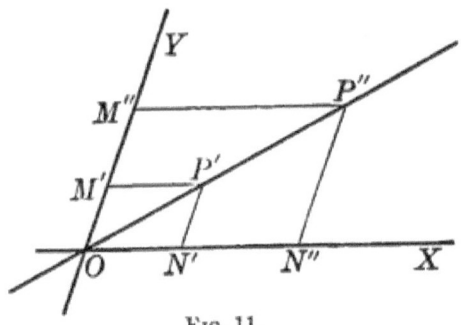

Fig. 11.

(0, 0) which satisfies the equation of the locus, is the origin O; therefore the locus passes through the origin. Next, give x any special values x' and x'', and let y' and y'' be the corresponding values of y; then plot the two points $P'(x'y')$ and $P''(x''y'')$, and draw the straight lines OP' and OP''. Since, by the given condition,

$$\frac{P'N'}{ON'} = \frac{P''N''}{ON''},$$

it follows that the triangles $OP'N'$ and $OP''N''$ are similar, and that the angle $P'ON' = P''ON''$; therefore the point P' falls on the straight line OP'', as would any other point whose abscissa equals its ordinate. We find, then, that the locus of $y = x$ is a straight line passing through the origin, and bisecting the angle XOY both for rectangular and oblique axes.

Again, suppose a point to move so that its ordinate y is in a constant ratio to its abscissa x. If we denote this constant ratio by m, then, by the given condition, $\frac{y}{x} = m$, or $y = mx$, is the equation of the locus.

Let us find the position and form of this locus when the axes are rectangular.

If in the equation of the locus we make $x = 0$, we also have $y = 0$; therefore $y = mx$ is satisfied by the coordinates (0, 0) of the origin, and the locus passes through this point. Next, let $P(xy)$ be any other point on the locus; then, since

$$\frac{PN}{ON} = \frac{y}{x} = \tan NOP = m, \text{ a constant,}$$

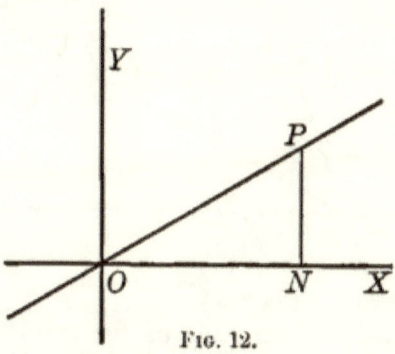

Fig. 12.

it follows that P must move in a constant direction, and therefore the locus represented by the equation $y = mx$ is a straight line passing through the origin.

For another simple case, suppose that the point P in the above figure must during all its motion remain at a constant distance from the origin O. Let this constant $OP = r$; then $x^2 + y^2 = r^2$ expresses the relation which the coordinates xy must always bear to each other, and is the equation of the locus, which is a circle. The position of this circle is determined by the coordinates of its centre (0, 0), and its magnitude by the length of its radius.

If the moving point xy must always remain at a constant distance r from a fixed point (d, e), then the condition is expressed (Art. 4) by the equation

$$(x - d)^2 + (y - e)^2 = r^2;$$

the locus is a circle fixed both in position and size.

It appears, then, that *a single equation between the coordinates xy of a moving point denotes the geometrical locus, or the path of this moving point.* Such an equation will be satisfied by an infinite number of sets of values of x and y, each set corresponding to a point on the locus; but a set corresponding to a point *not* on the locus will *not* satisfy the equation of the locus. If, then, we give x any special value, say a, and substitute this value in the equation of the locus, the resulting equation will contain but one unknown quantity, y. If this resulting equation be of the first degree, it will give a single value of $y = b$, say, corresponding to the value of $x = a$. This set (a, b) will determine one point on the locus. In the same way other points can be found. If the equation of the locus is of the second degree, substituting a for x in its equation will give an equation in y of the second degree; and by the solution of this equation we shall get two values of y, each corresponding to the value of $x = a$. Suppose these values of y are b and c; then (a, b) and (a, c) are two points on the locus. In this way we can find as many points as we wish, and by plotting them we can get an idea of the geometrical position and shape of the locus.

Classification of Loci.

Loci are classified according to the degree of the equation which represents them. An equation is of the *first* degree when the highest exponent of either variable is *unity*, such as

$$ax + by + c = 0.$$

As we shall see hereafter, an equation of the first degree always has a straight line for its locus. An equation is of the *second* degree, when, after freeing it from all negative and fractional powers of x and y, the sum of the exponents of x and y in any term does not exceed *two;* such as

$$ax^2 + 2hxy + by^2 + 2gx + 2fy + c = 0.$$

If any two of the terms containing x^2, xy, y^2 were missing, the equation would still be of the second degree. In the same way any equation, after being freed from negative and fractional powers of x and y, will be of the nth degree, when n is the highest sum of the exponents of x and y in any term.

EXAMPLES OF LOCI.

17. Rectangular coordinates are understood unless the contrary is stated.

1. A point xy so moves that its ordinate y always exceeds by a constant, b, a constant, m, times its abscissa x. Find the equation of the locus. *Ans.* $y = mx + b$.

2. A point xy so moves that the square of its ordinate y is always equal to a constant, n, times its abscissa. Find the equation of its path. *Ans.* $y^2 = nx$.

3. A point xy so moves that its direction from the point (a, b) always makes a constant angle θ with the axis of X. Find the equation of the locus. *Ans.* $y - b = (x - a)\tan\theta$.

4. A point xy so moves that its distances from the points $(-3, 2)$ and $(5, -1)$ are always equal. Find its equation.
Ans. $(x+3)^2 + (y-2)^2 = (x-5)^2 + (y+1)^2$, or $16x - 6y - 13 = 0$.

5. A point xy so moves that the sum of the squares of its distances from the two fixed points $(a, 0)$ and $(-a, 0)$ is a constant $2c^2$. Find its equation. *Ans.* $x^2 + y^2 = c^2 - a^2$.

6. A point xy so moves that the difference of the squares of its distances from the two fixed points $(a, 0)$ and $(-a, 0)$ is a constant c^2. Find the equation of its locus. *Ans.* $4ax = \pm c^2$.

7. A point xy so moves that its distance from the origin is always equal to its distance from the axis of Y, axes oblique. Find the equation of the locus. *Ans.* $x^2\cos^2\omega + y^2 + 2xy\cos\omega = 0$.

8. A point xy so moves that its distance from the axis of X is half its distance from the origin. Find the equation of its locus.
Ans. $3y^2 - x^2 = 0$.

9. A point xy so moves that its distance from the axis of X is always equal to its distance from the point $(1, 1)$. Find the equation of its locus.   *Ans.* $x^2 - 2x - 2y + 2 = 0$.

10. A point xy so moves that its distance from a fixed point $(a, 0)$ divided by its distance from the axis of Y is always equal to a constant e. Find the equation of its locus.
$$Ans.\ y^2 + (1 - e^2) x^2 - 2ax + a^2 = 0.$$

Construction of Loci.

18. *To construct the locus corresponding to a given equation when the axes are rectangular.*

It is always best to begin by finding the points in which the locus cuts the coordinate axes. This will give us some idea of its position.

For all points on the axis of X the ordinates are zero; that is, $y = 0$. If, then, in the equation of the locus we make $y = 0$, the resulting equation in x will give the abscissæ of all points whose ordinates are zero; that is, of all points in which the locus cuts the axis of X. In the same way, if we make $x = 0$ in the equation of the locus, the resulting equation in y will give the ordinates of all the points in which the locus cuts the axis of Y. If the values of x thus found for $y = 0$, or the values of y for $x = 0$ are *real*, then the locus cuts the axes in *real* or actual points; but if these values are *imaginary*, then the locus cuts the axes in *imaginary* points; that is, does not cut them at all. It is convenient, however, to say that a locus *always* cuts the axes, which we can do by introducing the idea of imaginary points.

The student will notice that here, as in other algebraic problems, impossible conditions are indicated by imaginary roots.

Again, suppose we wish to find the coordinates of other points through which the locus passes besides those on the coordinate axes. The problem may be stated in this way:

Does the locus pass through a point whose abscissa is any given quantity a, that is, through points for which $x = a$?

Make $x = a$ in the equation of the locus, and the resulting equation in y will give the ordinates of all the points on the locus having the abscissa a.

If these values of y are all real, then the points are all on the locus, or the locus passes through them all; but if the values of y are all imaginary, or some real and some imaginary, then the points having the given abscissa a are none of them, or only some of them, on the locus.

We see, then, that by giving to x a series of values, and computing the corresponding values of y, we can, by plotting the series of points thus obtained and drawing a line through them, get an approximation to the locus, which will be more or less exact, according to the less or greater distance apart of the computed points.

If, however, the equation happens to have a straight line or a circle for its locus, lines whose properties we already know, then the computation and the plotting of a few points will usually determine completely the position and magnitude of the locus.

In the following exercises the axes are rectangular.

EXERCISES IN THE CONSTRUCTION OF LOCI.

1. Construct the locus of the equation $y = 2x + 4$.

First find where the locus cuts the axes.

For $y = 0$, $x = -2$; therefore the locus cuts the axis of X in the point $(-2, 0)$.

For $x = 0$, $y = 4$; therefore the locus cuts the axis of Y in the point $(0, 4)$.

Next plot the point $(-2, 0)$ by laying off two units to the left on the axis of X, and find the point A; then plot the point $(0, 4)$ by laying off four units on the positive direction of the axis of Y. Through the points A, B so determined draw the straight line ABC; it now remains to show that this is the locus of the given equation.

EXERCISES IN THE CONSTRUCTION OF LOCI.

Give x any series of small values, $-2, -1, 0, 1, 2$; the corresponding values of y are $0, 2, 4, 6, 8$; plot the points $(-2, 0)$, $(-1, 2)$, $(0, 4)$, $(1, 6)$, $(2, 8)$; they all fall on the straight line ABC. This line must therefore be the required locus, since any other points obtained in the same way will also fall on the same line.

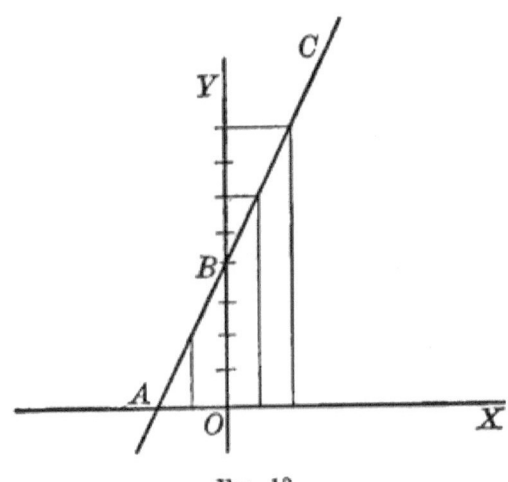

Fig. 13.

So far, then, as construction shows, the required locus is a straight line; but the definite proof that the locus of any equation of the first degree containing x and y, as $ax + by + c = 0$ is a straight line, must be reserved for a future chapter.

2. Given the equations

(1) $y + 2x = 4$;

(2) $3x + 2y = 6$;

(3) $\tfrac{2}{3}x - \tfrac{3}{4}y = 2$;

(4) $x - 2y - 5 = 0$;

(5) $\dfrac{y}{2} + \dfrac{x}{3} = 1$;

(6) $2y - 3x = 6$;

to construct their loci.

Find the points in which each locus cuts the axes; through these points draw a straight line, and then show that any other point computed from the equation will fall on this line.

36   PLANE ANALYTIC GEOMETRY.

3. Given the equation $x^2 + y^2 = 25$, to construct its locus.

(1) For $\quad y = 0, \quad x = \pm 5 = OA$ or $-OA'$.
For $\quad x = 0, \quad y = \pm 5 = OB$ or $-OB'$.

The locus cuts the axes in four equidistant points from the origin.

(2) Does the locus pass through any points for $x > \pm 5$ or $y > \pm 5$?

From the given equation we get

$$x = \pm \sqrt{5^2 - y^2}. \quad (1) \qquad y = \pm \sqrt{5^2 - x^2}. \quad (2)$$

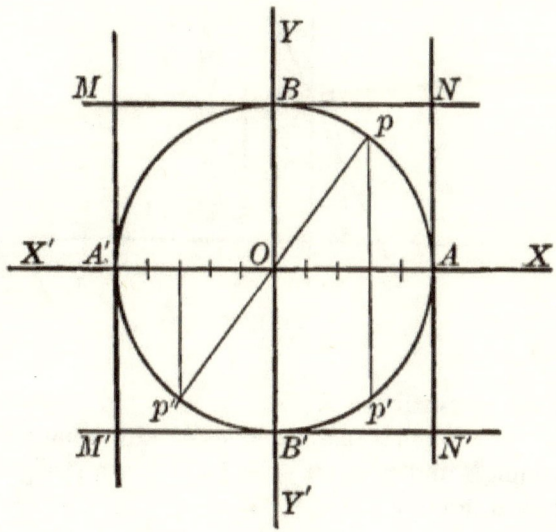

Fig. 14.

Any values of x greater than ± 5 substituted in (2) will make y imaginary; therefore all points on the locus must fall within the parallels NN' and MM'; any value of y greater than ± 5 substituted in (1) will make x imaginary; therefore all points on the locus will fall within the parallels NM and $N'M'$.

It only remains, then, to examine the locus for points within the above limits.

For $x = 3$ in (2), $y = \pm 4$, giving two points p and p' equidistant from the axis of X. In the same way we shall get two equal values of y with opposite signs for any value of x falling within the limits $x < \pm 5$.

The locus is therefore *symmetrical* with respect to the axis of X; that is, if the locus is folded along the axis of X, its upper half will fall upon and coincide with the lower half, since each point on the upper has its corresponding point on the lower.

In the same way it can be shown that the locus is symmetrical with respect to the axis of Y. Again the point p (3, 4) has a corresponding point $p''(-3, -4)$ in the third quadrant, by Art. 3, Ex. 12; but $(-3, -4)$ also satisfies the given equation. It appears, then, that any point on the locus in the first quadrant has a corresponding point in the third quadrant. But by (a'), Art. 4,

$$Op = \sqrt{(3)^2 + (4)^2} = \sqrt{(-3)^2 + (-4)^2} = 5 = Op'',$$

the same as the distances cut off the axes.

As the same can be proved true of any two corresponding points on the locus, all its points must be equidistant from O; therefore the locus is a circle whose centre is at the origin, and radius = 5.

4. Construct the locus of

$$x^2 + y^2 - 10x - 8y + 32 = 0.$$

For $y = 0$, $x^2 - 10x + 32 = 0.$ (1)

For $x = 0$, $y^2 - 8y + 32 = 0.$ (2)

But we find that the roots of (1) and (2) are imaginary; therefore the locus does not cut the axes.

Next, let us find between what limits values of x must lie which correspond to real values of y, and thus to points on the curve.

Solving the equation of the locus for y, we get

$$y = 4 \pm \sqrt{10x - 16 - x^2} = 4 \pm \sqrt{(x-2)(8-x)}.$$

This shows that values of x less than 2 or greater than 8 will give imaginary values of y.

Next, make $x = 2, 3, 4, 5, 6, 7, 8$, and find the corresponding values of y, denoting those which belong to each value of x by y' and y''. Also find $y = \dfrac{y' + y''}{2}$ for each value of x, which by (c), Art. 5, will be the y of the point midway between y' and y''. We have then the following table of values:

x	2	3	4	5	6	7	8
y'	4	$4+\sqrt{5}$	$4+\sqrt{8}$	$4+\sqrt{9}$	$4+\sqrt{8}$	$4+\sqrt{5}$	4
y''	4	$4-\sqrt{5}$	$4-\sqrt{8}$	$4-\sqrt{9}$	$4-\sqrt{8}$	$4-\sqrt{5}$	4
y	4	4	4	4	4	4	4

Now plot the lines AB and CD for $x = 2$ and $x = 8$, parallels to the axis of Y, between which the locus must line. Next plot the line $y = 4$, and divide the part MN which falls between the limiting

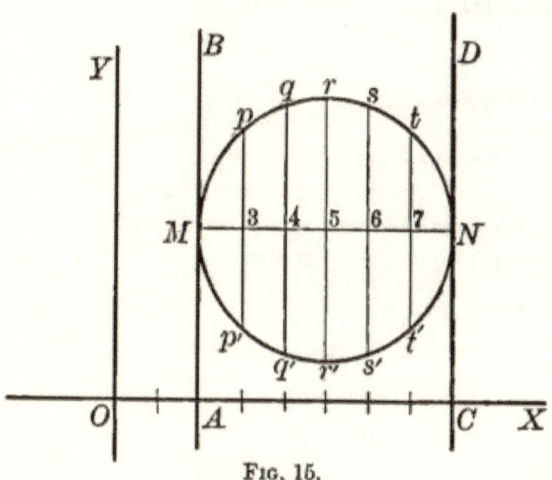

Fig. 15.

parallels AB and CD into 6 equal parts, and at the points of division 3, 4, 5, 6, 7 lay off above and below the line MN the equal radical parts of the values of y' and y'' which belong with each value of x. It appears then that the locus is symmetrical with reference to the line MN; that $MN = rr' = 6$; and it is easily shown that all the points found on the locus are equidistant from the point (5, 4).

The locus is therefore a circle with its centre at the point (5, 4), and radius = 3.

5. Construct the locus of the equation $y^2 = 8x$.

For $x = 0$, $y = 0$; therefore the locus passes through the origin. Since $y = \pm \sqrt{8x}$, we see that all negative values of x will make y

imaginary; therefore no part of the locus lies to the left of the axis of Y. Each positive value of x will give two equal values of y with opposite signs; therefore the locus is symmetrical with reference to the axis of X. As there is no limit to the positive values of x, there are no limits to the corresponding positive and negative values of y; both halves of the locus therefore extend indefinitely in the positive direction. It will be easy now to compute and plot a sufficient number of points to give an idea of the shape of the locus. We shall hereafter see that this locus is a parabola.

6. Construct the locus of the equation $y = x^2 - 3x - 2$. For $x = 0$, $y = -2$; for $y = 0$, $x^2 - 3x - 2 = 0$. Solving, we get

$$x = \frac{3 + \sqrt{17}}{2}, \quad \text{and} \quad x = \frac{3 - \sqrt{17}}{2}.$$

We see, then, that the locus cuts the axis of Y but once, in the point $(0, -2)$, and the axis of X twice, in the points $\left(\frac{3 + \sqrt{17}}{2}, 0\right)$, and $\left(\frac{3 - \sqrt{17}}{2}, 0\right)$. Between these points the locus lies below the axis of X; but for all values of x outside of these points the values of y are positive, and the locus extends indefinitely in the first and second quadrants. Compute and plot the following points:

$x = -1, \;-\frac{1}{2}, \;0, \;\frac{1}{2}, \;1, \;\frac{3}{2}, \;2, \;\frac{5}{2}, \;3, \;\frac{7}{2}, \;4, \;\cdots$

$y = \;\;2, \;-\frac{1}{4}, \;-2, \;-\frac{13}{4}, \;-4, \;-\frac{17}{4}, \;-4, \;-\frac{13}{4}, \;-2, \;-\frac{1}{4}, \;2, \;\cdots$

They will sufficiently indicate the form of the locus.

7. Construct the loci of the following six equations:

1. $x^2 + y^2 - 4x - 8y + 15 = 0$. 4. $3y = x^2 + x + 6$.
2. $4x^2 + 9y^2 = 36$. 5. $y^2 = 8x + 4$.
3. $x^2 - 16y^2 = 4$. 6. $y^2 = x^3 + x^2 - 6x$.

8. Construct the locus whose equation is $y = \sin x$. In this case values of the arc x are taken as abscissæ. For $x = 0$, $y = 0$; for $x = \pi$, $y = 0$; for all values of x between these limits y is positive, and the curve lies above the axis of X.

On the axis of X lay off $x = \pi = 3.1$, and call the unit one inch. It will be sufficient to divide π into twelve equal parts, and then compute and plot the corresponding values of y. For

$x = \dfrac{\pi}{12},\ \dfrac{\pi}{6},\ \dfrac{\pi}{4},\ \dfrac{\pi}{3},\ \dfrac{5\pi}{12},\ \dfrac{\pi}{2},\ \dfrac{7\pi}{12},\ \dfrac{2\pi}{3},\ \dfrac{3\pi}{4},\ \dfrac{5\pi}{6},\ \dfrac{11\pi}{12},\ \pi.$

$y = .26,\ \ .50,\ \ .71,\ \ .87,\ \ .97,\ \ 1.00,\ \ .97,\ \ .87,\ \ .71,\ \ .50,\ \ .26,\ \ 0.$

A table of natural sines, cosines, and tangents, is most convenient for computing these values of y. If this curve is continued between $x = \pi$ and $x = 2\pi$, all the values of y are negative, but have the same numerical values as those already computed. The second half of the curve is identical with the first half, and the like parts will repeat indefinitely.

9. Construct the loci of $y = \cos x$, $y = \tan x$, $y = \cot x$, $y = \sec x$, $y = \operatorname{cosec} x$, and $y = \operatorname{versin} x$.

Use the values of x given in Ex. 8. It is an interesting way of comparing these trigonometric curves to plot them in one figure on the same axes.

10. Construct the locus whose polar equation is $r = a \sin \theta$.

It is always best at first to get an approximate idea of the locus by examining it for a few special values of θ. Take O as the pole, and any line OX as the initial direction, and measure the angle from this line to the left, as in Trigonometry. For

$\theta = 0°,\quad 45°,\quad 90°,\quad 135°,\quad 180°,\quad 225°,\quad 270°,\quad 315°,\quad 360°.$

$r = 0,\quad \dfrac{a}{\sqrt{2}},\quad a,\quad \dfrac{a}{\sqrt{2}},\quad 0,\quad -\dfrac{a}{\sqrt{2}},\quad -a,\quad -\dfrac{a}{\sqrt{2}},\quad 0.$

Through O draw lines making the given angles with OX, and on these lines lay off the corresponding values of r, making a any convenient unit. This examination shows that the locus passes through the pole, and consists of two loops, each symmetrical about the 90° line, one above and the other below the initial line. Next compute points for as many values of θ between those already given as are needed to lay down the locus with any desired accuracy.

11. Construct the loci whose polar equations are

$$r = 2a \cos \theta, \quad r = a \tan \theta, \quad r = a(1 + \cos \theta), \quad r = a \sec^2 \left(\dfrac{\pi}{4} - \dfrac{\theta}{2} \right).$$

19. *The intersections of loci.*

If the two loci are expressed by equations of the first degree, as

$$ax + by + c = 0, \quad (1) \qquad a'x + b'y + c' = 0, \quad (2)$$

they can meet in only one point, since there is only one pair of roots or values of x and y which will satisfy (1) and (2) simultaneously, which is the condition that the point so determined must lie on both loci at the same time. Therefore loci of the first degree have only one point of intersection.

Solving (1) and (2), we get

$$x = \frac{bc' - b'c}{ab' - a'b}, \qquad y = \frac{a'c - ac'}{ab' - a'b}.$$

Since the elimination of x and y between two equations, one of the first and the other of the second degree, can give only two pairs of roots or values of x and y, it follows that their loci can only intersect in these two points.

And in general two loci will intersect in as many points, and no more, as there are pairs of roots found by the simultaneous solution of their equations, and these roots in pairs are the coordinates of the points of intersection.

But we know from Algebra that the roots of simultaneous equations fall in one of the following cases:

First, the roots may be real and unequal.
Second, the roots may be real and equal.
Third, the roots may be imaginary.

In the first case the corresponding loci intersect in *different* real points; in the second case in *coincident* real points, which is the *condition that the loci are tangent to each other at this point;* and in the third case in *imaginary* points; or, in other words, the loci do not intersect at all.

20. Suppose that the elimination of y between the equations of the two loci gives the quadratic equation

$$ax^2 + bx + c = 0, \quad \text{or} \quad x^2 + \frac{b}{a}x + \frac{c}{a} = 0. \qquad (1)$$

Denote the two roots of this equation by x' and x''; then

$$x' = \frac{-b + \sqrt{b^2 - 4ac}}{2a} = \frac{2c}{-b - \sqrt{b^2 - 4ac}}, \quad (2)$$

$$x'' = \frac{-b - \sqrt{b^2 - 4ac}}{2a} = \frac{2c}{-b + \sqrt{b^2 - 4ac}}. \quad (3)$$

I. When these roots are real and unequal, $b^2 > 4ac$.
II. When these roots are real and equal, $b^2 = 4ac$.
III. When these roots are imaginary, $b^2 < 4ac$.
IV. By adding (2) and (3) we get $\quad x' + x'' = -\dfrac{b}{a}$.

V. By multiplying (2) and (3) we get $\quad x'x'' = \dfrac{c}{a}$.

We see then that the sum of the roots of a quadratic is the *negative of the coefficient of the first power of the unknown quantity*, and their product *is the independent term*.

It sometimes happens that loci intersect in points at an *infinite* distance from the origin, which is the same as saying that they do not intersect at all, but only approach each other as they recede from the origin, becoming indefinitely near each other when the point considered is indefinitely distant from the origin; that is, the limit of approach is an *intersection* when the limit of distance is *absolute infinity*, neither of which, by the law of continuity, is ever reached.

Suppose now that of the two points of intersection given by equation (1), one is at a *finite* and the other at an *infinite* distance from the origin. Then one root must be finite, and the other infinite.

For $a = 0$, (2) and (3) become

$$x' = -\frac{c}{b}, \quad x'' = \infty;$$

that is, the condition that one root of a quadratic shall be finite and the other infinite is, *that the coefficient of the square of the unknown quantity shall equal zero.*

EXERCISES ON THE INTERSECTION OF LOCI.

Again, suppose that both roots are *infinite*, then $a=0$ and $b=0$, and the conditions are *that the coefficients of both the second and first powers of the unknown quantity are zero when both roots are infinite.*

The student is advised to become very familiar with all these properties of the quadratic roots, as he will have occasion to apply them hereafter.

EXERCISES ON THE INTERSECTION OF LOCI.

1. In what points do the loci expressed by the following sets of equations intersect?

   1. $3x + 5y = 13,$ $4x - y = 2.$
   2. $x^2 + y^2 = 25,$ $x - y = 1.$
   3. $x^2 + y^2 = 65,$ $3x + y = 25.$
   4. $x^2 + y^2 = 5,$ $xy = 2.$
   5. $x^2 - 5x + y + 3 = 0,$ $x^2 + y^2 - 5x - 3y + 6 = 0.$

Ans. 1. $(1, 2)$; 2. $(4, 3)$, $(-3, -4)$; 3. $(7, 4)$, $(8, 1)$; 4. $(1, 2)$, $(-1, -2)$, $(2, 1)$, $(-2, -1)$; 5. $(1, 1)$, $(2, 3)$, $(3, 3)$, $(4, 1)$.

2. Two loci are expressed by

$$x^2 + y^2 = r^2, \qquad y = mx + b. \qquad (1)$$

Find the value of b when the loci pass through two coincident points; that is, when they are tangent to each other. Eliminating y, we have

$$x^2 + (mx + b)^2 - r^2 = 0,$$

or $x^2(1 + m^2) + 2mbx + b^2 - r^2 = 0.$

By the condition for equal roots, II., Art. 20,

$$(1 + m^2)(b^2 - r^2) = b^2 m^2, \qquad (2)$$

from which we readily find $b = \pm r \sqrt{1 + m^2}.$

Therefore the two loci expressed by
$$y = mx + r\sqrt{1+m^2}, \quad y = mx - r\sqrt{1+m^2},$$
are both tangent to $x^2 + y^2 = r^2$ for all values of m.

Also from (2), $r^2 = \dfrac{b^2}{1+m^2}$;

and therefore the loci
$$x^2 + y^2 = \dfrac{b^2}{1+m^2}, \qquad y = mx + b,$$
are tangents.

3. Show that the loci expressed by
$$x^2 + y^2 = r^2, \qquad x\cos a + y\sin a = p,$$
are tangents when $r = p$.

4. Show that the loci expressed by
$$y^2 - 4ax = 0, \qquad y = mx + b,$$
are tangents when $b = \dfrac{a}{m}$.

5. Show that the loci expressed by
$$\dfrac{x^2}{a^2} + \dfrac{y^2}{b^2} = 1, \qquad y = mx + c,$$
are tangents when $c = \pm\sqrt{a^2m^2 + b^2}$.

6. Show that the loci expressed by
$$(x-c)^2 + (y-2c)^2 = 25c^2, \qquad 4x + 3y = b,$$
are tangents when $b = 35c$, or $-15c$.

7. In what points do the loci expressed by the following sets of equations intersect?

    1. $x^2 + y^2 = 25$, $y - 3x = 10$.

    2. $y^2 = 8x$, $3y - 9x = 2$.

    3. $\dfrac{x^2}{16} + \dfrac{y^2}{9} = 1$, $y - x = 5$.

    4. $x^2 + y^2 - 6x + 2y = 15$, $3y - 4x = 10$.

EXERCISES ON THE INTERSECTIONS OF LOCI.

5. $y^2 = 4ax,$ $\qquad y = 3x - a.$

6. $y^2 - 4ax = 0,$ $\qquad y = 2x + \dfrac{a}{2}.$

7. $y^2 - 7x - 8y + 14 = 0,$ $\qquad 7x + 6y = 13.$

8. $x^2 + y^2 = 2,$ $\qquad x^2 + y^2 - 6x - 6y + 10 = 0.$

9. $4x^2 - 3y^2 - 26 = 0,$ $\qquad 3y = x - 3.$

8. Under what conditions do the loci expressed by
$$x^2 + y^2 = r^2, \qquad y = mx + b,$$
intersect in imaginary points; that is, not at all.

9. Do the loci expressed by

1. $\dfrac{x^2}{25} - \dfrac{y^2}{16} = 1,$ $\qquad 3y - 4x = 0;$

2. $\dfrac{x^2}{a^2} - \dfrac{y^2}{b^2} = 1,$ $\qquad y^2 = \dfrac{b^2}{a^2}(2ax - x^2);$

3. $x^2 + y^2 = 16,$ $\qquad \dfrac{x^2}{9} + \dfrac{y^2}{4} = 1;$

intersect?

10. In what points does the locus expressed by
$$ax^2 + by^2 + 2gx + 2fy + c = 0$$
cut the coordinate axes?

$\qquad$ Ans. $\left[\dfrac{-g \pm \sqrt{g^2 - ac}}{a},\ 0\right]$ on the axis of X.

$\qquad\qquad\quad\ \left[0,\ \dfrac{-f \pm \sqrt{f^2 - bc}}{b}\right]$ on the axis of Y.

11. The equations of two loci are
$$y = mx + c. \quad (1) \qquad \dfrac{x^2}{a^2} - \dfrac{y^2}{b^2} = 1. \quad (2)$$

(1) What must the value of m be when one intersection is at a finite and the other at an infinite distance from the origin?
$\qquad\qquad\qquad\qquad\qquad\qquad\qquad$ Ans. $m = \pm \dfrac{b}{a}.$

Therefore $y = \pm \dfrac{b}{a} x + c$ are the equations of two loci, both fulfilling the given conditions.

(2) What must the values of m and c be when both roots are infinite, or when the loci intersect at infinity?

$$\text{Ans. } m = \pm \frac{b}{a}, \quad c = 0.$$

And $y = \pm \frac{b}{a} x$ are the equations of two loci, both of which intersect (2) at infinity.

12. Do the loci whose equations are

$$y = mx + c, \qquad \frac{x^2}{a^2} + \frac{y^2}{b^2} = 1,$$

intersect at infinity?

$\quad$ *Ans.* No; for the condition makes m imaginary.

Transformation of Coordinates.

21. It often happens that the equation of a locus can be simplified by referring the locus, or, what is the same thing, the coordinates of its generating point, to a new pair of coordinate axes. This change of reference is called the *transformation of coordinates*.

CASE I. — *To change to a new origin, the new pair of coordinate axes remaining parallel to the old pair.*

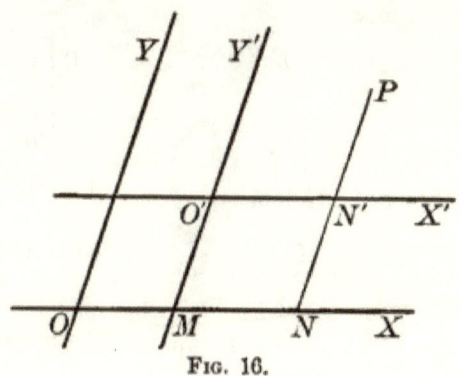

Fig. 16.

Let O' be the new origin, and $O'X'$, $O'Y'$ the new axes, respectively parallel to the old; also let (h, k) denote the coordinates of O' referred to the old axes. The coordinates of the point P are (ON, PN) referred to the old axes, and $(O'N', PN')$ referred to the new axes. Then

$$ON = OM + O'N', \qquad PN = O'M + PN';$$

or, $\qquad x = h + x', \qquad\qquad y = k + y';\qquad\qquad (a)$

which are the required equations for transformation.

Since (h, k) may have all values, the new origin may be anywhere on the plane of the coordinate axes; and also, since these values are *arbitrary* and *independent*, they may be used to impose upon the equation of the locus two independent conditions.

CASE II. — *To change from one pair of oblique axes to another, the origin remaining the same.*

Let OX, OY be the old axes, and OX', OY', the new ones; and also, let (ON, PN), (xy) be the coordinates of any point P referred to the old axes, and (ON', PN'), $(x'y')$, the coordinates of the same point referred to the new axes.

We are now to find the values of (xy) in terms of $(x'y')$ and the angles which determine the positions of the axes relatively to each other.

Denote these angles as follows:

$$XOY = \omega, \qquad X'OY' = \omega', \qquad XOX' = \theta,$$
$$X'OY = \omega - \theta, \qquad Y'OY = \omega - (\omega' + \theta).$$

As in Trigonometry, an angle measured from any axis to the left is *positive*, and to the right *negative*. Thus $XOX' = +\theta$; and $X'OX = -\theta$.

Let us now take as *axes of projection* PR and PS, respectively perpendicular to the old axes OX and OY. The projections of the line OP on the assumed axes are PR and PS;

and since OP is a side in the two triangles ONP and $ON'P$, it follows (Art. 15) that the sum of the projections of x and y

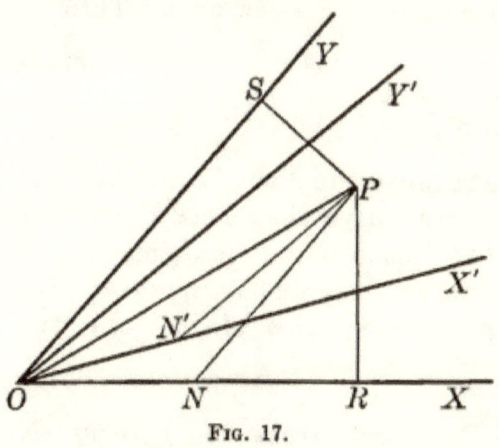

Fig. 17.

will equal the sum of the projections of x' and y' upon these axes. The coordinates

	$x,$	$y,$	$x',$	$y',$		
make angles	$0,$	$\omega,$	$\theta,$	$\omega'+\theta,$	with	$OX,$
and angles	$-\omega,$	$0,$	$-(\omega-\theta),$	$-[\omega-(\omega'+\theta)]$	with	$OY.$

Since the axes of projection PR and PS are perpendicular to OX and OY, we must project by the sine of the angle, Art. 13. Therefore

$$\left.\begin{array}{l} y\sin\omega = x'\sin\theta + y'\sin(\omega'+\theta), \\ x\sin\omega = x'\sin(\omega-\theta) + y'\sin[\omega-(\omega'+\theta)], \end{array}\right\} \quad (b)$$

are the equations of transformation from one pair of oblique axes to another.

SPECIAL CASE I. — *When both pairs are rectangular.*

In this case $\omega = 90°$, $\omega' = 90°$, and equations (b) become

$$y = x'\sin\theta + y'\cos\theta, \qquad x = x'\cos\theta - y'\sin\theta, \qquad (c)$$

the equations for changing from one pair of rectangular axes to another pair also rectangular.

SPECIAL CASE II. — *When the new axes are rectangular, and the new axis of X' coincides with the old one.*

In this case $\omega' = 90°$, $\theta = 0$, and equations (b) become

$$y \sin \omega = y', \qquad x \sin \omega = x' \sin \omega - y' \cos \omega. \qquad (d)$$

These two special cases are of most frequent use.

If the origin and directions of the axes are to be changed at the same time, then equations (a) and (b) are to be applied in succession; or, combining (a) and (b),

$$\left. \begin{array}{l} y = k + \dfrac{x' \sin \theta + y' \sin (\omega' + \theta)}{\sin \omega}, \\[1em] x = h + \dfrac{x' \sin (\omega - \theta) + y' \sin [\omega - (\omega' + \theta)]}{\sin \omega}. \end{array} \right\} \qquad (e)$$

EXERCISES ON TRANSFORMATION OF COORDINATES.

1. When the new pair of axes are rectangular, $\omega' = 90°$, and equations (b) become

$$\left. \begin{array}{l} y \sin \omega = x' \sin \theta + y' \cos \theta, \\ x \sin \omega = x' \sin (\omega - \theta) - y' \cos (\omega - \theta). \end{array} \right\} \qquad (f)$$

2. When the old pair are rectangular, $\omega = 90°$, and equations (b) become

$$\left. \begin{array}{l} y = x' \sin \theta + y' \sin (\omega' + \theta), \\ x = x' \cos \theta + y' \cos (\omega' + \theta). \end{array} \right\} \qquad (g)$$

3. When the new axes make the same angle with each other as the old, and the new axis of X bisects the angle made by the old, $\omega' = \omega$, $\theta = \frac{1}{2}\omega$, and equations (b) become

$$\left. \begin{array}{l} y \sin \omega = x' \sin \tfrac{1}{2}\omega + y' \sin \tfrac{3}{2}\omega, \\ x \sin \omega = x' \sin \tfrac{1}{2}\omega - y' \sin \tfrac{1}{2}\omega. \end{array} \right\} \qquad (h)$$

4. When both sets are rectangular, and the origin is changed at the same time,

$$\left. \begin{array}{l} y = k + x' \sin\theta + y'\cos\theta, \\ x = h + x'\cos\theta - y'\sin\theta. \end{array} \right\} \quad (i)$$

5. If, in Case II., OX and OY are taken as the axes of projection, we should have

$$\left. \begin{array}{l} x + y\cos\omega = x'\cos\theta + y'\cos(\omega' + \theta), \\ x\cos\omega + y = x'\cos(\omega - \theta) + y'\cos[\omega - (\omega' + \theta)]. \end{array} \right\} \quad (j)$$

Show that by eliminating x and y between these equations we shall get equations (b).

6. Deduce by elimination from equations (b) the equations

$$\left. \begin{array}{l} y'\sin\omega' = y\sin(\omega - \theta) - x\sin\theta, \\ x'\sin\omega' = x\sin(\omega' + \theta) - y\sin[\omega - (\omega' + \theta)], \end{array} \right\} \quad (k)$$

which give the new coordinates $(x'y')$ in terms of the old, (xy).

Also deduce these equations by projections, using perpendiculars from the point P on the axes OX' and OY'. Proceed as in Case II.

7. Show that the area of a triangle, as given in Art. 7, is not changed by transferring the coordinates of its vertices to any new origin (h, k), and a new set of axes parallel to the old.

8. A locus is expressed by $3x - 2y - 6 = 0$. What will the equation be if the origin is changed to the point $(4, 3)$, the new axes remaining parallel to the old?

By Case I., $y = 3 + y'$, $x = 4 + x'$. Substitute and reduce, leaving off the primes. *Ans.* $3x - 2y = 0$.

9. A locus referred to rectangular axes is

$$x^2 + y^2 - 4x - 6y = 18.$$

Transfer to a new origin $(2, 3)$, the axes remaining rectangular.

Ans. $x^2 + y^2 = 31$.

EXERCISES ON TRANSFORMATION OF COORDINATES.

10. What will the equations of transformation become if both sets of axes are rectangular, and the new axis of X bisects the angle of the old pair?
Ans. $y = \dfrac{1}{\sqrt{2}}(x' + y')$, $x = \dfrac{1}{\sqrt{2}}(x' - y')$.

11. Suppose that the old pair make an angle of $60°$, the new pair is rectangular, and the new axis of X bisects the angle of the old pair. What are the equations?
Ans. $y = \dfrac{x'}{\sqrt{3}} + y'$, $x = \dfrac{x'}{\sqrt{3}} - y'$.

12. The old axes make an angle of $60°$, the new axes are rectangular, and the new axis of X coincides with the old one. What are the equations?
Ans. $y = \dfrac{2y'}{\sqrt{3}}$, $x = x' - \dfrac{y'}{\sqrt{3}}$.

13. A locus referred to rectangular axes is expressed by $y^2 - x^2 = 6$. What will this equation become if the locus is referred to a new pair of rectangular axes which bisect the angles between the old ones?
Ans. $xy = 3$.

14. If the equation of a locus is $2x^2 - 5xy + 2y^2 = 4$ when referred to axes making $60°$ with each other, what will the equation become when the new axes bisect the angles between the old ones?
Ans. $x^2 - 27y^2 + 12 = 0$.

15. Transform the same equation to rectangular axes, retaining the old axis of X.
Ans. $3x^2 + 10y^2 - 7xy\sqrt{3} = 6$.

16. Show from equations (c) that $x^2 + y^2 = x'^2 + y'^2$.

17. Show from equations (b) that

$$x^2 + y^2 + 2xy\cos\omega = x'^2 + y'^2 + 2x'y'\cos\omega'.$$

Put
$$M = x'\cos\theta + y'\cos(\omega' + \theta), \qquad (1)$$
$$L = x'\sin\theta + y'\sin(\omega' + \theta); \qquad (2)$$

then equations (b) become

$$y\sin\omega = L, \quad (3) \qquad x\sin\omega = M\sin\omega - L\cos\omega. \quad (4)$$

Now square (3) and (4), multiply their product by $2\cos\omega$, and add the results; then

$$x^2 + y^2 + 2xy\cos\omega = L^2 + M^2.$$

But by squaring (1) and (2), and adding,

$$L^2 + M^2 = x'^2 + y'^2 + 2x'y' \cos \omega',$$

and the proposition is proved.

18. If a locus, referred to rectangular axes, is expressed by the equation

$$ax^2 + 2hxy + by^2 = 0,$$

to what new pair of rectangular axes, having the same origin, must the locus be referred in order that the term containing the product xy shall disappear? What will be the values of a' and b', the new coefficients of x^2 and y^2?

Ans. The angle which the new axis of X must make with the old one is

$$\theta = \tfrac{1}{2} \tan^{-1} \frac{2h}{a-b},$$

$$a' = \tfrac{1}{2}\left[(a+b) + \sqrt{(a-b)^2 + 4h^2}\right],$$

$$b' = \tfrac{1}{2}\left[(a+b) - \sqrt{(a-b)^2 + 4h^2}\right].$$

19. If a locus, expressed by the equation

$$Ax^2 + 2Hxy + By^2 + 2Gx + 2Fy + C = 0 \qquad (1)$$

is referred to a new origin and to a new pair of axes parallel to the old ones, what must be the values of (h, k), the coordinates of the new origin, in order that the transformed equation shall not contain the first powers of the new coordinates $(x'y')$?

By Case I., $\qquad x = h + x', \qquad y = k + y'.$

Substituting these values in (1), we get

$$Ax'^2 + 2Hx'y' + By'^2 + 2G'x' + 2F'y' + C' = 0,$$

in which

$$G' = Ah + Hk + G, \qquad F' = Bk + Hh + F,$$

$$C' = Ah^2 + 2Hhk + Bk^2 + 2Gh + 2Fk + C.$$

EXERCISES ON TRANSFORMATION OF COORDINATES.

If the terms containing the first powers of x' and y' must disappear, then

$$G' = Ah + Hk + G = 0, \quad F' = Bk + Hh + F = 0,$$

from which we get

$$h = \frac{BG - HF}{H^2 - AB}, \qquad k = \frac{AF - HG}{H^2 - AB},$$

the required values of (h, k).

20. If in equation (1) $H = 0$, what are the values of (h, k), what the value of C', and what does the transformed equation become? $\quad$ Ans. $Ax'^2 + By'^2 + \dfrac{ABC - AF^2 - BG^2}{AB} = 0.$

22. *The degree of an equation in respect to the coordinates is not altered by transformation.*

When a locus is referred to a new pair of axes, the degree of its equation is not changed, because the first powers of x and y referred to the old axes are expressed in terms of the first powers of x' and y', the coordinates of the same point referred to the new axes; therefore any powers of x and y will always be expressed in the same powers of x' and y'. The degree of the transformed equation cannot therefore be increased; nor can it be diminished; for if it could, by transforming back to the old axes, it would be increased, which is contrary to what has just been proved.

CHAPTER III.

THE STRAIGHT LINE.

Its Position on the Coordinate Axes.

23. The Position of a Straight Line. — As in the case of a point, so in the case of a straight line, *two* independent conditions are *necessary* and *sufficient* to determine its position with reference to the coordinate axes.

These two conditions are, a *distance* and a *direction*, or two *distances*.

The quantities which express these conditions are called *constants of position*, or *parameters*.

Distance parameters, also called *intercepts*, are the distances measured from the origin, which the given line cuts from the coordinate axes, and from a line passing through the origin perpendicular to the given line, which may be called the *normal axis* of the line.

Direction parameters are the angles which the given line and its normal axis make with the coordinate axes.

24. *To determine the position of a straight line in terms of its parameters.*

First, *when the axes are rectangular.*

Let AB be the given straight line; then the straight lines

$$OA = a, \quad OB = b, \quad OD = p,$$

are its distance parameters, and the angles

$$XAB = \theta, \quad OBA = \theta - 90°, \quad AOD = a, \quad BOD = 90° - a,$$

are its direction parameters.

THE POSITION OF A STRAIGHT LINE.

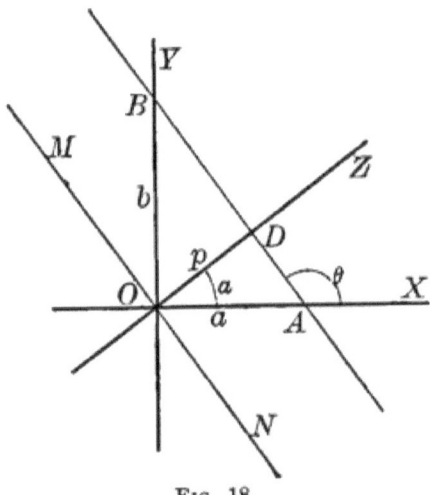

Fig. 18.

It is plain that any two of these parameters, provided one is a distance, will determine the position of the given line.

The pairs of parameters commonly used are

$$(a, b), \quad (a, \theta), \quad (b, \theta), \quad (p, a), \quad (p, \theta).$$

Construction.

I. *Given the parameters (a, b).*

Lay off $OA = a$, $OB = b$, and through the points A and B so determined draw the required line AB. Since a and b can have any positive or negative values, they will determine the positions of all possible straight lines, not passing through O.

That these distance parameters also involve the direction parameters of the line will be clearly seen by stating the problem as follows:

Through a given point on the axis of X draw a line in such a direction that it shall cut off a given distance on the axis of Y.

II. *Given the parameters (a, θ).*

Find the point A as before. Through the origin O draw MN, making the angle $AOM = \theta$, the given angle; then through A draw the required line AB parallel to MN.

The vertex of the angle θ is always the point where the line cuts the axis of X; its *initial* side is the axis of X measured in the positive direction, and its *terminal* side is the line itself. The angle is measured to the left, as in Trigonometry, from $0°$ to $180°$.

Since θ may have any values between $0°$ and $180°$, and a any positive or negative values, these parameters will determine the positions of all possible straight lines.

III. *Given the parameters* (b, θ).

Find the point B on the axis of Y, and the angle $AOM = \theta$, as before; then through B draw the required line parallel to MN.

These parameters will also determine the positions of all possible straight lines.

IV. *Given the parameters* (p, a).

Draw the normal axis OZ, making the angle $AOZ = a$, the given angle; lay off on OZ the given distance $OD = p$; then through D draw the required line AB perpendicular to OZ.

'The angle a is measured to the left from $0°$ to $360°$, and p is always positive. These parameters, since they have arbitrary values, will determine the position of any straight line.

V. *Given the parameters* (p, θ).

Construct the angle $AOM = \theta$, as before; draw OZ perpendicular to MN, and lay off $OD = p$; then through D draw the required line parallel to MN.

In this case, θ must be measured from $0°$ to $360°$, in order that p may always be positive.

25. *To find the relations which the parameters of a given line bear to each other.*

THE EQUATIONS OF A STRAIGHT LINE.

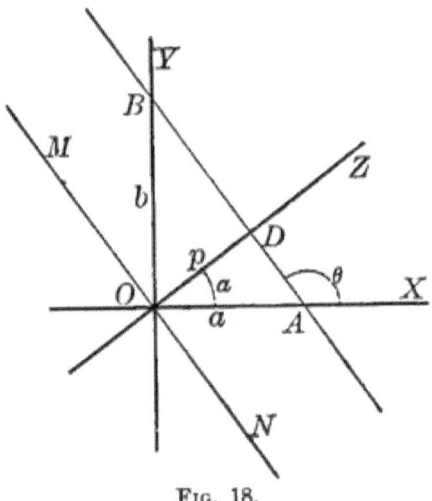

Fig. 18.

The three triangles AOB, AOD, BOD are similar; their sides are

$$OA = a, \quad OB = b, \quad OD = p, \quad AB = \sqrt{a^2 + b^2};$$

their angles are

$$OAB = \pi - \theta = BOD = 90° - a; \quad AOD = a = OBD.$$

By Trigonometry,

$$\frac{p}{a} = \cos a \qquad = \sin(\pi - \theta) = \sin\theta \quad = \frac{b}{\sqrt{a^2 + b^2}};$$

$$\frac{p}{b} = \cos(90° - a) = \cos(\pi - \theta) = \sin a \quad = -\cos\theta = \frac{a}{\sqrt{a^2 + b^2}};$$

$$\frac{b}{a} = \tan(\pi - \theta) \quad = -\tan\theta \qquad = \cot a;$$

$$\therefore p = a \cos a \qquad = a \sin\theta \qquad = b \sin a = -b \cos\theta = \frac{ab}{\sqrt{a^2 + b^2}}.$$

The Equations of a Straight Line.

26. *To find the equations of the straight line AB in terms of each pair of parameters.*

58 PLANE ANALYTIC GEOMETRY.

DEFINITION. — *The equation of a straight line expresses the relation which exists between the coordinates (xy) of every point on the line and the two parameters which determine the position of the line.*

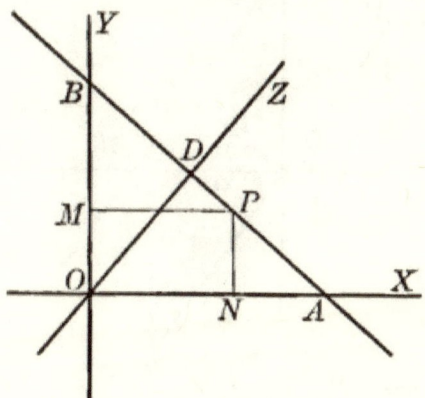

FIG. 19.

Let $P\,(ON,\,PN)\,(xy)$ be any point on the given line AB. Draw PM parallel to the axis of x; then

$$BM = b - y, \quad PM = x, \quad PN = y, \quad NA = a - x.$$

The similar triangles ANP, PMB, AOB give

$$\frac{PN}{NA} = \frac{BM}{PM} = \frac{OB}{OA} = \tan OAB = \cot AOD,$$

or

$$\frac{y}{a-x} = \frac{b-y}{x} = \frac{b}{a} = \tan(\pi - \theta) = -\tan\theta = \cot\alpha.$$

I. *Equation in terms of the parameters* (a, θ).

$$\frac{y}{a-x} = -\tan\theta = -m, \text{ for brevity};$$

$$\therefore y = m(x - a) \qquad\qquad (a)$$

is the required equation.

Note. — If we put $n = \dfrac{1}{m} = \cot\theta$, equation (a) may be written $x = ny + a$, a form sometimes useful.

II. *Equation in terms of parameters (b, θ).*

$$\frac{b-y}{x} = -\tan\theta = -m;$$

$$\therefore y = mx + b \qquad (b)$$

is the required equation.

III. *Equation in terms of the parameters (a, b).*

$$\frac{y}{a-x} = \frac{b}{a}, \text{ or } \frac{y}{b} = \frac{a-x}{a} = 1 - \frac{x}{a};$$

$$\therefore \frac{x}{a} + \frac{y}{b} = 1 \qquad (c)$$

is the required equation.

Note. — If we put $l = \frac{1}{a}$, $n = \frac{1}{b}$, equation (c) may be written $lx + ny = 1$, a simpler form useful in some cases.

IV. *Equation in terms of the parameters (p, a).*

$$\frac{y}{a-x} = \cot a = \frac{\cos a}{\sin a},$$

or $\qquad y \sin a = (a-x)\cos a = a\cos a - x\cos a = p - x\cos a;$

$$\therefore x\cos a + y\sin a = p \qquad (d)$$

is the required equation, called the *normal equation*.

V. *Equation in terms of the parameters (p, θ).*

$$\frac{y}{a-x} = -\tan\theta = -\frac{\sin\theta}{\cos\theta},$$

or $\qquad -y\cos\theta = (a-x)\sin\theta = a\sin\theta - x\sin\theta = p - x\sin\theta;$

$$\therefore x\sin\theta - y\cos\theta = p \qquad (e)$$

is the required equation.

It will be noticed that the form of the equation of the straight line depends upon the pair of parameters chosen to determine its position, and that the variable point $P(xy)$ can only vary by moving on the line.

The General Equation of the First Degree
$$Ax + By + C = 0.$$

27. In deducing the preceding equations of the straight line, we have in each case assumed the *position* and *form* of the locus, and have found that its equation is of the *first degree* with reference to the coordinates x and y.

We will now prove the proposition

That the locus of the general equation of the first degree
$$Ax + By + C = 0 \qquad (f)$$
is a straight line.

It will be sufficient to show the truth of the proposition for rectangular axes, since by Art. 22 the degree of an equation is not changed when its locus is referred to any new origin and pair of coordinate axes.

Let us define a straight line as

The path of a point which always moves in the same direction.

Now, whatever the locus of (f) may be, we may assume $P'(x'y')$ as any fixed point on the locus; the coordinates of this point must satisfy the equation of the locus; therefore
$$Ax' + By' + C = 0. \qquad (g)$$

Next, let $P(xy)$ be any other point on the locus; then, by subtracting (g) from (f), we get
$$A(x - x') + B(y - y') = 0,$$
or
$$-\frac{A}{B} = \frac{y - y'}{x - x'} = \text{a constant}$$
for all points (xy) on the locus.

But by Art. 4 this constant ratio is the tangent of the angle which the straight line connecting the two points (xy) and $(x'y')$ makes with the axis of X. Therefore, by the definition, the locus is a straight line.

RELATIONS OF PARAMETERS.

It must be possible, therefore, to put the general equation (f) into each of the forms already found by finding the values of the parameters in each case in terms of (A, B, C).

But we have seen that only two independent conditions, or parameters, are needed to determine the position of a straight line; so that, of the three parameters (A, B, C), any one may be regarded as arbitrary, and used in determining the form of the equation of the locus.

28. *To determine the parameters* a, b, p, a, θ, *in terms of* A, B, C.

Let us use the arbitrary parameter in

$$Ax + By + C = 0 \qquad (f)$$

as a divisor, and take the two resulting ratios as the parameters of the line.

Dividing by B and transposing terms, we get

$$y = -\frac{A}{B}\left(x + \frac{C}{A}\right); \qquad (a')$$

also
$$y = -\frac{A}{B}x - \frac{C}{B}; \qquad (b')$$

which are the forms of (a) and (b), Art. 26;

$$\therefore m = \tan\theta = -\frac{A}{B}, \quad a = -\frac{C}{A}, \quad b = -\frac{C}{B}.$$

Again, divide (f) by $-C$, change the coefficients of x and y into divisors, and it may be written

$$\frac{x}{-\frac{C}{A}} + \frac{y}{-\frac{C}{B}} = 1, \qquad (c')$$

which is of the form of (c), Art. 26; and

$$\therefore a = -\frac{C}{A}, \quad b = -\frac{C}{B},$$

as before.

By division we get
$$\frac{b}{a} = \frac{A}{B};$$
from which, and Art. 25, it follows that
$$\cos \alpha = \sin \theta = \frac{b}{\sqrt{a^2+b^2}} = \frac{A}{\sqrt{A^2+B^2}},$$
$$\sin \alpha = -\cos \theta = \frac{a}{\sqrt{a^2+b^2}} = \frac{B}{\sqrt{A^2+B^2}},$$
$$p = a \cos \alpha = -\frac{C}{A} \cdot \frac{A}{\sqrt{A^2+B^2}} = -\frac{C}{\sqrt{A^2+B^2}}.$$

By substituting these values in equations (d) and (e), they both become
$$\frac{A}{\sqrt{A^2+B^2}} x + \frac{B}{\sqrt{A^2+B^2}} y + \frac{C}{\sqrt{A^2+B^2}} = 0, \quad (d'), (e')$$

which is got from (f) by simply dividing each term by $\sqrt{A^2+B^2}$; that is, *by the square root of the sum of the squares of the coefficients of x and y.* Since p must always be positive when written in the second member of the equation, it appears that C and the radical $\sqrt{A^2+B^2}$ must have opposite signs.

It also appears that the direction parameters in equation (f) depend upon A and B, and are independent of C.

If any equation of a line whose parameters are referred to the coordinate axes is given, and its equation is required when the parameters of the line are referred to the normal axis, apply the following rule:

Divide each term of the given equation by the square root of the sum of the squares of its coefficients of x and y, and the result will be the required equation.

Relations of Parameters.

29. We may find the relations of the parameters as follows:

Put all the equations of Art. 26, including the general equation (f), in the form of any one of them, as (b); then

I. $y = mx - ma.$ IV. $y = -\cot a \cdot x + \dfrac{p}{\sin a}.$

II. $y = mx + b.$ V. $y = \tan\theta \cdot x - \dfrac{p}{\cos\theta}.$

III. $y = -\dfrac{b}{a}x + b.$ (f) $y = -\dfrac{A}{B}x - \dfrac{C}{B}.$

Since these equations represent the same line, they are identical, and therefore,

$$m = -\frac{b}{a} = -\cot a = \tan\theta = -\frac{A}{B},$$

$$b = -ma = \frac{p}{\sin a} = -\frac{p}{\cos\theta} = -\frac{C}{B},$$

as we have already found in Art. 25.

Variable Parameters.

30. In the preceding forms of the equation of the straight line we have seen that given values of the parameters determine the position of a single and definite line, and that the point $P(xy)$ can vary only by moving along this line. If, however, the value of one or both of the parameters vary, the position of the line will also vary; but as the line carries with it the variable point $P(xy)$, that is, since $P(xy)$ always remains on the varying line, the equations already found will still be true for the varying line in all its new positions.

For example, suppose that one or more of the parameters (A, B, C) vary; then the position of the line

$$Ax + By + C = 0 \qquad (f)$$

will vary; but the expression $Ax + By + C$ will *not* vary if the point $P(xy)$ is always on the varying line; that is, $Ax + By + C$ will remain equal to zero.

On the other hand, suppose that the parameters (A, B, C) are constant, and that $P(xy)$ is any variable point *not* on the line (f); then for this point the expression $Ax + By + C$ will *not* reduce to zero, but to some variable quantity, which we will denote by L; then for this point (f) becomes

$$Ax + By + C = L. \qquad (g)$$

But for any given value of L, (g) is the equation of a straight line, because it is of the first degree in x and y, and it is parallel to line (f), because it has the same direction parameters (A, B).

It follows, then, that if the parameters (A, B, C) are constant, and L is a new variable parameter, equation (g) represents a system of lines parallel to and including the type line (f) which (g) becomes for $L = 0$.

When the position of a line is referred to the coordinate axes; that is, when its equation is expressed in terms of the parameters (a, θ), (b, θ), or (a, b) or (A, B, C), we shall sometimes, for the sake of brevity in writing and facility in use, denote such equation by a single letter of the set $L, M, N, \ldots$; but if the position of the line is referred to the normal axis; that is, if its equation is expressed in terms of (p, a) or (p, θ), we shall denote it by a single letter of the set $P, Q, R, \ldots$.

The equations of these two systems are connected by the relation

$$L = P\sqrt{A^2 + B^2}.$$

For example, if $y - mx - b = L$, then

$$\frac{y - mx - b}{\sqrt{1 + m^2}} = \frac{L}{\sqrt{1 + m^2}} = x \cos a + y \sin a - p = P.$$

$$\therefore y - mx - b = L = (x \cos a + y \sin a - p) \csc a = P \csc a.$$

$$\therefore x \cos a + y \sin a - p = P = (y - mx - b) \sin a = L \sin a.$$

The idea and use of variable parameters will be sufficiently illustrated for the present in the following elementary exercises.

EXERCISES ON THE STRAIGHT LINE.

I. *Equation* (a). $\qquad y = m(x - a)$. $\qquad$ *Parameters* (a, θ).

1. If the intercept $a = -5$, and $\theta = 45°$, what is the equation of the line? $\qquad$ *Ans.* $y - x = 5$.

2. For $a = 3$ and $\theta = 135°$, what is the equation of the line?

3. If $a = 0$, through what point does the line pass?

4. When the line passes through the origin, and $\theta = 120°$, what is its equation?

5. For $\theta = 0$, what does equation (a) become, and what line does it represent? $\qquad$ *Ans.* $y = 0$; axis of X.

6. For $\theta = 90°$, what does equation (a) become, and what the line? Put (a) in the form $\frac{y}{m} = x - a$. For $m = \infty$, $x = a$, a parallel to the axis of Y.

7. If a is constant and m variable, what system of lines does equation (a) represent?

> *Ans.* A system of lines passing through the point A $(a, 0)$, and called a *pencil of lines*. See figure, Art. 26.

8. If a is variable and m constant, what system of lines does (a) represent?

> *Ans.* A system of parallels, because all the lines of the system have the *same direction parameter m*.

9. If a and m are both constant, and k an arbitrary factor, what system of lines does $y = m(x - ka)$ represent? What is the value of k for that line of the system which passes through the origin?

10. What is the value of k for the line of the system which passes through the point $(x'y')$? $\qquad$ *Ans.* $k = \dfrac{mx' - y'}{am}$.

11. For what values of k do lines of this system and the origin lie on opposite sides of line (a)? *Ans.* $k > 1$.

12. Show that equation (a) can be written

$$-\frac{y}{\sqrt{1+m^2}} + \frac{mx}{\sqrt{1+m^2}} = \frac{ma}{\sqrt{1+m^2}},$$

$$x \cos \alpha + y \sin \alpha = a \cos \alpha,$$

or $\qquad x \sin \theta - y \cos \theta = a \sin \theta,$

which are the forms of (d) and (e).

II. *Equation* (b). $\qquad y = mx + b. \qquad$ *Parameters* (b, θ).

1. When $b = -5$, and $\theta = 120°$, what is the equation of the line? What when $b = 10$ and $\theta = 30°$? What when $b = -8$ and $\theta = \frac{3\pi}{4}$? What when $b = -2$ and $\theta = \pi$?

2. When $b = 0$, through what point will the line pass?

3. When $\theta = 0$, what do the equation and line become?

4. When $\theta = \frac{\pi}{2}$, what do the equation and line become? Put (b) in the form $\frac{y}{m} = x + \frac{b}{m}$, and then make m infinite.

5. When b is constant and m varies, what system of lines does equation (b) represent? $\qquad$ *Ans.* A pencil through the point $(0, b)$.

6. If m is constant and b varies, what system of lines does equation (b) represent?

7. If m and b are constants and L variable, what system of lines does the equation $y - mx - b = L$ represent?

8. What is the value of L for the line of this system which passes through the origin. $\qquad$ *Ans.* $L = -b$.

9. For what values of L do lines of this system and the origin lie on opposite sides of line (b)?
$\qquad$ *Ans.* All values having the same sign as b.

10. What is the value of L for the line of the system which passes through the given point $(x'y')$? \qquad Ans. $L' = y' - mx' - b$.

11. Do the point $(x'y')$ and the origin lie on the same side or on opposite sides of line (b)?

12. Find the equation of the line passing through the point $(x'y')$ parallel to (b).

Ans. $y - mx - b = L' = y' - mx' - b$, or $y - y' = m(x - x')$.

13. Show that equation (b) can be written

$$\frac{y}{\sqrt{1+m^2}} - \frac{mx}{\sqrt{1+m^2}} = \frac{b}{\sqrt{1+m^2}},$$

$$x \cos a + y \sin a = b \sin a,$$

or $\qquad x \sin \theta - y \cos \theta = -b \cos \theta,$

which are the forms of (d) and (e).

III. *Equation* (c). $\qquad \dfrac{x}{a} + \dfrac{y}{b} = 1.$ \qquad *Parameters* (a, b).

1. If $a = -5$ and $b = 3$, what is the equation of the line?

Ans. $\dfrac{x}{-5} + \dfrac{y}{3} = 1$, or $3x - 5y + 15 = 0$.

2. If the intercepts are equal and have the same sign, what angle does the line (c) make with the axis of x? \qquad Ans. $135°$.

3. What is the angle when the intercepts are equal and have opposite signs? \qquad Ans. $45°$.

4. Write the equations of the four lines whose intercepts are (a, b), $(-a, b)$, $(-a, -b)$, $(a, -b)$.

5. What figure do these four lines form?

6. Put their equations in the form of (d), and show that the distance between parallel lines is $\dfrac{2ab}{\sqrt{a^2+b^2}}$.

7. In Fig. 19, the area of the triangle AOB equals the sum of the areas of AOP and BOP; that is,

$$AOB = AOP + BOP.$$

From this equation deduce (c).

8. Show by a figure that the lines $\frac{x}{a}+\frac{y}{b}=1$, $\frac{x}{b}-\frac{y}{a}=1$ are perpendicular to each other.

9. If a is constant and b varies, what system of lines does equation (c) represent? What are the equation and line for $b=0$? what for $b=$ infinity?

10. If k is an arbitrary factor, what system of lines does
$$\frac{x}{ka}+\frac{y}{kb}=1, \text{ or } \frac{x}{a}+\frac{y}{b}=k,$$
represent?

11. What relation do the lines of this system bear to line (c)?

12. What is the value of k for the line of the system which passes through the origin?

13. Show that the distances between (c) and the parallels $\frac{x}{a}+\frac{y}{b}=k$, measured on the axes of X and Y, are $(k-1)a$ and $(k-1)b$, and on the normal axis $\frac{(k-1)ab}{\sqrt{a^2+b^2}}$.

14. What does the equation $\frac{x}{a}+\frac{y}{b}-1=L$ represent? Find the equation of the line passing through the point $(x'y')$ parallel to (c).

Ans. $\frac{x}{a}+\frac{y}{b}-1=L'=\frac{x'}{a}+\frac{y'}{b}-1;$ or $y-y'=-\frac{b}{a}(x-x')$.

15. Given a right triangle ABC. Take the right angle C at the origin, and the base b and the altitude a in the positive directions of the axes of X and Y respectively.

(1) The equation of the hypothenuse is $\frac{x}{b}+\frac{y}{a}=1$.

(2) The line passing through the middle points of a and b is
$$\frac{x}{b}+\frac{y}{a}=\frac{1}{2}.$$

(3) The line through the vertex B and the middle of the opposite side is $\frac{2x}{b}+\frac{y}{a}=1$.

(4) The line through the vertex A and the middle of the opposite side is $\frac{x}{b}+\frac{2y}{a}=1$.

(5) The last two lines intersect in the point $(\tfrac{1}{3}b, \tfrac{1}{3}a)$.

IV. *Equation* (d). $x \cos a + y \sin a = p$.   *Parameters* (p, a).

1. If $p = 3$ and $a = \dfrac{\pi}{4}$, what does (d) become?

   If $p = 5$ and $a = \tfrac{2}{3}\pi$, what does (d) become?

   If $p = 2$ and $a = \tfrac{7}{8}\pi$, what does (d) become?

2. If a varies from $0°$ to $360°$ and p is constant, what locus will p describe, and what relation will (d) in all its positions bear to this locus?

3. If a is constant and p varies, what system of lines will (d) represent?

4. The equation of a line parallel to (d) is
$$x \cos(a + \pi) + y \sin(a + \pi) = p',$$
which reduces to $-x \cos a - y \sin a = p'$.

5. The equation of a line perpendicular to (d) is
$$x \cos\left(a + \frac{\pi}{2}\right) + y \sin\left(a + \frac{\pi}{2}\right) = p',$$
which reduces to $y \cos a - x \sin a = p'$.

6. Find the equations of two lines, each making an angle of $45°$ with (d), and having the p of (d). *Ans.* $x \cos\left(a \pm \dfrac{\pi}{4}\right) + y \sin\left(a \pm \dfrac{\pi}{4}\right) = p$.

Show that these equations may be reduced to
$$x \cos\left(a \pm \frac{\pi}{2}\right) + y \sin\left(a \pm \frac{\pi}{2}\right) = p \tan \frac{\pi}{8},$$
and now represent two parallel lines perpendicular to (d).

7. If $a = 30°$, and the line (d) passes through the point $(4, 3)$, what is its distance from the origin?   *Ans.* $p = 2\sqrt{3} + \dfrac{3}{2}$.

8. If a and p are constant and P variable, what system of lines does the equation
$$x \cos a + y \sin a - p = P$$
represent?

9. For what value of P will a line of this system pass through the origin? What value of P gives (d)?

10. What is the perpendicular distance between (d) and any line of this system? *Ans.* P.

11. What is the value of P for the line which passes through a given point $(x'y')$? *Ans.* $P' = x' \cos a + y' \sin a - p$.

Show that P' is the length of the perpendicular dropped from the point $(x'y')$ on line (d).

12. For what values of P will lines of this system and the origin lie on the same side of (d), and for what values on opposite sides?

13. If P increases without limit, where will the corresponding line lie?

V. *Equation* (e). $\qquad x \sin \theta - y \cos \theta = p$. $\qquad$ *Parameters* (p, θ).

1. Through what point does the line pass when $p = 0$, and what does the equation become? *Ans.* $y = mx$.

2. If a line makes an angle of $45°$ with the axis of X, and $p = 10$, what is its equation? What if $\theta = 150°$ and $p = 5$?

3. When will the line (e) be perpendicular, and when parallel, to the axis of X? What does equation (e) become for these cases?

4. Show that $x \cos \theta + y \sin \theta = p'$ is the equation of a line perpendicular to (e).

5. Show that $y \cos \theta - x \sin \theta = p'$ is parallel to (e).

6. Show that $x \cos \theta + y \sin \theta = p \tan \dfrac{\pi}{8}$ makes an angle of $90°$ with (e).

7. If a line makes an angle of $60°$ with the axis of X, and passes through the point $(-1, -4)$, what is its distance from the origin?
Ans. $p = 2 - \tfrac{1}{2}\sqrt{3}$.

8. The direction of a force F makes an angle of $120°$ with the axis of X, and it acts at the point $(8, 6)$ in this direction; what is its moment about an axis through the origin perpendicular to the plane of the coordinate axes?
Ans. $Fp = F(8 \cdot \tfrac{1}{2}\sqrt{3} + 6 \cdot \tfrac{1}{2}) = F(4\sqrt{3} + 3)$.

EXERCISES ON THE STRAIGHT LINE.

9. Find the equations of the two straight lines which pass through the given point $(x'y')$ and make a given angle β with line (e).

Ans.
$$x\sin(\theta \pm \beta) - y\cos(\theta \pm \beta) - p' = P' = x'\sin(\theta \pm \beta) - y'\cos(\theta \pm \beta) - p';$$
$$\text{or } y - y' = \tan(\theta \pm \beta)(x - x').$$

VI. Equation (f). $\quad Ax + By + C = 0. \quad$ **Parameters** (A, B, C).

1. In this trinomial equation C is called the independent term, and the others the x and y terms. If the independent term $C = 0$, through what point does the line pass? In the equation $Ax + By = 0$, what is the tangent of the angle which the line makes with the axis of X? Upon which of the parameters does the direction of line (f) depend?

2. If $A = 0$, $By + C = 0$; if $B = 0$, $Ax + C = 0$. What lines do these equations represent?

3. If $x = 0$, $By + C = 0$; if $y = 0$, $Ax + C = 0$. What do these equations mean?

4. If $A = B$, what angle does line (f) make with the axis of X? What angle if $A = -B$?

5. If the normal axis passes through the point (A, B), then
$$\cos \alpha = \sin \theta = \frac{A}{\sqrt{A^2 + B^2}}, \quad \sin \alpha = -\cos \theta = \frac{B}{\sqrt{A^2 + B^2}}.$$
$$p = a \cos \alpha = -\frac{C}{A} \cdot \frac{A}{\sqrt{A^2 + B^2}} = -\frac{C}{\sqrt{A^2 + B^2}},$$
and equation (f) may be written in the normal form
$$-\frac{A}{\sqrt{A^2 + B^2}} x - \frac{B}{\sqrt{A^2 + B^2}} y = \frac{C}{\sqrt{A^2 + B^2}}.$$

6. If $A = 0$, $B = 0$, and C is finite, where will the line (f) lie? Where for A and B finite and C infinite?

> *Ans.* The intercepts $a = -\dfrac{C}{A}$, $b = -\dfrac{C}{B}$, both become infinite for both cases; therefore the line lies at an infinite distance from the origin; that is, all points on the line are *at infinity.*

Show the same from equation of Exercise 5.

7. What system of lines is represented by the equation
$$Ax + By + C = L$$
when the parameters (A, B, C) are constant and L variable? What relation do the lines of this system bear to (f)?

8. For what value of L will a line of this system pass through the origin? *Ans.* $L = C$.

9. What will the value of L be for the line which passes through the point $(x'y')$, and what its equation?
Ans. $Ax + By + C = L' = Ax' + By' + C$;
or, $A(x-x') + B(y-y') = 0$ is the required equation.

10. Find the equation of the line of the system which passes through the two given points $(x'y')$, $(x''y'')$.
Ans. $L = L' = L''$;
or $L - L' = L' - L'' = 0$;
or $A(x-x') + B(y-y') = A(x'-x'') + B(y'-y'') = 0$;
or $-\dfrac{A}{B} = \dfrac{y-y'}{x-x'} = \dfrac{y'-y''}{x'-x''}$, is the required equation.

11. If $Ax + 3y - 10 = 0$, what will the value of A be when the line passes through the point $(2, -4)$? *Ans.* $A = 11$.

12. If $Ax + By - 14 = 0$, what will the values of A and B be when the line passes through the points $(2, 3)$, $(-1, 2)$?
Ans. $A = -2$, $B = 6$.

VII. *Numerical Equations of Straight Lines.*

1. $2x - 7y - 2 = 0$.
2. $x - 2y + 6 = 0$.
3. $3x - 4y - 2 = 0$.
4. $4x - 3y - 18 = 0$.
5. $7x + 5y - 35 = 0$.
6. $3y - 20x + 1 = 0$.
7. $2y - \tfrac{2}{3}x - 8 = 0$.
8. $x + \tfrac{3}{4}y - 2 = 0$.
9. $\tfrac{2}{3}x - \tfrac{3}{4}y - 6 = 0$.
10. $\tfrac{1}{8}x - \tfrac{3}{8}y + \tfrac{2}{3} = 0$.
11. $3x + \tfrac{1}{4}y - 5 = 0$.
12. $10x + 9y - 90 = 0$.
13. $3x - 2y\ \ \ \ = 0$.
14. $2x + 3y\ \ \ \ = 0$.
15. $3x - 5y + 30 = 0$.
16. $7x - 4y - 8\tfrac{1}{2} = 0$.
17. $5x + 6y - 2 = 0$.
18. $8x + 2y - 10 = 0$.

(1) Find the points in which each line intersects the axes, and draw the line; that is, find a and b.

(2) Find the values of the parameters θ, a, and p.

(3) When $C = 0$, through what point does the line pass?

Next let us find what changes the forms of the foregoing equations of the straight line, found in Art. 26, undergo when the axes are oblique.

31. *To find the relations which the parameters of a straight line bear to each other when the axes are oblique.*

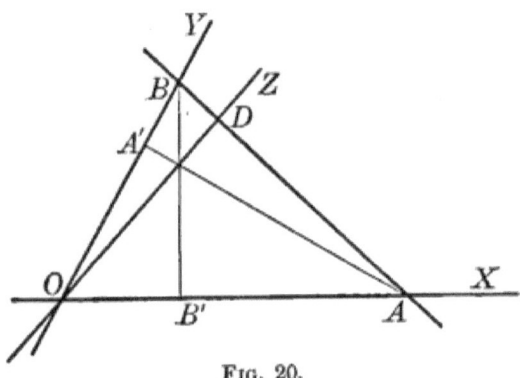

Fig. 20.

Let OX, OY, OZ be the oblique and normal axes, and AB the straight line. Drop perpendiculars from A and B on the axes of Y and X. The distance parameters, as before, are

$$OA = a, \quad OB = b, \quad OD = p;$$

the direction parameters are

$$AOB = \omega, \quad XAB = \theta, \quad OAB = \pi - \theta, \quad OBA = \theta - \omega,$$
$$AOD = ABB' = \alpha, \quad BOD = BAA' = \omega - \alpha.$$

Then, by Trigonometry,

$$AB = \sqrt{a^2 + b^2 - 2ab \cos \omega} = l, \quad \text{for brevity};$$
$$AA' = a \sin \omega, \quad OA' = a \cos \omega,$$

$$BB' = b \sin \omega, \qquad OB' = b \cos \omega,$$

$$AB' = a - b \cos \omega, \qquad A'B = b - a \cos \omega.$$

$$\frac{p}{a} = \cos \alpha = \sin(\pi - \theta) = \sin \theta = \frac{BB'}{AB} = \frac{b \sin \omega}{l}.$$

$$\frac{p}{b} = \cos(\omega - \alpha) = -\sin(\omega - \theta) = \frac{AA'}{AB} = \frac{a \sin \omega}{l}.$$

$$\frac{b}{a} = -\frac{\sin \theta}{\sin(\omega - \theta)} = \frac{\cos \alpha}{\cos(\omega - \alpha)}.$$

$$\sin \alpha = -\cos \theta = \frac{AB'}{AB} = \frac{a - b \cos \omega}{l}.$$

$$\tan \alpha = -\cot \theta = \frac{AB'}{BB'} = \frac{a - b \cos \omega}{b \sin \omega}.$$

$$\therefore p = a \cos \alpha = a \sin \theta = b \cos(\omega - \alpha) = -b \sin(\omega - \theta) = \frac{ab \sin \omega}{l}.$$

32. *To find the equations of the straight line AB in terms of each pair of parameters.*

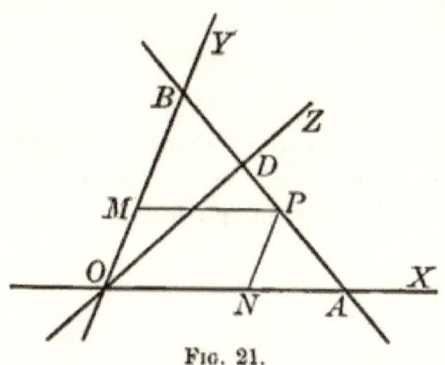

Fig. 21.

Let $P\ (ON, PN)(xy)$ be any point on the given line AB, and draw PM parallel to the axis of X. Then

$$PN = y, \quad PM = x, \quad BM = b - y, \quad NA = a - x.$$

The similar triangles ANP, PMB, AOB, give

$$\frac{PN}{NA} = \frac{BM}{PM} = \frac{OB}{OA},$$

or
$$\frac{y}{a-x} = \frac{b-y}{x} = \frac{b}{a} = -\frac{\sin\theta}{\sin(\omega-\theta)} = \frac{\cos a}{\cos(\omega-a)}.$$

I. *Equation in terms of the parameters* (a, θ).

$$\frac{y}{a-x} = -\frac{\sin\theta}{\sin(\omega-\theta)} = -m, \text{ for brevity.}$$

$$\therefore y = m(x-a) \qquad (a')$$

is the required equation.

II. *Equation in terms of the parameters* (b, θ).

$$\frac{b-y}{x} = -\frac{\sin\theta}{\sin(\omega-\theta)} = -m.$$

$$\therefore y = mx + b \qquad (b')$$

is the required equation.

III. *Equation in terms of the parameters* (a, b).

$$\frac{y}{a-x} = \frac{b}{a}, \text{ or } \frac{y}{b} = \frac{a-x}{a} = 1 - \frac{x}{a}.$$

$$\therefore \frac{x}{a} + \frac{y}{b} = 1 \qquad (c')$$

is the required equation.

IV. *Equation in terms of the parameters* (p, a).

$$\frac{y}{a-x} = \frac{\cos a}{\cos(\omega-a)},$$

or
$$y \cos(\omega-a) = (a-x)\cos a = a\cos a - x\cos a.$$

$$\therefore x \cos a + y \cos(\omega-a) = p \qquad (d')$$

is the required equation.

V. *Equation in terms of the parameters* (p, θ).

$$\frac{y}{a-x} = -\frac{\sin\theta}{\sin(\omega-\theta)},$$

or $\qquad -y\sin(\omega-\theta) = (a-x)\sin\theta = a\sin\theta - x\sin\theta.$

$$\therefore\ x\sin\theta - y\sin(\omega-\theta) = p \qquad (e')$$

is the required equation.

Equation (c') is independent of ω, and is identical with (c) for rectangular axes; (a') and (b') are of the same form as the corresponding ones for rectangular axes, but the difference in the values of m must not be forgotten. If in all these equations we make $\omega = 90°$, we shall get the corresponding ones for rectangular axes.

33. *To determine the parameters* a, b, p, a, θ, *in terms of the parameters* (A, B, C), *axes oblique.*

As in Art. 28, we can at once put

$$Ax + By + C = 0 \qquad (f)$$

in the forms

$$y = -\frac{A}{B}\left(x + \frac{C}{A}\right), \qquad (a'')$$

$$y = -\frac{A}{B}x - \frac{C}{B}, \qquad (b'')$$

$$\frac{x}{-\frac{C}{A}} + \frac{y}{-\frac{C}{B}} = 1, \qquad (c'')$$

which are those of (a'), (b'), (c'), Art. 32; therefore

$$m = -\frac{\sin\theta}{\sin(\omega-\theta)} = -\frac{A}{B}, \quad a = -\frac{C}{A}, \quad b = -\frac{C}{B}.$$

Also, by Art. 31, substituting the values of a and b just found,

$$l = \sqrt{a^2 + b^2 - 2ab\cos\omega}$$

by putting
$$= -\frac{C}{AB}\sqrt{A^2 + B^2 - 2AB\cos\omega} = -\frac{CL}{AB},$$

$\sqrt{A^2 + B^2 - 2AB\cos\omega} = L$, for brevity;

$$\cos a = \sin\theta = \frac{b\sin\omega}{l} = \frac{A\sin\omega}{L},$$

$$\cos(\omega - a) = -\sin(\omega - \theta) = \frac{a\sin\omega}{l} = \frac{B\sin\omega}{L},$$

$$p = \frac{ab\sin\omega}{l} = -\frac{C\sin\omega}{L}.$$

Substituting these values in equations (d') and (e'), they both become

$$\frac{A\sin\omega}{L}x + \frac{B\sin\omega}{L}y + \frac{C\sin\omega}{L} = 0, \qquad (d''), (e'')$$

which is (f) multiplied by the factor $\frac{\sin\omega}{L}$.

C and L must have opposite signs to make p positive.

We have thus found the values of a, b, p, a, θ, in terms of the parameters (A, B, C), when the axes are oblique.

34. Relation of Parameters. — If we put equations (a'), (b'), (c'), (d'), (e'), Art. 32, and the general equation (f), in the form of (b'), they become

I. $y = mx - ma$.

II. $y = mx + b$.

III. $y = -\frac{b}{a}x + b$.

IV. $y = -\frac{\cos a}{\cos(\omega - a)}x + \frac{p}{\cos(\omega - a)}.$

V. $y = \frac{\sin\theta}{\sin(\omega - \theta)}x - \frac{p}{\sin(\omega - \theta)}.$

(f) $y = -\frac{A}{B}x - \frac{C}{B}.$

78   PLANE ANALYTIC GEOMETRY.

Since these equations represent the same line, they are identical, and

$$m = -\frac{b}{a} = -\frac{\cos a}{\cos(\omega - a)} = \frac{\sin\theta}{\sin(\omega - \theta)} = -\frac{A}{B},$$

$$b = -ma = \frac{p}{\cos(\omega - a)} = -\frac{p}{\sin(\omega - \theta)} = -\frac{C}{B};$$

from which we also get

$$\tan\theta = \frac{m\sin\omega}{1 + m\cos\omega} = -\cot a = -\frac{b\sin\omega}{a - b\cos\omega} = -\frac{A\sin\omega}{B - A\cos\omega}.$$

These relations are the same as found in Art. 33.

EXERCISES ON THE STRAIGHT LINE, AXES OBLIQUE.

1. Given the equation $3x - 2y - 5 = 0$, to find the intercepts on the coordinate axes, and the angle which the line makes with the axis of X.

Ans. $a = \frac{5}{3}$; $b = -\frac{5}{2}$; $m = \frac{\sin\theta}{\sin(\omega-\theta)} = \frac{3}{2}$; $\tan\theta = \frac{3\sin\omega}{2 + 3\cos\omega}$.

2. Put the equation $3x - 2y - 5 = 0$ in the forms of (d') and (e'), and determine the corresponding parameters.

Ans. $\dfrac{(3x - 2y - 5)\sin\omega}{\sqrt{13 + 12\cos\omega}} = 0$;

$\therefore \cos a = \sin\theta = \dfrac{3\sin\omega}{\sqrt{13 + 12\cos\omega}}$;

$\therefore \cos(\omega - a) = -\sin(\omega - \theta) = -\dfrac{2\sin\omega}{\sqrt{13 + 12\cos\omega}}$.

$\therefore p = \dfrac{5\sin\omega}{\sqrt{13 + 12\cos\omega}}$.

For other numerical equations of straight lines see p. 72.

3. Given the triangle AOB and notation of Fig. 20. Find the equations of the following lines:

(1) AB, $\dfrac{x}{a} + \dfrac{y}{b} = 1$.

(2) AA', $x\cos\omega + y = a\cos\omega$.

(3) BB', $\qquad y\cos\omega + x = b\cos\omega.$

(4) OD, $\qquad x(a - b\cos\omega) - y(b - a\cos\omega) = 0.$

(5) $A'B'$, $\qquad \dfrac{x}{b} + \dfrac{y}{a} = \cos\omega.$

(6) Medial through A, $\qquad \dfrac{x}{a} + \dfrac{2y}{b} = 1.$

(7) Medial through B, $\qquad \dfrac{y}{b} + \dfrac{2x}{a} = 1.$

(8) Medial through O, $\qquad \dfrac{x}{a} - \dfrac{y}{b} = 0.$

(9) A line cuts off the mth part of a and b; its equation is
$$\frac{x}{a} + \frac{y}{b} = \frac{1}{m}.$$

(10) Show that this line is parallel to AB.

(11) The perpendiculars AA', BB', OD meet in the point
$$y = \frac{(a - b\cos\omega)\cos\omega}{\sin^2\omega}, \qquad x = \frac{(b - a\cos\omega)\cos\omega}{\sin^2\omega}.$$

(12) The medials meet in the point $\left(\dfrac{a}{3}, \dfrac{b}{3}\right).$

(13) Show that the equations of the lines AA', BB', are already in the form of (d').

(14) Put the equation of $A'B'$ in the form of (d').
$$\frac{a\sin\omega\, x}{\sqrt{a^2 + b^2 - 2ab\cos\omega}} + \frac{b\sin\omega\, y}{\sqrt{a^2 + b^2 - 2ab\cos\omega}} = \frac{ab\sin 2\omega}{2\sqrt{a^2 + b^2 - 2ab\cos\omega}}$$

(15) Show that this equation may be written
$$x\sin A + y\sin B = b\sin A\cos\omega = a\sin B\cos\omega.$$

4. Two opposite sides AA' and BB' of a quadrilateral are produced till they meet in O. Take OA', OB' as axes of X and Y, and let $OA = 2a$, $OA' = 2a'$, $OB = 2b$, $OB' = 2b'$; then

(1) The equations of the sides AB and $A'B'$ are
$$\frac{x}{a} + \frac{y}{b} = 2, \qquad \frac{x}{a'} + \frac{y}{b'} = 2.$$

(2) The equations of the diagonals AB', $A'B$ are

$$\frac{x}{a} + \frac{y}{b'} = 2, \qquad \frac{x}{a'} + \frac{y}{b} = 2.$$

(3) The middle points of the diagonals are (a, b'), (a', b).

(4) The diagonals meet in the point

$$x = \frac{2aa'(b-b')}{ab - a'b'}, \qquad y = \frac{2bb'(a-a')}{ab - a'b'}.$$

(5) The sides AB, $A'B'$ meet in the point O',

$$x = \frac{2aa'(b-b')}{a'b - ab'}, \qquad y = \frac{2bb'(a'-a)}{a'b - ab'}.$$

(6) The equation of the line OO' is

$$aa'(b-b')y = bb'(a'-a)x.$$

(7) The middle point of the line OO' is

$$x = \frac{a'b \cdot a - ab' \cdot a'}{a'b - ab'}, \qquad y = \frac{a'b \cdot b' - ab' \cdot b}{a'b - ab'}.$$

(8) Divide the line passing through the points (a, b') and (a', b) externally in the ratio $a'b : ab'$, and show that it is the point found in (7).

The Polar Equation of the Straight Line.

35. *To find the polar equation of the straight line.*

Let the pole be at the origin of rectangular axes, and the axis of X the initial direction from which the vectorial angle is measured.

If the position of the line is determined by the parameters (p, a), then its equation is

$$x \cos a + y \sin a = p. \qquad (d)$$

But for any point (r, θ) on the line,

$$x = r \cos \theta, \quad y = r \sin \theta.$$

By substituting these values of x and y in (d) it becomes

$$r \cos(\theta - a) = p, \qquad (a)$$

the required polar equation of the line.

If the equation of the given line is

$$Ax + By + C = 0, \qquad (f)$$

then $\qquad r(A \cos \theta + B \sin \theta) + C = 0 \qquad (b)$

is the required polar equation, which may be reduced to the form (a) by dividing each term by $\sqrt{A^2 + B^2}$, Art. 28.

EXERCISES ON POLAR EQUATIONS OF THE STRAIGHT LINE.

1. What is the polar equation of the straight line which is perpendicular to the axis of X? *Ans.* $r \cos \theta = p$.

2. What is the polar equation of the straight line which is parallel to the axis of X? *Ans.* $r \sin \theta = p$.

3. If the normal axis is taken as the initial direction of the vectorial angle, what is the equation of the line? *Ans.* $r \cos \theta = p$.

4. If the parameters of the given line are (p, θ'), what is its polar equation? *Ans.* $r \sin(\theta' - \theta) = p$.

5. If the given line passes through the point $(x'y')$, and makes an angle θ' with the axis of X, what is its polar equation?
 Ans. $r \sin(\theta - \theta') = y' \cos \theta' - x' \sin \theta'$.

6. Find the polar equation of a straight line passing through the two given points (r', θ'), (r'', θ''). Suggestion: The area of the triangle made by these two points and any other point on the line must equal zero. See Art. 11.
 Ans. $r[r \sin(\theta' - \theta) + r'' \sin(\theta - \theta'')] = r'r'' \sin(\theta' - \theta'')$.

7. Deduce this result from the equation

$$(y - y')(x' - x'') - (x - x')(y' - y'') = 0.$$

8. Reduce the polar equation $r = 2a \sec\left(\theta + \dfrac{\pi}{6}\right)$ to rectangular coordinates.

9. Find the polar coordinates of the point of intersection of the lines $r\cos\left(\theta-\frac{\pi}{2}\right)=2a$, $r\cos\left(\theta-\frac{\pi}{6}\right)=a$; and also the angle between them. *Ans.* Intersection, $\left(2a,\frac{\pi}{2}\right)$; angle $=\frac{\pi}{3}$.

Lines Subject to Given Conditions.

36. *To find the equation of a straight line which passes through a given point and makes a given angle with the axis of X.*

In this case the parameters are the given point $(x'y')$ and the given angle θ. The equation

$$y = mx + b \qquad (a)$$

already contains one of the given parameters, θ, and it only remains to eliminate b, and replace it by the given point $(x'y')$. The value of b is determined by the condition

$$y' = mx' + b, \quad \text{or} \quad b = y' - mx', \qquad (b)$$

since the required line must pass through $(x'y')$. Substituting this value of b in (a), we get

$$y - y' = m(x - x'), \qquad (c)$$

which is the required equation.

If m is variable, (c) represents a system of lines all passing through the same point $(x'y')$. This equation is true both for rectangular and oblique axes, by using the proper value of m.

Second Solution. — If $-\dfrac{A}{B} = \tan\theta$, then the line whose equation is

$$Ax + By + C = L = 0 \qquad (a)$$

makes the given angle θ with the axis of X. But the equation

$$Ax + By + C = L, \qquad (b)$$

or, more briefly L, is any parallel to $L = 0$, Art. 30, and the equation of the line which passes through the point $(x'y')$ is

$$L = L', \text{ or } L - L' = 0.$$

If this equation is expressed directly in terms of x and y, then

$$Ax + By + C = L' = Ax' + By' + C,$$

or
$$y - y' = -\frac{A}{B}(x - x'). \tag{c}$$

Since, by Art. 29,

$$-\frac{A}{B} = m = -\frac{b}{a} = -\cot\alpha = \tan\theta,$$

this equation can be written in terms of either of the direction parameters.

Third Solution. — Let $P'(x'y')$ be the given point, and $OAB = PP'R = \theta$, the given angle; also let $P(xy)$ be any other point on the line.

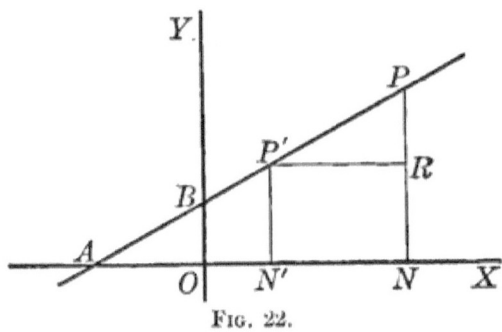

Fig. 22.

Then
$$PR = y - y', \qquad P'R = x - x',$$

$$\tan\theta = \frac{PR}{P'R} = \frac{y - y'}{x - x'} = m.$$

$$\therefore y - y' = m(x - x') \tag{c}$$

is the required equation for rectangular axes.

This equation may be written

$$y - y' = \frac{\sin\theta}{\cos\theta}(x - x'), \quad \text{or} \quad \frac{y - y'}{\sin\theta} = \frac{x - x'}{\cos\theta} = r,$$

in which $r = P'P$. Therefore the coordinates of any point $P(xy)$, at a given distance r from $P'(x'y')$, are

$$y = y' + r \sin \theta, \quad x = x' + r \cos \theta. \tag{d}$$

For oblique axes, the oblique triangle $PP'R$ gives

$$\frac{y-y'}{x-x'} = \frac{\sin \theta}{\sin(\omega - \theta)} = m.$$

$$\therefore y - y' = m(x - x') \tag{c'}$$

is the required equation for oblique axes.

37. From the oblique triangle $PP'R$ we also get

$$\frac{y-y'}{\sin \theta} = \frac{x-x'}{\sin(\omega - \theta)} = \frac{r}{\sin \omega}.$$

If for brevity we put

$$l = \frac{\sin \theta}{\sin \omega}, \quad n = \frac{\sin(\omega - \theta)}{\sin \omega},$$

then

$$\frac{y-y'}{l} = \frac{x-x'}{n} = r.$$

$$\therefore y = y' + lr, \quad x = x' + nr. \tag{d'}$$

The constants l and n are not independent; find their relation.

By (c), Art. 4, and (d') just found,

$$(y-y')^2 + (x-x')^2 + 2(y-y')(x-x')\cos \omega$$
$$= l^2 r^2 + n^2 r^2 + 2 l n r^2 \cos \omega = r^2.$$

$$\therefore l^2 + n^2 + 2 l n \cos \omega = 1$$

is the required relation.

By substituting the values of l and n, this relation becomes

$$\sin^2 \theta + \sin^2(\omega - \theta) + 2 \sin \theta \sin(\omega - \theta) \cos \omega = \sin^2 \omega;$$

which might have been inferred, since the sines of the angles in any triangle are proportional to their opposite sides.

Also, since, by Art. 31,

$$\sin\theta = \cos a, \quad \sin(\omega - \theta) = -\cos(\omega - a);$$

$$\therefore \cos^2 a + \cos^2(\omega - a) - 2\cos a \cos(\omega - a)\cos\omega = \sin^2\omega.$$

38. *To find the equation of a straight line passing through a given point parallel to a given line.*

The equation of any line passing through a given point $(x'y')$ is, by Art. 36,

$$y - y' = m(x - x'); \tag{a}$$

and if this line must be parallel to the given line,

$$y = m'x + b', \tag{b}$$

then it must have the same direction parameter m';

$$\therefore y - y' = m'(x - x') \tag{c}$$

passes through $(x'y')$ and is parallel to (b).

Second Solution. — Let the equation of the given line be

$$Ax + By + C = L = 0, \tag{a}$$

then for the line L parallel to $L = 0$, which passes through the given point $(x'y')$, the equation is

$$L = L', \text{ or } L - L' = 0;$$

which may be expressed directly in terms of x and y, as in Art. 36.

39. *To find the equation of a straight line which passes through two given points.*

In this case the parameters m and b, in the equation

$$y = mx + b, \tag{a}$$

must be so determined that the line shall pass through the two given points $(x'y')(x''y'')$.

The equations of condition for determining m and b are, therefore,
$$y' = mx' + b, \qquad (a')$$
$$y'' = mx'' + b; \qquad (a'')$$

from which, by elimination, we readily get
$$m = \frac{y' - y''}{x' - x''}, \quad b = \frac{x'y'' - x''y'}{x' - x''},$$

and (a) becomes
$$y = \frac{y' - y''}{x' - x''} x + \frac{x'y'' - x''y'}{x' - x''},$$

which is the required equation.

A more useful form of this equation is found as follows.

By Art. 36,
$$y - y' = m(x - x')$$

is the equation of any line passing through the point $(x'y')$, and a second point $(x''y'')$ gives m as above; therefore,
$$y - y' = \frac{y' - y''}{x' - x''}(x - x'), \quad \text{or} \quad \frac{y - y'}{y' - y''} = \frac{x - x'}{x' - x''}, \qquad (c)$$

is the required equation.

Second Solution. — Let $P'(x'y')$ and $P''(x''y'')$ be the two given points, and $P(xy)$ any other point on the line AB. Then from the similar triangles $P''P'S$, $P''PT$, $PP'R$, we get
$$\frac{P''S}{P'S} = \frac{P''T}{PT} = \frac{PR}{P'R},$$

or
$$\frac{y'' - y'}{x'' - x'} = \frac{y - y''}{x - x''} = \frac{y - y'}{x - x'}, \qquad (c)$$

which are three equivalent forms of the required equation.

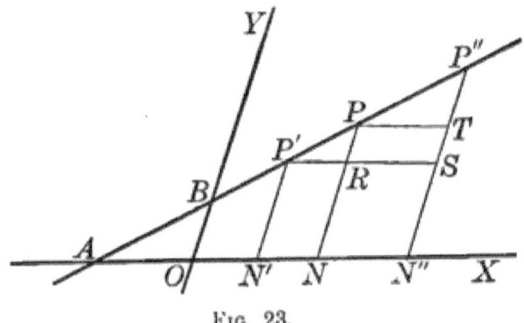

Fig. 23.

The third of these forms,

$$(y - y'')(x - x') = (y - y')(x - x''),$$

is not a quadratic as it appears to be, since the products xy found in both members cancel.

All the forms reduce to

$$(y' - y'')x - (x' - x'')y + x'y'' - x''y' = 0.$$

Third Solution. — Let it be required to so determine the parameters (A, B, C) that the line

$$Ax + By + C = L = 0 \qquad (a)$$

shall pass through the two given points $(x'y')$ and $(x''y'')$.

The equations of the parallel L, which passes through the two given points, are

$$L = L' = L'', \quad \text{or} \quad L - L' = L - L'' = L' - L'' = 0;$$

from which, by eliminating A, B, C, we get

$$-\frac{A}{B} = \frac{y - y'}{x - x'} = \frac{y - y''}{x - x''} = \frac{y' - y''}{x' - x''},$$

as before. These equations are the same for both rectangular and oblique axes.

40. *To find the condition that three given points shall lie on the same straight line.*

If the line found in Art. 39 must also pass through the point $(x''' y''')$, then the condition is

$$\frac{y''' - y'}{x''' - x'} = \frac{y''' - y''}{x''' - x''} = \frac{y' - y''}{x' - x''};$$

which reduces to

$$y'(x'' - x''') + y''(x''' - x') + y'''(x' - x'') = 0,$$

a form easily remembered from the symmetry of the notation.

EXERCISES ON LINES PASSING THROUGH GIVEN POINTS.

1. Find the equation of a straight line which passes through the given point $(2, -5)$ and makes an angle of $45°$ with the axis of X.
   Ans. $y + 5 = x - 2$, or $y - x + 7 = 0$.

2. If the axes are oblique and $\omega = 75°$, what is the equation?
   Ans. $y = \sqrt{2}\,x - 5 - 2\sqrt{2} = 0$.

3. Find the equation of a line passing through the points $(3, 5)$ and $(-2, 3)$.   *Ans.* $\dfrac{y-5}{5-3} = \dfrac{x-3}{3-(-2)}$, or $5y - 2x - 19 = 0$.

4. Find the equation of a line passing through the point $(3, -5)$ parallel to a given line $4x - 7y + 2 = 0$.
   Ans. For the given line $m = \tfrac{4}{7}$. $\therefore y + 5 = \tfrac{4}{7}(x - 3)$,
   or $4x - 7y - 47 = 0$.
   By Art. 38, second solution,
   $4x - 7y + 2 = L' = 4 \cdot 3 - 7(-5) + 2$, or $4x - 7y - 47 = 0$.

5. Find the equations of the sides of a triangle the coordinates of whose vertices are $(2, 1)$, $(3, -2)$, $(-4, -1)$.
   Ans. $x + 7y + 11 = 0$, $3y - x - 1 = 0$, $3x + y - 7 = 0$.

6. Given the vertices $(2, 3)$, $(4, -5)$, $(-3, -6)$ of a triangle:

(1) Find the equations of the opposite sides.

(2) Find the equations of the medial lines.

(3) Find the equations of the bisectors of the adjacent sides.

(4) Find the equations of the lines through the vertices parallel to the opposite sides.

(5) Find the equations of the lines through the vertices parallel to the medial lines.

$$(2, 3) \qquad (4, -5) \qquad (-3, -6)$$

(1) $x - 7y - 39 = 0; \quad 9x - 5y - 3 = 0; \quad 4x + y - 11 = 0.$
(2) $17x - 3y - 25 = 0; \quad 7x + 9y + 17 = 0; \quad 5x - 6y - 21 = 0.$
(3) $x - 7y - 10 = 0; \quad 9x - 5y - 32 = 0; \quad 8x + 2y + 7 = 0.$
(4) $x - 7y + 19 = 0; \quad 9x - 5y - 61 = 0; \quad 4x + y + 18 = 0.$
(5) $7x + 9y - 41 = 0; \quad 17x - 3y - 83 = 0; \quad 17x - 3y + 33 = 0.$
$\ 5x - 6y + 8 = 0; \quad 5x - 6y - 50 = 0; \quad 7x + 9y + 75 = 0.$

7. Find the equation of the line passing through the points $(x'y')$ and $(0, b)$. *Ans.* $(b - y')x + x'y = bx'.$
Find the same result by eliminating m from (a) and (a'), Art. 39.

8. Given the vertices $(x'y')$, $(x''y'')$, $(x'''y''')$; find the equation of the medial line passing through the vertex $(x'y')$.
Ans. $(y'' + y''' - 2y')x - (x'' + x''' - 2x')y + x''y' - x'y'' + x'''y' - x'y''' = 0.$

Write the equations of the remaining medials by increasing the primes by one for each successive medial, noting that $'''$ becomes $'$.

9. Find the equation of the line passing through the points $(a, 0)$ and $(0, b)$. *Ans.* $bx + ay = ab.$

10. Find the equation of the line passing through the points $(-ae, 0)$ and $\left(ae, \dfrac{b^2}{a}\right)$. *Ans.* $2a^2ey - b^2x = ab^2e.$

41. *To find the condition that a given point and the origin shall lie on the same or opposite sides of a given straight line.*

Let the equation of the given line be

$$Ax + By + C = L = 0. \qquad (a)$$

Since L, the equation of any parallel to $L = 0$, is a continuous function of (xy), it follows that L must have opposite signs for points on opposite sides of the line $L = 0$. But for the line passing through the origin $(0, 0)$, $L = C$, and for the line passing through the given point $(x'y')$,

$$L' = Ax' + By' + C.$$

If, then, L' and C have the same sign, the given point $(x'y')$ and the origin lie on the same side of $L = 0$, and on opposite sides of $L = 0$ if L' and C have opposite signs. Two given parallel lines will lie on the same side of the origin if the signs of their corresponding intercepts on the coordinate axes are the same; and on different sides of the origin if the signs of their corresponding intercepts are unlike.

42. *To find the length of the perpendicular dropped from a given point to a given line.*

Let
$$x \cos a + y \sin a - p = P = 0 \qquad (a)$$

be the equation of the given line AB, and let $N'(x'y')$ be the given point; then

$$x \cos a + y \sin a - p = P$$

is the equation of any line parallel to AB, in which P is the distance measured on the normal axis between any line P and the given line $P = 0$. For the line $A'B'$, passing through the given point $N'(x'y')$,

$$P' = x' \cos a + y' \sin a - p, \qquad (b)$$

which is the length $D'D = N'N$ of the required perpendicular.

It will be noticed that the length P' of the required perpendicular is obtained directly from the equation of the given line by simply substituting for x and y the coordinates x' and y' of the given point.

LENGTHS OF PERPENDICULARS.

We have taken the distance $OD = p$ as positive; therefore $DO = -p$, and we may consider the origin as being on the negative side of the line, if we measure from the line to the origin. It follows, then, that values of $(x'y')$ which make P' negative show that this point and the origin lie on the same side of the line, and on opposite sides if they make P' positive.

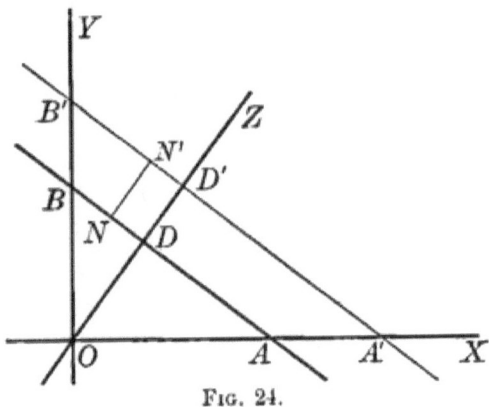

Fig. 24.

If the equation of the given line is

$$x \sin \theta - y \cos \theta - p = 0, \quad \text{or} \quad Ax + By + C = 0,$$

then

$$P' = x' \sin \theta - y' \cos \theta - p = \frac{Ax' + By' + C}{\sqrt{A^2 + B^2}}$$

for rectangular axes. If the axes are oblique, then

$$P' = x' \cos a + y' \cos(\omega - a) - p = \frac{(Ax' + By' + C)\sin \omega}{\sqrt{A^2 + B^2 - 2AB \cos \omega}}.$$

43. *To find the locus of a point (xy) whose perpendicular distance from a given line is constant.*

Let the equation of the given line be

$$Ax + By + C = L = 0, \tag{a}$$

and let P denote the length of the constant perpendicular; then $P = $ a constant is the equation of the required locus.

To express P directly in terms of x and y we have

$$\frac{Ax + By + C}{\sqrt{A^2 + B^2}} = \frac{L}{\sqrt{A^2 + B^2}} = P,$$

or $\qquad Ax + By + C = L = P\sqrt{A^2 + B^2}.$ \hfill (b)

This equation may be written

$$x + \frac{By}{A} + \frac{C}{A} = \frac{L}{A} = \frac{P\sqrt{A^2 + B^2}}{A} = P \sec a = P \csc \theta,$$

or $\qquad y + \dfrac{Ax}{B} + \dfrac{C}{B} = \dfrac{L}{B} = \dfrac{P\sqrt{A^2 + B^2}}{B} = P \csc a = -P \sec \theta,$

in which $\dfrac{L}{A}$ and $\dfrac{L}{B}$, or their equivalents, are the distances between the parallels L and $L = 0$ measured on the axes of X and Y. For oblique axes these distances are $P \sec a = P \csc \theta$ on the axis of X, and $P \sec(\omega - a) = -P \csc(\omega - \theta)$ on the axis of Y.

EXERCISES ON THE RELATIVE POSITIONS OF POINTS AND LINES.

1. Do the origin and the point $(2, 3)$ lie on the same or on opposite sides of the line $3x + y - 5 = 0$? \hfill *Ans.* Opposite.

2. Does the origin lie in the acute or in the obtuse angle made by the lines $x - 2y + 7 = 0$, $3x + y - 4 = 0$? \hfill *Ans.* Acute.

3. Does the origin lie within or without the triangle formed by the lines $3x - 2y - 5 = 0$, $x + 4y + 7 = 0$, $2x - 3y + 6 = 0$?
\hfill *Ans.* Within.

4. Find the length of the perpendicular from the origin on the line $3x + 4y + 20 = 0$. \hfill *Ans.* 4.

5. Find the length of the perpendicular from the point $(2, 3)$ on the line $3x - 4y - 4 = 0$.
\hfill *Ans.* 2. Point and origin are on the same side of the line.

6. Find the length of the perpendicular from the point $(2, 3)$ on the line $2x + y - 4 = 0$.
\hfill *Ans.* $\dfrac{3}{\sqrt{5}}$. Point on opposite side from origin.

7. Find the length of the perpendicular from the point $(3, -4)$ on the line $4x + 2y = 7$, when $\omega = 60°$.

Ans. $\frac{3}{4}$. Point and origin are on the same side of the line.

8. Find the lengths of the perpendiculars from each vertex $(2,1)$, $(3,-2)$, $(-4,-1)$ of a triangle to the opposite side.

Ans. $2\sqrt{2}$, $\sqrt{10}$, $2\sqrt{10}$, and the origin is within the triangle.

9. Find the perpendicular distance between the two lines which are parallel to the line $3x + 4y - 2 = 0$, and pass respectively through the points $(2,3)$, $(-3,-4)$. *Ans.* $8\frac{3}{5}$.

10. Find the distance between two parallels to $L = 0$ measured on the axis of X, one of which passes through the point $(x'y')$, and the other through $(x''y'')$. *Ans.* $\dfrac{L' - L''}{A}$, or $x' - x'' + \dfrac{B}{A}(y' - y'')$.

11. Find the length of the perpendicular from the point (A, B) on the line $Ax + By + C = 0$. *Ans.* $\sqrt{A^2 + B^2} + \dfrac{C}{\sqrt{A^2 + B^2}}$.

12. Find the length of the perpendicular from the point (b, a) on the line $\dfrac{x}{a} + \dfrac{y}{b} = 2$. *Ans.* $\dfrac{(b-a)^2}{\sqrt{a^2 + b^2}}$.

13. Given the vertices $(x'y')$, $(x''y'')$, $(x'''y''')$ of a triangle.

(1) Find the length of the perpendicular from $(x'y')$ on the opposite side.

(2) Find the length of the side opposite $(x'y')$.

(3) Find the area of the triangle.

Ans. (1) $\dfrac{(y' - y'')(x'' - x''') - (x' - x'')(y'' - y''')}{\sqrt{(x'' - x''')^2 + (y'' - y''')^2}}$.

(2) $\sqrt{(x'' - x''')^2 + (y'' - y''')^2}$.

(3) $\frac{1}{2}[y'(x'' - x''') + y''(x''' - x') + y'''(x' - x'')]$.

14. Find the length of the perpendicular from the point $(p, 0)$ on the line $y = mx + \dfrac{p}{m}$. *Ans.* $\dfrac{p}{m}\sqrt{1 + m^2} = p \csc \theta$.

15. Find the length of the perpendicular from the origin on the line $y = mx + r\sqrt{1 + m^2}$. *Ans.* r

16. Find the length of the perpendicular from the point $(ae, 0)$ on the line $y = mx + \sqrt{a^2m^2 + b^2}$. Ans. $ae \sin \theta + \sqrt{a^2 \sin^2 \theta + b^2 \cos^2 \theta}$.

17. Find the length of the perpendicular from the point $(r \cos \alpha, r \sin \alpha)$ on the line $x \cos \alpha + y \sin \alpha - p = 0$. Ans. $r - p$.

Angles between Given Lines.

44. *To find the angle between two given straight lines.*

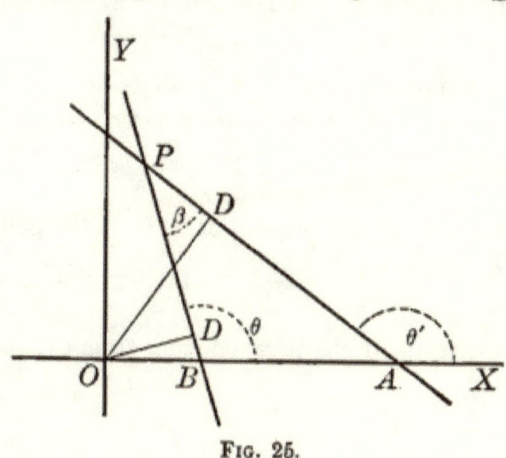

Fig. 25.

First, *when the axes are rectangular.*

Let the direction parameters of the two given lines AP and BP be denoted by θ' and θ, or a' and a, and let β denote the required angle APB; then $\beta = \theta' - \theta = a' - a$, and, by Trigonometry, Art. 63,

$$\tan \beta = \tan (\theta' - \theta) = \frac{\tan \theta' - \tan \theta}{1 + \tan \theta' \tan \theta}.$$

But by Art. 29,

$$\tan \theta' = m' = -\frac{b'}{a'} = -\cot a' = -\frac{A'}{B'},$$

$$\tan \theta = m = -\frac{b}{a} = -\cot a = -\frac{A}{B},$$

and therefore the corresponding forms of the value of $\tan \beta$ are

ANGLES BETWEEN GIVEN LINES. 95

$$\tan \beta = \frac{m'-m}{1+m'm} = \frac{a'b-ab'}{aa'+bb'} = \tan(a'-a) = \frac{AB'-A'B}{AA'+BB'}. \quad (a)$$

45. Parallel Lines. — In this case, $\beta = \theta' - \theta = a' - a = 0$, and $\tan \beta = 0$; therefore,

$m' - m = 0$,   or $m' = m$;

$a'b - ab' = 0$,   or $\dfrac{b'}{a'} = \dfrac{b}{a}$;

$\tan(a' - a) = 0$,   or $a' = a$, or $= a + \pi$;

$AB' - A'B = 0$,   or $\dfrac{A'}{B'} = \dfrac{A}{B}$;

which shows *that the direction parameters of parallel lines are either equal or proportional.* The condition of the parallelism of two given lines may also be stated as follows:

When two straight lines are parallel, the ratios of the coefficients of x and y in their equations are equal.

46. Perpendicular Lines. — In this case,

$$\beta = \theta' - \theta = a' - a = \frac{\pi}{2},$$

and $\tan \beta = $ infinity; therefore,

$1 + m'm = 0$,   or $m' = -\dfrac{1}{m}$;

$aa' + bb' = 0$,   or $\dfrac{b'}{a'} = -\dfrac{a}{b}$;

$\tan(a' - a) = \infty$,   or $a' = a + \dfrac{\pi}{2}$, or $= a + \dfrac{3\pi}{2}$;

$AA' + BB' = 0$,   or $\dfrac{A'}{B'} = -\dfrac{B}{A}$.

Hence, two lines are mutually perpendicular, *if the ratio of the coefficients of x and y in the equation of one is minus the reciprocal of this ratio in the equation of the other.*

The condition of perpendicularity may also be stated as follows:

Two lines are mutually perpendicular when the tangents of the angles which they make with the axis of x are each the minus reciprocal of the other.

47. Second, *when the axes are oblique.*

In this case, Art. 34,

$$\frac{\sin \theta'}{\sin(\omega - \theta')} = m' = -\frac{b'}{a'} = -\frac{A'}{B'};$$

$$\frac{\sin \theta}{\sin(\omega - \theta)} = m = -\frac{b}{a} = -\frac{A}{B}.$$

$$\tan \theta' = \frac{m' \sin \omega}{1 + m' \cos \omega}, \quad \tan \theta = \frac{m \sin \omega}{1 + m \cos \omega}.$$

$$\therefore \tan \beta = \tan(\theta' - \theta) = \frac{\tan \theta' - \tan \theta}{1 + \tan \theta' \tan \theta} = \tan(\alpha' - \alpha)$$

$$= \frac{(m' - m) \sin \omega}{1 + m'm + (m' + m) \cos \omega}$$

$$= \frac{(a'b - ab') \sin \omega}{aa' + bb' - (a'b + ab') \cos \omega}$$

$$= \frac{(AB' - A'B) \sin \omega}{AA' + BB' - (AB' + A'B) \cos \omega}.$$

48. Parallel Lines. — In this case, $\beta = \theta' - \theta = \alpha' - \alpha = 0$, and $\tan \beta = 0$; therefore,

$$(m' - m) \sin \omega = 0, \quad \text{or} \quad m' = m;$$

$$(a'b - ab') \sin \omega = 0, \quad \text{or} \quad \frac{b'}{a'} = \frac{b}{a};$$

$$(AB' - A'B) \sin \omega = 0, \quad \text{or} \quad \frac{A'}{B'} = \frac{A}{B};$$

$$\alpha' = \alpha \text{ or } = \pi + \alpha;$$

which are the same conditions as for rectangular axes, Art. 45.

49. Perpendicular Lines. — In this case,

$$\beta = \theta' - \theta = \alpha' - \alpha = \frac{\pi}{2},$$

and $\tan \beta = $ infinity; therefore,

$$1 + m'm + (m'+m) \cos \omega = 0,$$
$$aa' + bb' - (a'b + ab') \cos \omega = 0,$$
$$AA' + BB' - (AB' + A'B) \cos \omega = 0,$$

from which we get for mutually perpendicular lines,

$$m' = -\frac{1 + m \cos \omega}{m + \cos \omega},$$

$$\frac{b'}{a'} = -\frac{a - b \cos \omega}{b - a \cos \omega},$$

$$\frac{A'}{B'} = -\frac{B - A \cos \omega}{A - B \cos \omega},$$

which reduce to the corresponding results for rectangular axes by making $\omega = \frac{\pi}{2}$.

50. *To find the equation of a straight line which passes through a given point, and is perpendicular to a given line.*

First, *when the axes are rectangular.*

The equation of any line through the given point $P(x'y')$ is, Art. 36,

$$y - y' = m'(x - x'), \qquad (a)$$

and it only remains to so determine m' that this line shall be perpendicular to the given line.

For the given line let the direction parameters be

$$m = -\frac{b}{a} = -\cot \alpha = -\frac{A}{B},$$

according to the form of its equation. Therefore, for perpendicularity, Art. 46,

$$m' = -\frac{1}{m} = \frac{a}{b} = \tan\alpha = \frac{B}{A},$$

and (*a*) becomes

$$y - y' = -\frac{1}{m}(x - x') = \frac{a}{b}(x - x') = \tan\alpha(x - x') = \frac{B}{A}(x - x'), \quad (b)$$

according to the direction parameters of the given line.

Second Solution. — If the given line is

$$Ax + By + C = L = 0, \tag{a}$$

then $\quad Bx - Ay + C = M = 0 \tag{b}$

is perpendicular to $L = 0$, Art. 46.

But the equation of the line M passing through the given point $(x'y')$ is

$$M = M',$$

or $\quad Bx - Ay + C = M' = Bx' - Ay' + C,$

or $\quad y - y' = \frac{B}{A}(x - x'),$

which is the required equation.

Second, *when the axes are oblique.*

If the axes are oblique, then, by Art. 49,

$$m' = -\frac{1 + m\cos\omega}{m + \cos\omega} = \frac{a - b\cos\omega}{b - a\cos\omega} = \frac{B - A\cos\omega}{A - B\cos\omega},$$

and the required line is

$$y - y' = -\frac{1 + m\cos\omega}{m + \cos\omega}(x - x');$$

and so for the other parameters of the given line.

LINES MAKING GIVEN ANGLES WITH GIVEN LINES. 99

51. *To find the equations of the two straight lines which pass through a given point and make a given angle with a given line.*

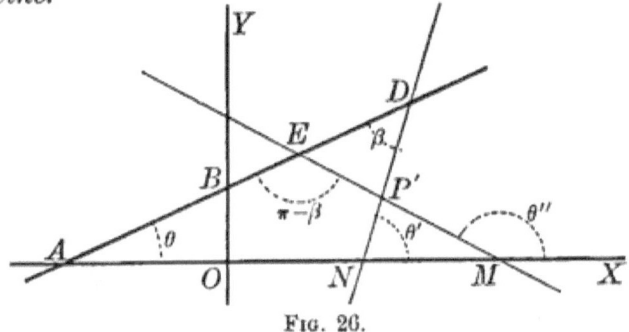

Fig. 26.

First, *when the axes are rectangular.*

Let AB be the given line, $P'(x'y')$ the given point, and β the given angle which the required lines ND and ME make with the given line.

For the line ND, $\theta' = \theta + \beta$, and for the line ME, $\theta'' = \theta + \pi - \beta$; therefore, by Trigonometry, Art. 63, for ND,

$$\tan\theta' = \tan(\theta+\beta) = \frac{\tan\theta + \tan\beta}{1 - \tan\theta\tan\beta} = \frac{m + \tan\beta}{1 - m\tan\beta};$$

for ME, $\quad \tan\theta'' = \tan(\theta+\pi-\beta) = \dfrac{\tan\theta - \tan\beta}{1 + \tan\theta\tan\beta} = \dfrac{m - \tan\beta}{1 + m\tan\beta}.$

But the equation of any line through $P'(x'y')$ is

$$y - y' = m'(x - x');$$

therefore the equation of ND is

$$y - y' = \frac{m + \tan\beta}{1 - m\tan\beta}(x - x'),$$

and of ME is

$$y - y' = \frac{m - \tan\beta}{1 + m\tan\beta}(x - x').$$

If $\beta = 90°$, the equations of both ND and ME reduce to

$$y - y' = -\frac{1}{m}(x - x'),$$

as they should, since the lines coincide, and are then perpendicular to the given line AB.

Second Solution. — If the equation of the given line AB is

$$x \sin \theta - y \cos \theta - p = P = 0, \tag{a}$$

then $\quad x \sin(\theta \pm \beta) - y \cos(\theta \pm \beta) - p' = P$

is the equation of two systems of parallels, each making the given angle β with the given line $P = 0$.

For the line of each system passing through the given point $(x'y')$,

$$x \sin(\theta \pm \beta) - y \cos(\theta \pm \beta) - p' = P' = x' \sin(\theta \pm \beta) - y' \cos(\theta \pm \beta) - p',$$

or, $\quad y - y' = \tan(\theta \pm \beta)(x - x'),$

which are the equations of ND and ME as already found.

Second, *when the axes are oblique.*

In this case, we get from Art. 47, using $\pm \tan \beta$,

$$m' = \frac{m \sin \omega \pm (1 + m \cos \omega) \tan \beta}{\sin \omega \mp (m + \cos \omega) \tan \beta}$$

$$= -\frac{b \sin \omega \mp (a - b \cos \omega) \tan \beta}{a \sin \omega \pm (b - a \cos \omega) \tan \beta}$$

$$= -\frac{A \sin \omega \mp (B - A \cos \omega) \tan \beta}{B \sin \omega \pm (A - B \cos \omega) \tan \beta}.$$

$$\therefore \ y - y' = \frac{m \sin \omega \pm (1 + m \cos \omega) \tan \beta}{\sin \omega \mp (m + \cos \omega) \tan \beta}(x - x')$$

are the equations of ND and ME, and so for other values of the parameter m'.

For $\omega = 90°$, these equations reduce to those given for rectangular axes.

For $\beta = 90°$, the lines ND and ME are perpendicular to AB, and their equations reduce to
$$y - y' = -\frac{1 + m \cos \omega}{m + \cos \omega}(x - x'),$$
as found in Art. 50.

Intersection of Lines.

52. *To find the coordinates of the point of intersection of two given straight lines.*

Let the equations of the two given lines be
$$Ax + By + C = 0, \quad A'x + B'y + C' = 0.$$

Let $(x'y')$ be the point in which these lines intersect; then $(x'y')$ is on both lines and must satisfy both equations, and therefore
$$Ax' + By' + C = 0, \quad A'x' + B'y' + C' = 0.$$

By solving these equations, we get
$$x' = \frac{BC' - B'C}{AB' - A'B}, \quad y' = \frac{A'C - AC'}{AB' - A'B},$$
the required coordinates.

If $AB' - A'B = 0$, or $\frac{A}{B} = \frac{A'}{B'}$, then x' and y' are both infinite, and the lines intersect at *infinity;* or, in other words, are parallel, since their direction parameters are proportional (Art. 45).

If $AB' - A'B = 0$, and also $BC' - B'C = 0$, $A'C - AC' = 0$; that is, if $\frac{A}{A'} = \frac{B}{B'} = \frac{C}{C'}$,

then the values of x' and y' are both $\frac{0}{0}$, or indeterminate, and the two lines meet in all points on each, or are coincident lines.

Therefore, *two straight lines are coincident when their corresponding parameters have the same ratio.*

EXERCISES ON ANGLES BETWEEN GIVEN LINES, AND ON THEIR INTERSECTIONS.

1. Find the angle which the lines $3x - y = 5$, $2y - x = 8$ make with each other. *Ans.* $45°$.

2. Show that the lines $4y - 3x = 8$, $6x + 8y = 25$ make supplementary angles with the axis of x; and the angle $\tan^{-1} \frac{24}{7}$ with each other.

3. The sides of a triangle are $x + 7y + 11 = 0$, $3y - x = 1$, $3x + y = 7$; find the acute angles which they make with each other, and the coordinates of the vertices.

Ans. $90°$, $\tan^{-1} 2$, $\tan^{-1} \frac{1}{2}$; $(2, 1)$, $(3, -2)$, $(-4, -1)$.

4. Do the lines $2x + 3y = 13$, $5x - y = 7$, $x - 4y + 10 = 0$ meet in a point? Find the acute angles which they make with each other.

Ans. $\tan^{-1} \frac{17}{7}$, $\tan^{-1} \frac{11}{10}$, $\tan^{-1} \frac{19}{9}$.

5. Find the equations of the two lines which pass through $(-3, 1)$ and each make an angle $\tan^{-1} \frac{3}{4}$ with the line $2x - 3y + 7 = 0$.

Ans. $6y - 17x - 57 = 0$, $18y + x - 15 = 0$.

6. Find the equations of the lines passing through the point $(3, -1)$, and making angles of $45°$ with $2y + 3x = 6$.

Ans. $5y + x + 2 = 0$, $y - 5x + 16 = 0$.

7. Find the equations of the two straight lines which pass through the point $(2, 3)$ and make angles of $30°$ with the given line $3x + 5y - 2 = 0$, when $\omega = 60°$.

Ans. $8y + x - 26 = 0$, $7y + 8x - 37 = 0$.

8. Find the equation of a line passing through $(3, -1)$ parallel to $2y + 3x - 6 = 0$.

Ans. $2y + 3x - 6 = L' = -2 + 9 - 6$, or $2y + 3x - 7 = 0$.

9. Find the equation of a line passing through $(-3, 5)$ perpendicular to $3y - 2x - 2 = 0$.

Ans. $2y + 3x - 2 = L' = 2 \cdot 5 - 3 \cdot 3 - 2$, or $2y + 3x - 1 = 0$.

EXERCISES ON ANGLES AND INTERSECTIONS.

10. When are the lines
$$x + (a+b)y + c = 0 \text{ and } a(x + ay) + b(x - by) + d = 0$$
parallel? When perpendicular? *Ans.* For $b = 0$; for $a^2 - b^2 + 1 = 0$.

11. A point xy so moves that its perpendicular distance from the line $5x + 12y - 2 = 0$ is always ± 5; find the equations of the loci.
Ans. $5x + 12y - 67 = 0$, or $5x + 12y + 63 = 0$.

12. A point xy so moves that its perpendicular distances from the two lines $5x + 12y + 5 = 0$, $15x - 8y - 3 = 0$ are always equal; find the equations of the loci.
Ans. $110x - 308y - 124 = 0$, $280x + 100y + 46 = 0$.

Show that these loci are perpendicular to each other.

Show that these loci pass through the intersection of the given lines.

13. A point xy so moves that its perpendicular distances from the lines $P = 0$, $Q = 0$ are always equal; find the equations of the loci.
Ans. $P = Q$, $P = -Q$, or $P - Q = 0$, $P + Q = 0$.

Show that these loci pass through the intersection of $P = 0$, $Q = 0$.

14. Find the points on the line $x + 2y - 15 = 0$ whose perpendicular distances from the line $3x - 4y + 5 = 0$ are 2.
Ans. $(7, 4)$, $(3, 6)$.

15. Find the point equidistant from the three lines
$$4x + 3y - 7 = 0, \quad 5x + 12y - 20 = 0, \quad 3x + 4y - 8 = 0,$$
and its distance from each. *Ans.* $(2, 3)$; 2.

16. Given the vertices $(2, 1)$, $(-4, 3)$, $(6, -5)$ of a triangle.

(1) Find the equations of the sides.

(2) Find the equations of the medial lines.

(3) Find the equations of the lines passing through the vertices perpendicular to the opposite sides.

(4) Find the equations of the lines perpendicular to the three sides through their middle points.

(5) Show that the lines of each set (2), (3), (4), meet in a point, and find the coordinates of these points.

(6) Show that the three points thus found lie on a straight line, and find its equation.

Ans. (1) $5y+4x+1=0$; $2y+3x-8=0$; $3y+x-5=0$.

(2) $y-2x+3=0$; $8y+5x-4=0$; $y+x-1=0$.

(3) $4y-5x+6=0$; $3y-2x-17=0$; $y-3x+23=0$.

(4) $4y-5x+9=0$; $3y-2x+14=0$; $y-3x-5=0$.

(5) $(\frac{4}{3}, -\frac{1}{3})$; $(\frac{86}{7}, \frac{97}{7})$; $(-\frac{29}{7}, -\frac{52}{7})$.

(6) $805y-1043x+1659=0$.

17. Given the vertices $(x'y')$, $(x''y'')$, $(x'''y''')$ of a triangle.

(1) Find the tangents of the angles which the sides make with the axis of X.

(2) Find the condition that the angle at $(x'y')$ shall be a right angle.

(3) Find the condition that the sides through $(x'y')$ shall make supplementary angles with the axis of X.

(4) Find the equations of the lines through the vertices parallel to the opposite sides.

(5) Find the equations of the lines through the vertices perpendicular to the opposite sides.

(6) Find the equations of the lines perpendicular to the sides through their middle points.

Ans.

(2) $(x'-x'')(x'-x''') + (y'-y'')(y'-y''') = 0$.

(3) $(x'-x''')(y'-y'') + (x'-x'')(y'-y''') = 0$.

(4) $y(x''-x''') - x(y''-y''') - y'(x''-x''') + x'(y''-y''') = 0$;
$y(x'''-x') - x(y'''-y') - y''(x'''-x') + x''(y'''-y') = 0$;
$y(x'-x'') - x(y'-y'') - y'''(x'-x'') + x'''(y'-y'') = 0$.

EXERCISES ON ANGLES AND INTERSECTIONS. 105

(5) $y(y'' - y''') + x(x'' - x''') - y'(y'' - y''') - x'(x'' - x''') = 0$;

$y(y''' - y') + x(x''' - x') - y''(y''' - y') - x''(x''' - x') = 0$;

$y(y' - y'') + x(x' - x'') - y'''(y' - y'') - x'''(x' - x'') = 0$.

(6) $y(y'' - y''') + x(x'' - x''') = \frac{1}{2}(x''^2 - x'''^2) + \frac{1}{2}(y''^2 - y'''^2)$;

$y(y''' - y') + x(x''' - x') = \frac{1}{2}(x'''^2 - x'^2) + \frac{1}{2}(y'''^2 - y'^2)$;

$y(y' - y'') + x(x' - x'') = \frac{1}{2}(x'^2 - x''^2) + \frac{1}{2}(y'^2 - y''^2)$.

18. Take the side through $(x''y'')$, $(x'''y''')$ as the axis of X, and the perpendicular through $(x'y')$ as the axis of Y; then the coordinates of the vertex are $(0, y')$, and those of the base angles are $(-x'', 0)$ and $(x''', 0)$.

(1) Find the equations of the three sides.

(2) Find the tangents of the three angles of the triangle.

(3) Find the equations of the three medial lines, and the point in which they meet.

(4) Find the equations of the three perpendiculars through the vertices on the opposite sides, and the point in which they meet.

(5) Find the equations of the perpendiculars to the sides through their middle points, and the point in which they meet.

(6) Show that the points found in (3), (4), (5) lie on the same straight line; find its equation.

Ans. (1) $xy' + yx''' - y'x''' = 0$; $xy' - yx'' + y'x''' = 0$; $y = 0$.

(2) $\dfrac{y'}{x''}$; $\dfrac{y'}{x'''}$; $\dfrac{y'(x'' + x''')}{x''x''' - y'^2}$.

(3) $(x''' - x'')y + 2y'x - y'(x''' - x'') = 0$;

$(x'' + 2x''')y + y'x - y'x''' = 0$;

$(x''' + 2x'')y - y'x - y'x'' = 0$; $\left(\dfrac{x''' - x''}{3}, \dfrac{y'}{3}\right)$.

(4) $y'y - x'''x - x''x''' = 0$; $y'y + x''x - x''x''' = 0$;

$x = 0$; $\left(0, \dfrac{x''x'''}{y'}\right)$.

(5) $2y'y + 2x''x + x''^2 - y'^2 = 0$; $2y'y - 2x'''x + x'''^2 - y'^2 = 0$;

$2x + x'' - x''' = 0$; $\left(\dfrac{x''' - x''}{2}, \dfrac{y'^2 - x''x'''}{2y'}\right)$.

(6) $y'(x''' - x'')y + (3x''x''' - y'^2)x - x''x'''(x''' - x'') = 0$.

Lines passing through the Intersection of Two Given Lines.

53. *To find the equation of any straight line which passes through the point of intersection of two given straight lines.*

In Art. 52 we found the coordinates $(x'y')$ of the point in which the two given lines

$$Ax + By + C = L = 0, \quad A'x + B'y + C' = M = 0$$

intersect; then, by Art. 36,

$$y - y' = m(x - x')$$

is the required equation of any line passing through the point $(x'y')$.

Second Solution. — Let l and m be any two arbitrary factors; then

$$lL + mM = 0, \qquad (a)$$

or $\qquad l(Ax + By + C) + m(A'x + B'y + C') = 0$,

is the required equation; first, because it is of the first degree, in x and y, and thus represents some straight line; and second, because $(L = 0, M = 0)$, or $(x'y')$ the coordinates of the point in which the two given lines intersect, reduce both of its terms to zero, and thus satisfy it.

If we put $m : l = k$, then equation (a) can be written $L + kM = 0$. Since k is arbitrary, we may find the line fulfilling one additional condition, such as passing through a given point, or making a given angle with the axis of X.

LINES THROUGH INTERSECTIONS OF GIVEN LINES. 107

The lines $L=0$ and $M=0$ make two angles with each other which are supplementary; the one in which the origin lies is called the *internal angle*, and the other the *external angle*. If $L+kM=0$ represents all lines lying in the external angle, then $L-kM=0$ will represent all lines lying in the internal angle, since for $k=0$ we have $L=0$, and k must change signs in passing through zero.

For example, if the line, besides passing through the point $(L=0, M=0)$, or $(x'y')$, must also pass through the given point $(x''y'')$, then the condition for determining k is

$$L''+kM''=0, \text{ or } k=-\frac{L''}{M''}=-\frac{Ax''+By''+C}{A'x''+B'y''+C'},$$

and the required equation is

$$M''L - L''M = 0,$$

or $\quad Ax+By+C-\dfrac{Ax''+By''+C}{A'x''+B'y''+C'}(A'x+B'y+C')=0.$

If the line passing through the point $(L=0, M=0)$ must also pass through the origin, then $(x''y'')$ becomes $(0,0)$, and the equation just found reduces to

$$(AC'-A'C)x+(BC'-B'C)y=0.$$

54. *To find the equations of the two lines which bisect the angles made by two given lines.*

Let the equations of the two given lines MC and $M'C$ be

$$x\cos a + y\sin a - p = P = 0, \tag{a}$$

$$x\cos a' + y\sin a' - p' = Q = 0. \tag{b}$$

The line BC which bisects the angle MCM' in which the origin lies is called the *internal bisector;* the line BC which bisects the angle NCN', supplementary to MCM', is called the **external bisector.**

The origin and any point $B(xy)$ on the internal bisector always lie on the same or on opposite sides of both MC and $M'C$, and therefore the perpendiculars BM and BM' always have the same sign (Art. 42); while the origin and any point $B(xy)$ on the external bisector always lie on the same side of one of the lines and on the opposite sides of the other, and therefore the perpendiculars BN and BN' always have opposite signs.

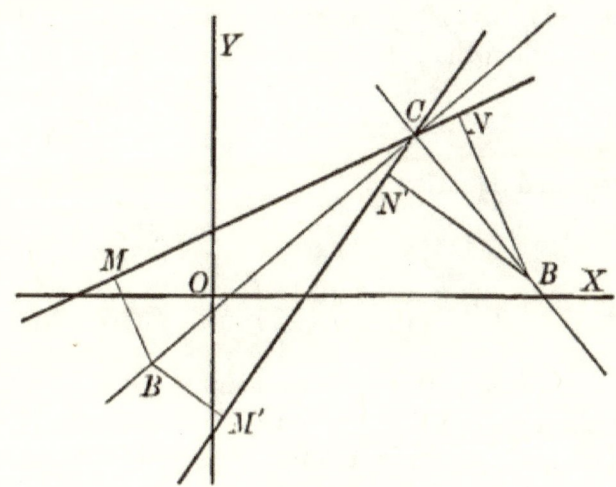

Fig. 27.

The length of the perpendicular from any point $B(xy)$ to the line $MC = P$, and to the line $MC' = Q$, Art. 42. Since these perpendiculars are equal for all points $B(xy)$ on the bisectors, we have

$$P = Q, \text{ or } P - Q = 0,$$

for the equation of the internal bisector, and

$$P = -Q, \text{ or } P + Q = 0,$$

for the equation of the external bisector.

If the perpendiculars P and Q are not equal, but proportional to the two arbitrary constants p and q, then

$$\frac{P}{p} = \frac{Q}{q}, \text{ or } P - \frac{p}{q}Q = 0, \text{ or } P - kQ = 0.$$

is the equation of any line lying in the internal angle MCM', and

$$\frac{P}{p} = -\frac{Q}{q}, \text{ or } P + \frac{p}{q}Q = 0, \text{ or } P + kQ = 0,$$

is the equation of any line lying in the external angle NCN'.

Of the two lines $P - kQ = 0$ and $Q - kP = 0$, which correspond to reciprocal values of k, $P - kQ = 0$ makes the same angle with $P = 0$ that $Q - kP = 0$ makes with $Q = 0$, since their bisectors are

$$P - kQ \mp (Q - kP) = P \mp Q = 0,$$

the same as the bisectors of $P = 0$, $Q = 0$.

If the equations of the given lines are

$$Ax + By + C = L = 0, \quad A'x + B'y + C' = M = 0,$$

then the equations of the bisectors are

$$\frac{Ax + By + C}{\sqrt{A^2 + B^2}} \mp \frac{A'x + B'y + C'}{\sqrt{A'^2 + B'^2}} = 0.$$

If the axes are oblique, then

$$\frac{Ax + By + C}{\sqrt{A^2 + B^2 - 2AB\cos\omega}} \mp \frac{A'x + B'y + C'}{\sqrt{A'^2 + B'^2 - 2A'B'\cos\omega}} = 0$$

or $\quad x\cos a + y\cos(\omega - a) - p \mp [x\cos a' + y\cos(\omega - a') - p'] = 0.$

are the equations of the bisectors.

The equations of these bisectors may be written

$$\frac{L}{A}\cos a \mp \frac{M}{A'}\cos a' = 0, \text{ or } \frac{L}{A}\sin\theta \mp \frac{M}{A'}\sin\theta' = 0,$$

both for rectangular and oblique axes.

110 PLANE ANALYTIC GEOMETRY.

55. *To construct the equations* $P \mp kQ = 0$ *and* $L \mp kM = 0$.

I. The equations $P \mp kQ = 0$. Take the two lines CP_0 and CQ_0 (Fig. 28), whose equations are

$$x \cos a + y \sin a - p = P = 0, \quad x \cos a' + y \cos a' - p' = Q = 0,$$

as coordinate axes, and their intersections $C\,(P = 0,\ Q = 0)$ as the origin, a point through which the required lines must pass.

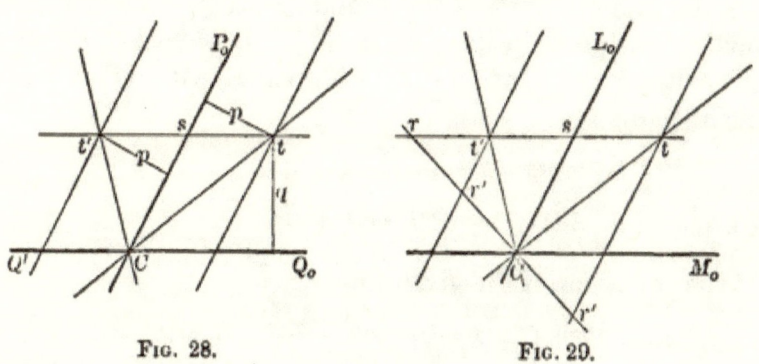

Fig. 28. Fig. 29.

The perpendicular coordinates of the point of intersection of any two lines P and Q, respectively, parallel to the axes, are (P, Q), Art. 42. If $k = \pm p : q$, the points $t\,(p, q)$ and $t'(-p, q)$ are the intersections of the parallels p and $-p$ to $P = 0$, by q the parallel to $Q = 0$. Draw the required lines Ct and Ct'.

If (P, Q) are the corresponding coordinates of any other points on these lines, then, for Ct,

$$\frac{P}{Q} = k = \frac{p}{q} = \frac{p}{Ct} : \frac{q}{Ct} = \frac{\sin tCP_0}{\sin tCQ_0},$$

which shows that k is the ratio of the sines of the two angles into which the line Ct divides the internal angle P_0CQ_0.

For Ct',

$$\frac{P}{Q} = -k = -\frac{p}{q} = -\frac{p}{Ct'} : \frac{q}{Ct'} = \frac{\sin t'CP_0}{\sin t'CQ_0},$$

which shows that $-k$ is the ratio of the sines of the two angles into which the line Ct' divides the external angle $P_0CQ'_0$. If we suppose the line Ct to revolve into the position Ct', we see that p equals zero and changes sign as Ct coincides with and passes $P=0$, and that the angles tCP_0 and $t'CP_0$ are measured in opposite directions. It also appears that $ts = st'$, and that $P=0$ bisects the portion of any parallel to $Q=0$, intercepted between Ct and Ct'.

When Ct and Ct' are bisectors, $q = \pm p$, $k = \pm 1$, and their equations are $P \mp Q = 0$.

II. The equations $L \mp kM = 0$. Take the two lines CL_0 and CM_0 (Fig. 29), whose equations are

$$Ax + By + C = L = 0, \quad A'x + B'y + C' = M = 0,$$

as axes, and their intersection $C(L=0, M=0)$ as the origin, a point through which the lines must pass. Let $k = \pm m : l$, instead of $\pm l : m$, to conform to the notation of Art. 53. The ratio of the distances, measured on the axis of X, of the parallels L and m from $L=0$, and of M and l from $M=0$, are $L:m$ and $M:l$, Art. 43. To find the points ($\pm m : l$). Draw through the origin a parallel rr' to the axis of X, on which lay off $Cr' = \pm m$ and $Cr = l$. Through the points r' draw parallels to $L=0$, and through r a parallel to $M=0$. The points t and t' thus found are on the required lines Ct and Ct', whose equations are

$$\frac{L}{\pm m} = \frac{M}{l}, \text{ or } L \mp kM = 0.$$

As before, the intercept tt' on the parallel to $M=0$ is bisected by $L=0$. These equations may be written

$$\frac{L}{A} q \sin \theta \mp \frac{M}{A'} p \sin \theta' = 0.$$

When Ct and Ct' are bisectors, $q = \pm p$, and their equations are

$$\frac{L}{A} \sin \theta \mp \frac{M}{A'} \sin \theta' = 0.$$

56. *To find the condition that three given straight lines shall intersect in the same point.*

Let their equations be

$$Ax + By + C = L = 0, \qquad (a)$$

$$A'x + B'y + C' = M = 0, \qquad (b)$$

$$A''x + B''y + C'' = N = 0. \qquad (c)$$

The coordinates of the point of intersection of either two must satisfy the third, if all three pass through the same point. Substitute the coordinates $(x'y')$ of the point of intersection of (a) and (b), Art. 52, in (c), and

$$\frac{A''(BC'-B'C)}{AB'-A'B} + \frac{B''(A'C-AC')}{AB'-A'B} + C'' = 0, \qquad (d)$$

or $\quad A''(BC'-B'C) + B''(A'C-AC') + C''(AB'-A'B) = 0,$

or $\quad A(B'C''-B''C') + B(C'A''-C''A') + C(A'B''-A''B') = 0,$

or $\quad A(B'C''-B''C') + A'(B''C-BC'') + A''(BC'-B'C) = 0,$

are different forms of the required equation of condition.

This is the relation which must exist between the nine parameters of the three given lines if they pass through the same point.

Second Solution. — We already know, Art. 53, that

$$lL + mM = 0,$$

or $\quad (lA + mA')x + (lB + mB')y + lC + mC' = 0,$

represents any line passing through the point $(L=0, M=0)$. If now the third line

$$A''x + B''y + C'' = N = 0$$

is coincident with one of the lines $lL + mM = 0$, it will also pass through the point $(L=0, M=0)$.

But for coincidence, Art. 52,

$$\frac{lA + mA'}{A''} = \frac{lB + mB'}{B''} = \frac{lC + mC'}{C''} = -n,$$

$-n$ being any arbitrary constant. It follows then that if values of l, m, n can be found which satisfy the identity $lL + mM + nN = 0$, or the equivalent equations of condition,

$$lA + mA' + nA'' = 0,$$
$$lB + mB' + nB'' = 0,$$
$$lC + mC' + nC'' = 0,$$

the three given lines will pass through the same point.

If we eliminate l, m, n from these equations of condition, the resulting equation will express the required relation (d) already found between the nine parameters of the three given lines.

If, however, values of l, m, n cannot be found which satisfy the above equations of condition, then the line $N = 0$ will not pass through the point ($L = 0$, $M = 0$), but the three given lines $L = 0$, $M = 0$, $N = 0$ will form a triangle.

In this case the line $lL + mM = 0$ passing through the vertex ($L = 0$, $M = 0$) will intersect the opposite side $N = 0$ in the point*($lL + mM. N$), and the equation

$$lL + mM + nN = 0,$$

instead of being indeterminate, will represent some straight line passing through this point of intersection.

But the symmetry of the equation shows that this line, which is called a *transversal*, will also intersect the side $L = 0$, in the point ($mM + nN. L$), and the side $M = 0$ in the point ($nN + lL. M$). If, then, we construct the triangle given by the lines $L = 0$, $M = 0$, $N = 0$, and then for given values of l, m, n construct the lines $lL + Mm = 0$, $mM + nN = 0$, $nN + lL = 0$,

* The point ($lL + mM = 0$, $N = 0$). This notation is used for brevity.

they will meet the opposite sides; the three points thus found will lie on the transversal whose equation is

$$lL + mM + nN = 0.$$

Assuming, as we may, that the origin of the xy coordinates is within the triangle, and its angles are internal, then for all positive values of l, m, n the corresponding lines

$$lL + mM = 0, \quad mM + nN = 0, \quad nN + lL = 0,$$

through the vertices of the triangle will all lie in the external angles, and the transversal

$$lL + mM + nN = 0$$

will cut all the sides produced, or externally; while the transversals

$$lL + mM - nN = 0, \quad lL - mM + nN = 0, \quad -lL + mM + nN = 0,$$

will each cut the respective sides $N = 0$, $M = 0$, $L = 0$ externally, and the other two internally.

If the equations of the sides of the triangle are given in their normal forms $P = 0$, $Q = 0$, $R = 0$, then the above transversal may be written

$$lP + mQ + nR = 0,$$

which is called the *trilinear equation** of the straight line. In this case the three sides of the triangle are taken as axes, and (P, Q, R) are the perpendicular coordinates of any point on the line, since, by Art. 42, P, Q, R respectively represent the lengths of the perpendiculars from any point (xy) on the three axes whose equations are $P = 0$, $Q = 0$, $R = 0$.

---

* The student who wishes to see a full discussion of the method of trilinear coordinates, with applications, may consult a work on "Trilinear Coordinates," by N. M. Ferrers; or "Trilinear Coordinates and Other Methods of Modern Analytic Geometry of Two Dimensions," by William Allen Whitworth; or Salmon's "Conic Sections."

57. *To find the ratio of the segments into which the line $Ax + By + C = L = 0$ divides the line joining the two given points $(x'y')$ and $(x''y'')$.*

The distances between the parallels to $L = 0$ through the given points $(x'y')$, $(x''y'')$, measured on the axis of X, are $\dfrac{L'}{A}$ and $\dfrac{L''}{A}$, Art. 43. Since $L = 0$ divides the distance between the given points $(x'y')$, $(x''y'')$ into segments proportional to $\dfrac{L'}{A}$ and $\dfrac{L''}{A}$, the required ratio is $\pm \dfrac{L'}{L''}$, according as the section is internal or external, Art. 5.

58. *A transversal cuts the sides of a triangle opposite to the angles $A(x'y')$, $B(x''y'')$, $C(x'''y''')$ in the points A', B', C'. Prove that $\dfrac{AC' \cdot BA' \cdot CB'}{C'B \cdot A'C \cdot B'A} = -1.$*

Let $T = 0$ be the equation of the transversal; then, by Art. 57,

$$\frac{AC'}{C'B} = \pm \frac{T'}{T''}, \quad \frac{BA'}{A'C} = \pm \frac{T''}{T'''}, \quad \frac{CB'}{B'A} = \pm \frac{T'''}{T'};$$

and since the number of external sections must be odd, we have

$$\frac{AC' \cdot BA' \cdot CB'}{C'B \cdot A'C \cdot B'A} = -\frac{T'}{T''} \cdot \frac{T''}{T'''} \cdot \frac{T'''}{T'} = -1.$$

59. *Lines passing through any assumed point $O(x^{\mathrm{IV}} y^{\mathrm{IV}})$ and the vertices $A(x'y')$, $B(x''y'')$, $C(x'''y''')$ of a triangle cut the opposite sides in the points A', B', C'. Prove that $\dfrac{AC' \cdot BA' \cdot CB'}{C'B \cdot A'C \cdot B'A} = +1.$*

The line through $O(x^{\mathrm{IV}} y^{\mathrm{IV}})$ and $A(x'y')$ is, Art. 39,

$$(y' - y^{\mathrm{IV}})x - (x' - x^{\mathrm{IV}})y + x'y^{\mathrm{IV}} - x^{\mathrm{IV}}y' = 0;$$

and it cuts the opposite side BC in the ratio

$$\frac{BA'}{A'C} = \pm \frac{(y'-y^{\text{iv}})x'' - (x'-x^{\text{iv}})y'' + x'y^{\text{iv}} - x^{\text{iv}}y'}{(y'-y^{\text{iv}})x''' - (x'-x^{\text{iv}})y''' + x'y^{\text{iv}} - x^{\text{iv}}y'}$$

$$= \pm \frac{x'(y''-y^{\text{iv}}) + x''(y^{\text{iv}}-y') + x^{\text{iv}}(y'-y'')}{x'(y^{\text{iv}}-y''') + x^{\text{iv}}(y'''-y') + x'''(y'-y^{\text{iv}})}.$$

In like manner,

$$\frac{CB'}{B'A} = \pm \frac{x''(y'''-y^{\text{iv}}) + x'''(y^{\text{iv}}-y'') + x^{\text{iv}}(y''-y''')}{x'(y''-y^{\text{iv}}) + x''(y^{\text{iv}}-y') + x^{\text{iv}}(y'-y'')},$$

$$\frac{AC'}{C'B} = \pm \frac{x'(y^{\text{iv}}-y''') + x^{\text{iv}}(y'''-y') + x'''(y'-y^{\text{iv}})}{x''(y'''-y^{\text{iv}}) + x'''(y^{\text{iv}}-y'') + x^{\text{iv}}(y''-y''')}.$$

Since the number of external sections is even, we have

$$\frac{AC' \cdot BA' \cdot CB'}{C'B \cdot A'C \cdot B'A} = +1.$$

60. In the following elementary propositions relating to the triangle, let

A, B, C, denote the angles;

a, b, c, the lengths of the opposite sides;

$P = 0, Q = 0, R = 0$, the normal equations of these sides;

(PQR), the trilinear coordinates of any point (xy) in the plane of the triangle, all positive when the point is within the triangle; when the point, in moving from within to without the triangle crosses a side, the corresponding coordinate will change sign to minus.

$\triangle$, the area of the triangle.

I. *To find the area of the triangle ABC in terms of the trilinear coordinates (PQR) of any point in the plane of the triangle.*

Join the point with each vertex. The algebraic sum of the three triangles thus formed is equal to the given triangle.

The double areas of these triangles are aP, bQ, cR, and therefore

$$aP + bQ + Cr = 2\Delta = 2\sqrt{s(s-a)(s-b)(s-c)} \quad \text{(Trig., Art. 150)}$$

$$= y'(x''-x''') + y''(x'''-x') + y'''(x'-x''), \text{(Art. 7)}$$

which expresses the constant relation between the trilinear coordinates of any point.

Since $\dfrac{a}{\sin A} = \dfrac{b}{\sin B} = \dfrac{c}{\sin C} = k,$ (Trig., Art. 144)

this relation may be written

$$aP + bQ + cR = 2\Delta = k(P\sin A + Q\sin B + r\sin C).$$

II. *To find the radii and the centres of the inscribed, circumscribed, and escribed circles, in terms of trilinear coordinates.*

Let r_i and R_c denote the radii of the inscribed and circumscribed circles, and r_a, r_b, r_c, the radii of the escribed circles which are tangents externally to the sides a, b, c, respectively.

By Trigonometry, Arts. 151, 152, 153,

$$r_i = \frac{\Delta}{s}, \quad R_c = \frac{abc}{4\Delta}, \quad r_a = \frac{\Delta}{s-a}, \quad r_b = \frac{\Delta}{s-b}, \quad r_c = \frac{\Delta}{s-c},$$

in which $2s = a + b + c$. For the inscribed circle

$r_i = P$, centre is $[P = Q = R.]$

For the circumscribed circle the radius

$$R_c = P\sec A = Q\sec B = R\sec C.$$

For the escribed circles

$r_a = -P$, centre is $[-P = Q = R]$;

$r_b = -Q$, centre is $[P = -Q = R]$;

$r_c = -R$, centre is $[P = Q = -R]$.

III. *To find the equations of the internal and external bisectors of the angles A, B, C.*

The required equations are

$$P \pm Q = 0, \quad Q \pm R = 0, \quad R \pm P = 0. \qquad \text{(Art. 54)}$$

A simple solution is as follows:

By Geometry, the centre of the inscribed circle is at the intersection of the three internal bisectors of the angles; and the centres of the escribed circles are at the intersections of the external bisectors of two of the angles and the internal bisector of the third angle.

But by II., the coordinates of the centre of the inscribed circle are $P = Q = R$; or

$$P - Q = 0, \quad Q - R = 0, \quad R - P = 0,$$

are the equations of the internal bisectors, since each is satisfied by the coordinates of a vertex and of this centre.

The coordinates of the centre of the escribed circle r_a are $-P = Q = R$; or

$$P + Q = 0, \quad R + P = 0, \quad Q - R = 0,$$

are the equations of two external bisectors and an internal bisector, since each is satisfied by the coordinates of a vertex and of this centre.

For the centres, r_b and r_c, the equations are

$$P + Q = 0, \quad Q + R = 0, \quad R - P = 0,$$
$$Q + R = 0, \quad R + P = 0, \quad P - Q = 0.$$

IV. *Each internal bisector bisects the angle between the other two external bisectors.*

The internal bisector of the angle between the external bisectors $P + Q = 0$ and $R + P = 0$ is

$$P + Q - (R + P) = Q - R = 0,$$

the third internal bisector; and so for the others, which proves the proposition.

V. *To find the equations of the lines drawn through the vertices and the centre of the circumscribed circle.*

For the centre, $P \sec A = Q \sec B = R \sec C$, by II., and the required equations are

$$P \sec A - Q \sec B = 0, \quad Q \sec B - R \sec C = 0, \quad R \sec C - P \sec A = 0,$$

since each is also satisfied by the coordinates of a vertex.

VI. *To find the equations of the medial lines in trilinear coordinates.*

The medials divide the triangle into two equal parts. The coordinates of the middle points of the sides a, b, c are $(0, Q, R)$, $(P, 0, R)$, $(P, Q, 0)$, and the equality of the triangles give

$$bQ - cR = 0, \quad cR - aP = 0, \quad aP - bQ = 0,$$

the required equations of the medials.

These equations may be written (Trig., Art. 144)

$$Q \sin B - R \sin C = 0, \quad R \sin C - P \sin A = 0, \quad P \sin A - Q \sin B = 0.$$

These medials pass through the same point, since the sum of their equations is zero (Art. 56).

VII. *To find the equations of the lines drawn through the vertices perpendicular to the opposite sides.*

By Art. 55,

$$\frac{P}{Q} = \frac{\sin BCC''}{\sin ACC''} = \frac{\cos B}{\cos A}.$$

Therefore

$P \cos A - Q \cos B = 0$ is the equation of CC'';

similarly

$Q \cos B - R \cos C = 0$ is the equation of AA',

and $R \cos C - P \cos A = 0$ is the equation of BB'.

These lines pass through the same point since the sum of their equations is zero (Art. 56).

VIII. *To find equations of the transversals which pass through the points in which the bisectors of the angles meet the opposite sides.*

First, the external bisectors. The external bisector $P+Q=0$ meets the opposite side $R=0$ in the point $(P+Q, R)$, and the equation
$$P+Q+R=0$$
represents a line passing through this point.

But the symmetry of this equation shows that the line also passes through the points $(Q+R, P)$ and $(R+P, Q)$, and is therefore the required equation. All three points are points of external section.

Second, the internal bisectors. In this case the transversals are
$$P-Q+R=0, \quad Q-R+P=0, \quad R-P+Q=0,$$
each of which passes through two points of internal and one of external section.

IX. *To find the equations of the transversals which pass through the points in which the lines drawn through the vertices and the centre of the circumscribed circle meet the opposite sides.*

The line $Q \sec B - R \sec C = 0$, through $A(Q=0, R=0)$ and the centre of the circumscribed circle meets the opposite side in the point $(Q \sec B - R \sec C, P)$, and
$$Q \sec B - R \sec C + P \sec A = 0 \qquad (a)$$
is the equation of some line passing through this point. But the equation shows that the line also passes through $(P \sec A - R \sec C, Q)$, the point in which the line through $B(P=0, R=0)$ and the centre meets the opposite side. Therefore (a)

ELEMENTARY PROPOSITIONS ON TRIANGLE. 121

is the equation of one of the transversals; the equations of the other two are readily found to be

$$Q\sec B + R\sec C - P\sec A = 0, \quad -Q\sec B + R\sec C + P\sec A = 0.$$

X. *To find the equations of the transversals which pass through the points in which the medials meet the opposite sides.*

The medial $aP - bQ = 0$ meets the opposite side $R = 0$ in the point $(aP - bQ, R)$, and the equation

$$aP - bQ + cR = 0$$

represents a line passing through this point. But this equation shows that the line also passes through the point $(cR - bQ, P)$ of internal section, and through the point $(aP + cR, Q)$ of external section. Therefore

$$aP - bQ + cR = 0$$

is the required equation of the transversal.

The other two are

$$aP + bQ - cR = 0, \quad -aP + bQ + cR = 0.$$

These equations may be written (Trig., Art. 144)

$$P\sin A + Q\sin B - R\sin C = 0,$$
$$P\sin A - Q\sin B + R\sin C = 0,$$
$$-P\sin A + Q\sin B + R\sin C = 0.$$

XI. *To find the equations of the transversals which pass through the points in which the perpendiculars meet the opposite sides.*

By VII., the equation of the perpendicular through ($P = 0$, $Q = 0$) is $P\cos A - Q\cos B = 0$. This perpendicular meets the opposite side in the point ($P\cos A - Q\cos B, R$), and

$$P\cos A - Q\cos B + R\cos C = 0$$

is the equation of some line passing through this point. But this equation shows that the line also passes through the point

($R \cos C - Q \cos B, P$) in which the perpendicular through ($Q = 0, R = 0$) intersects the opposite side, and is therefore the required equation of one of the transversals.

The other two are

$P \cos A + Q \cos B - R \cos C = 0, \quad -P \cos A + Q \cos B + R \cos C = 0.$

61. *Given the equations of the sides of a complete quadrilateral, to find the equations of its three diagonals.*

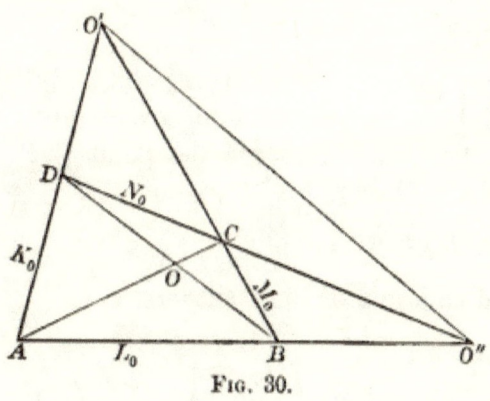

Fig. 30.

Let $ABCD$ be a quadrilateral; produce the opposite sides till they meet in the points O' and O''; then the lines AO', AO'' and BO', DO'' form the complete quadrilateral; AC, BD, $O'O''$ are the three diagonals whose equations we are to find.

Let the equations of AO', AO'', BO', DO'' be $K = 0$, $L = 0$, $M = 0$, $N = 0$, as in the figure. Since the diagonals pass through the intersections of the given sides, we must find values of the arbitrary factors k, l, m, n which will make the equations

$kK + lL = 0$, and $mM + nN = 0$, represent AC;

$lL + mM = 0$, and $kK + nN = 0$, represent BD;

$kK + mM = 0$, and $lL + nN = 0$, represent $O'O''$.

By Art. 56,

$kK + lL + mM = 0$

is the equation of the transversal DO'', which must be coincident with $N=0$. Therefore, the values of k, l, m, n which satisfy the identity

$$kK + lL + mM + nN = 0,$$

or the equivalent equations of condition,

$$kA + lA' + mA'' + nA''' = 0,$$
$$kB + lB' + mB'' + nB''' = 0,$$
$$kC + lC' + mC'' + nC''' = 0,$$

for coincidence, will, by substitution, give the required equations of the diagonals AC, BD, $O'O''$.

These three equations of condition will determine any three of the four arbitraries, as k, l, m, in terms of the fourth, n, to which we can assign such a value as will make k, l, m, n the least integers.

If, however, the transversal $kK + lL + mM = 0$ is not coincident with $N=0$, then the equation

$$kK + lL + mM + nN = 0$$

is not an identity, but represents a straight line passing through the point $(kK+lL, mM+nN)$ in which the two lines drawn respectively through the opposite vertices $A(K=0, L=0)$ and $C(M=0, N=0)$ intersect; also through the point $(kK+nN, mM+lL)$, in which the two lines drawn respectively through the two opposite vertices $D(K=0, N=0)$ and $B(M=0, L=0)$ intersect; and also through the point $(kK+mM, nN+lL)$, in which the two lines drawn respectively through the two opposite vertices $O'(K=0, M=0)$ and $O''(N=0, L=0)$ intersect.

If, then, we construct the quadrilateral given by the equations $K=0$, $L=0$, $M=0$, $N=0$, and then for given values of k, l, m, n construct the lines through the six vertices, we shall find that the intersections of the lines through opposite vertices all lie on the line $kK + lL + mM + nN = 0$.

EXERCISES ON LINES THROUGH THE INTERSECTIONS OF LINES.

1. Find the equation of any straight line passing through the intersection of $2x + 3y + 1 = 0$, $3x - 4y - 5 = 0$.

   Ans. $(2 + 3k)x + (3 - 4k)y = 5k - 1$.

2. Find the equation of a straight line passing through the origin and the intersection of $x + 3y - 2 = 0$, $2x - 3y + 1 = 0$.

   Ans. $5x - 3y = 0$.

3. Find the equation of a straight line passing through the point $(2, -5)$ and the intersection of $2x - y - 7 = 0$, $x + 3y + 2 = 0$.

   Ans. $24x - 5y - 73 = 0$.

4. Find the equation of a straight line passing through the intersection of $2x - 3y + 1 = 0$, $x + 2y - 5 = 0$, and parallel to the line $x + 4y - 2 = 0$.   Ans. $7x + 28y - 57 = 0$.

5. Find the equation of a straight line passing through the intersection of $3x + 2y + 7 = 0$, $2x - 5y - 3 = 0$, and perpendicular to the line $x - 3y - 2 = 0$.   Ans. $57x + 19y + 110 = 0$.

6. Find the value of k in terms of the tangent of the angle which any line passing through the intersection of $3x - 2y + 7 = 0$, $x + y - 5 = 0$ makes with the axis of X.

   Ans. $k = \dfrac{2 \tan \theta - 3}{\tan \theta + 1}$.

7. Find the length of the perpendicular dropped from the origin on any line passing through the intersection of

   $x \cos a + y \sin a - p = 0$, $x \cos \beta + y \sin \beta - p' = 0$.

   Ans. $\dfrac{p + kp'}{\sqrt{1 + k^2 + 2k \cos(a - \beta)}}$

8. Find the equations of the bisectors of the angles between $3x + 4y - 9 = 0$, $12x + 5y - 3 = 0$.

   Ans. $7x - 9y + 34 = 0$, $9x + 7y - 12 = 0$.

   Show that these bisectors are perpendicular to each other.

9. Find the equations of the bisectors of the angles between $4y + 3x - 12 = 0$, $3y + 4x - 24 = 0$.

   Ans. $y - x + 12 = 0$, $7y + 7x - 36 = 0$.

LINES THROUGH INTERSECTIONS OF LINES.   125

10. Find the equations of the bisectors of the angles between $x\cos a + y\sin a - p = 0$, $x\cos\beta + y\sin\beta - p' = 0$.

Ans. $x\cos\left[\tfrac{1}{2}(a+\beta) + \dfrac{\pi}{2}\right] + y\sin\left[\tfrac{1}{2}(a+\beta) + \dfrac{\pi}{2}\right] = \dfrac{p - p'}{2\sin\tfrac{1}{2}(a - \beta)}.$

$x\cos\tfrac{1}{2}(a+\beta) + y\sin\tfrac{1}{2}(a+\beta) = \dfrac{p + p'}{2\cos\tfrac{1}{2}(a - \beta)}.$

11. Find the equations of the bisectors of the angles between $y = mx + b$, $y = m'x + b'$.

Ans. $y = x\tan\left[\tfrac{1}{2}(\theta + \theta') + \dfrac{\pi}{2}\right] + \dfrac{b\cos\theta - b'\cos\theta'}{\cos\theta - \cos\theta'}.$

$y = x\tan\tfrac{1}{2}(\theta + \theta') + \dfrac{b\cos\theta + b'\cos\theta'}{\cos\theta + \cos\theta'}.$

12. Do the lines $5x + 3y - 7 = 0$, $3x - 4y - 10 = 0$, $7x + 10y - 4 = 0$ meet in the same point? Apply second solution of Art. 56.

13. Do the following lines meet in a point?

$17x - 3y - 25 = 0$, $7x + 9y + 17 = 0$, $5x - 6y - 21 = 0$.

14. Do the sets of lines in Ex. 8, Art. 40, and Exs. 17, (4), (5), (6), Art. 52, meet in a point?

15. For what values of A, B, C will the line $Ax + By + C = 0$ pass through the intersection of $2x - 5y + 7 = 0$, $4y - 5x - 2 = 0$?
Ans. $A = 3$, $B = 1$, $C = -5$.

16. For what values of a and p will the line $x\cos a + y\sin a - p = 0$ pass through the intersection of $3x + 4y = 12$, $5x - 12y = 3$?
Ans. $a = \tan^{-1}(-\tfrac{1}{8})$, $p = \dfrac{171}{8\sqrt{65}}.$

17. The equations of two straight lines APB and CPD are $x + 3y - a = 0$, $y - x + a = 0$. Find the equations of two straight lines passing through P, such that the ratio of the sines of their inclination to AB and CD may be as $1 : \sqrt{5}$. *Ans.* $y = 0$, $y + x = a$.

18. Given the equations of the sides of a triangle,

$3y - 2x - 6 = L = 0$, $4y + 3x - 12 = M = 0$, $5y - 2x + 10 = N = 0$.

When $l = 1$, $m = -2$, $n = 3$, find the equations of the lines passing through the vertices of the triangle, and also the equation of the corresponding transversal.   *Ans.* Transversal, $5y - 7x + 24 = 0$.

19. Given the equations of the sides of a quadrilateral,
$$2x + 3y - 4a = K = 0, \quad x + 6y - 7a = M = 0,$$
$$2x + y - a = L = 0, \quad 3x - 2y + 2a = N = 0.$$
Find the equations of the three diagonals.

Ans. $4x + 4y - 5a = 0, \quad x - 5y + 6a = 0, \quad x - 3y + 3a = 0.$

Also show that for $k = 1, l = 2, m = 3, n = 4$, the pairs of lines

$$\left.\begin{array}{l}6x + 5y - 6a = 0 \\ 15x + 10y - 13a = 0\end{array}\right\}, \quad \left.\begin{array}{l}7x + 20y - 23a = 0 \\ 14x - 5y + 4a = 0\end{array}\right\}, \quad \left.\begin{array}{l}8x - 3y + 3a = 0 \\ 5x + 21y - 25a = 0\end{array}\right\},$$

which pass through the external angles of each pair of opposite vertices, will intersect on the line $21x + 15y - 19a = 0$.

20. If $P = 0, Q = 0, R = 0, S = 0$, are the normal equations of the sides of a quadrilateral, then the intersection of the bisectors of the external angles of each pair of opposite vertices will lie on the line $P + Q + R + S = 0$.

21. Let $L = 0, M = 0, N = 0$ be the equations of the sides of a triangle. Draw any parallel to the side $L = 0$, make $rs = st$, and $r's' = s't'$, and complete the construction as in the figure. Find the equations of the lines $CO', CO''', BO', BO''', OO', O''O'''$, and show that OO' and $O''O'''$ pass through the vertex A.

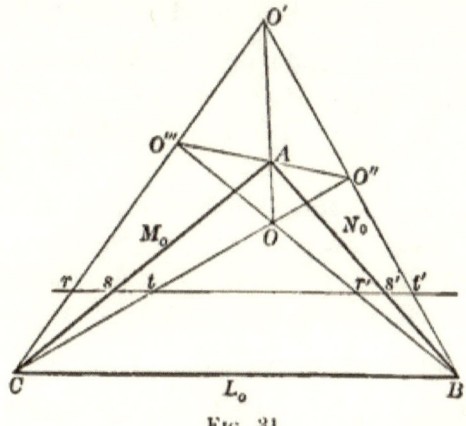

Fig. 31.

Equations above the First Degree which represent Straight Lines.

62. We already know that an equation of the first degree represents a straight line. If, then, any equation above the first degree can be resolved into first degree factors, it will represent the straight lines determined by these factors.

63. *A homogeneous equation of the nth degree in x and y will represent n straight lines passing through the origin.*

Let the equation be

$$Ay^n + By^{n-1}x + Cy^{n-2}x^2 + \cdots + Lx^n = 0,$$

or $\quad A\left(\dfrac{y}{x}\right)^n + B\left(\dfrac{y}{x}\right)^{n-1} + C\left(\dfrac{y}{x}\right)^{n-2} + \cdots + L = 0.$

Denote the roots of this equation by $m_1, m_2, m_3, \ldots, m_n$; then, by the theory of equations, the equation

$$A\left(\frac{y}{x} - m_1\right)\left(\frac{y}{x} - m_2\right)\left(\frac{y}{x} - m_3\right)\cdots\left(\frac{y}{x} - m_n\right) = 0$$

is identical with the given equation, and is satisfied by the n equations of the first degree,

$$y - m_1 x = 0, \quad y - m_2 x = 0, \quad \cdots, \quad y - m_n x = 0,$$

and by no others. The given equation therefore represents n straight lines passing through the origin.

For the given equation

$$A(y-b)^n + B(y-b)^{n-1}(x-a) + \cdots + L(x-a)^n = 0,$$

or $\quad A\left(\dfrac{y-b}{x-a}\right)^n + B\left(\dfrac{y-b}{x-a}\right)^{n-1} + \cdots + L = 0,$

the n first degree equations are

$$y - b = m_1(x-a), \quad y - b = m_2(x-a), \quad \cdots, \quad y - b = m_n(x-a),$$

and the given equation represents n straight lines passing through the point (a, b).

64. *To find the condition that the general equation of the second degree*
$$Ax^2 + 2Hxy + By^2 + 2Gx + 2Fy + C = 0 \qquad (a)$$
shall represent two straight lines.

Suppose that the required first degree factors are $lx + my + n$ and $l'x + m'y + n'$, then
$$(lx + my + n)(l'x + m'y + n') = 0,$$
or $ll'x^2 + (lm' + l'm)xy + mm'y^2 + (ln' + l'n)x + (mn' + m'n)y + nn' = 0.$

must be identical with the given equation (a).

Therefore, by equating coefficients,
$$A = ll', \quad B = mm', \quad C = nn',$$
$$2H = lm' + l'm, \quad 2G = ln' + l'n, \quad 2F = mn' + m'n.$$

The product of the last three equations is
$$8FGH = 2\,ll'mm'nn' + ll'(m^2n'^2 + m'^2n^2)$$
$$+ nn'(l^2m'^2 + l'^2m^2) + mm'(l^2n'^2 + l'^2n^2)$$
$$= 2ABC + A(4F^2 - 2BC) + B(4G^2 - 2AC) + C(4H^2 - 2AB),$$

which reduces to
$$ABC - AF^2 - BG^2 - CH^2 + 2FGH = 0,$$

the required condition.

A simple way of obtaining this condition is to solve the general quadratic (a) for x or y. Solving for y, we get
$$By + Hx + F = \pm\sqrt{[(H^2 - AB)x^2 + 2(HF - BG)x + F^2 - BC]}, \qquad (b)$$

which reduces to two equations of the first degree, if the quantity under the radical is a perfect square.

The condition for this is

$$(H^2 - AB)(F^2 - BC) = (HF - BG)^2, \qquad (c)$$

which readily reduces to the required condition found above.

The factors $(H^2 - AB)$ and $(F^2 - BC)$ must have the same sign, and if either is zero, $HF - BG$ will be zero also.

By introducing condition (c) into equation (b) it becomes

$$By + Hx + F$$
$$= \pm \sqrt{[(H^2 - AB)x^2 + 2x\sqrt{(H^2 - AB)(F^2 - BC)} + F^2 - BC]},$$

and now represents two straight lines which are

Real and intersecting for $H^2 - AB > 0$;

Real and parallel for $H^2 - AB = 0$;

Imaginary and intersecting for $H^2 - AB < 0$.

If $H^2 - AB$ and $F^2 - BC$ are zero at the same time, then the two parallels are coincident.

65. *To find the equation of the two straight lines represented by the equation* $Ax^2 + 2Hxy + By^2 = 0$.

Solving for $y : x$, we get

$$\frac{y}{x} = -\frac{H \pm \sqrt{H^2 - AB}}{B}.$$

Therefore,

$$y = -\frac{H + \sqrt{H^2 - AB}}{B} x, \quad y = -\frac{H - \sqrt{H^2 - AB}}{B} x,$$

or $\qquad y - mx = 0, \qquad\qquad y - m'x = 0,$

are the required equations.

These lines are real if $H^2 - AB > 0$; coincident if $H^2 - AB = 0$; and imaginary if $H^2 - AB < 0$, but pass through the real point $(0, 0)$.

66. *To find the angle between the two straight lines given by the equation* $Ax^2 + 2Hxy + By^2 = 0$.

Write the equation in the form

$$\left(\frac{y}{x}\right)^2 + \frac{2H}{B}\frac{y}{x} + \frac{A}{B} = 0.$$

If m and m' are its roots, then, by Art. 20,

$$m' + m = -\frac{2H}{B}, \qquad m'm = \frac{A}{B},$$

from which we get

$$m' - m = \frac{2\sqrt{H^2 - AB}}{B}.$$

If β denotes the required angle, then, by Art. 44,

$$\tan\beta = \frac{m' - m}{1 + m'm} = \frac{2\sqrt{H^2 - AB}}{A + B}.$$

If the two lines are perpendicular to each other, then $A + B = 0$, or the coefficients of x^2 and y^2 are equal with opposite signs.

If the axes are oblique, then, by Art. 47,

$$\tan\beta = \frac{(m' - m)\sin\omega}{1 + m'm + (m' + m)\cos\omega} = \frac{2\sqrt{H^2 - AB}\sin\omega}{A + B - 2H\cos\omega};$$

and the lines are perpendicular to each other if

$$A + B - 2H\cos\omega = 0.$$

67. *To find the equations of the bisectors of the angles between the lines represented by* $Ax^2 + 2Hxy + By^2 = 0$.

If the two lines given by this quadratic are

$$y - mx = 0, \quad y - m'x = 0,$$

EQUATIONS WHICH REPRESENT STRAIGHT LINES.

then, by Art. 54,

$$\frac{y-mx}{\sqrt{1+m^2}} = \pm \frac{y-m'x}{\sqrt{1+m'^2}}, \quad \text{or} \quad \frac{(y-mx)^2}{1+m^2} = \frac{(y-m'x)^2}{1+m'^2},$$

is the equation of the two bisectors, which reduces to

$$Hy^2 + (A-B)xy - Hx^2 = 0,$$

by substituting the values of m and m' found in Art. 65.

Second Solution. — Suppose that

$$y = \tan\gamma \cdot x \quad \text{and} \quad y = \tan\left(\gamma + \frac{\pi}{2}\right) \cdot x$$

are the equations of the bisectors of the angles between $y - mx = 0$ and $y - m'x = 0$. Then

$$\gamma = \frac{\theta' + \theta}{2}, \quad \gamma + \frac{\pi}{2} = \frac{\theta' + \theta}{2} + \frac{\pi}{2};$$

and in both cases

$$\tan 2\gamma = \tan(\theta' + \theta),$$

or

$$\frac{2\tan\gamma}{1 - \tan^2\gamma} = \frac{\tan\theta' + \tan\theta}{1 - \tan\theta'\tan\theta} = \frac{m' + m}{1 - m'm},$$

which reduces to

$$Hy^2 + (A-B)xy - Hx^2 = 0,$$

since $\quad \tan\gamma = \frac{y}{x}, \quad m' + m = -\frac{2H}{B}, \quad m'm = \frac{A}{B}.$

These bisectors are real lines, even when the lines $y - mx = 0$, $y - m'x = 0$ are imaginary.

EXERCISES ON EQUATIONS ABOVE THE FIRST DEGREE WHICH REPRESENT STRAIGHT LINES.

1. What loci do the following equations represent?

(1) $xy = 0$.
(2) $y^2 - x^2 = 0$.
(3) $y^2 + x^2 = 0$.
(4) $x^2 - 5xy + 6y^2 = 0$.
(5) $x^2 + xy = 0$.
(6) $xy - ax = 0$.
(7) $y^3 - x^3 = 0$.
(8) $y^4 - x^4 = 0$.

Ans. (1) $x = 0$, $y = 0$, the coordinate axes.

(2) $y - x = 0$, $y + x = 0$, the bisectors of the angles between the axes.

(3) $y = \pm x\sqrt{-1}$, a pair of imaginary lines through the origin.

(4) The lines $x - 3y = 0$, $x - 2y = 0$.

(7) The real line $y - x = 0$, and the two imaginaries $y = (-\tfrac{1}{2} \pm \tfrac{1}{2}\sqrt{-3})\, x$.

2. What loci are represented by the following equations?

(1) $(x - a)(y - b) = 0$.
(2) $(x - a)^2 + (y - b)^2 = 0$.
(3) $(x - y + a)^2 + (x + y - a)^2 = 0$.
(4) $(y - 3x + 3)(x + 3y - 9) = 0$.
(5) $y^2 - 2xy \sec \theta + x^2 = 0$.
(6) $x^2 + 2xy \cot 2\theta - y^2 = 0$.

Ans. (1) A parallel to each axis through (a, b).

(2) A pair of imaginaries through (a, b).

(3) A pair of imaginaries through $(0, a)$.

(4) A pair of perpendiculars through $(\tfrac{9}{5}, \tfrac{12}{5})$.

(5) $x - y \tan\left(\dfrac{\pi}{4} \pm \tfrac{1}{2}\theta\right) = 0$.

(6) $y = x \cot \theta$; $y = - x \tan \theta$.

3. Find the angles between the lines
$$x^2 + xy - 6y^2 = 0, \text{ and } x^2 - 2xy \sec \theta + y^2 = 0,$$
and the equations of their bisectors.

Ans. $\dfrac{\pi}{4}$; θ; $y^2 + 14xy - x^2 = 0$; $y^2 - x^2 = 0$.

EXERCISES ON STRAIGHT LINES.

4. Find the equation of the bisectors between the pair of straight lines $x^2 - 5xy + 6y^2 = 0$. *Ans.* $5x^2 - 10xy - 5y^2 = 0$.

5. Show that the pair $6x^2 + 5xy - 6y^2 = 0$ intersect at right angles.

6. Show that $6x^2 + 5xy - 6y^2 = 0$ is the equation of the bisectors of the angles between the pair $2x^2 + 12xy + 7y^2 = 0$.

7. Find the lines represented by $x^2 - 5xy + 4y^2 + x + 2y - 2 = 0$.
Ans. $y - x + 1 = 0$, $4y - x - 2 = 0$.

8. Show that $4x^2 - 12xy + 9y^2 - 4x + 6y - 12 = 0$ represents a pair of parallel lines, and find their equations.
Ans. $3y - 2x + 1 \pm \sqrt{13} = 0$.

9. Show that $4x^2 - 12xy + 9y^2 - 4x + 6y + 1 = 0$ denotes two coincident straight lines, and find their equation.
Ans. $3y - 2x + 1 = 0$.

10. For what value of C does the equation
$$12x^2 - 10xy + 2y^2 + 11x - 5y + C = 0$$
represent two straight lines? *Ans.* $C = 2$.

Show that the angle between them is $\tan^{-1} \frac{1}{7}$.

11. For what values of H does the equation
$$12x^2 + Hxy + 2y^2 + 11x - 5y + 2 = 0$$
represent straight lines? Find the lines.
Ans. -10 or $-\frac{35}{2}$.
$y - 3x - 2 = 0$, $2y - 4x - 1 = 0$.
$y - 8x - 2 = 0$, $4y - 3x - 2 = 0$.

12. For what value of B does the equation
$$12x^2 + 36xy + By^2 + 6x + 6y + 3 = 0$$
represent two straight lines? Are they real or imaginary?
Ans. 28.

13. For what value of H does the equation $Hxy + 5x + 3y + 2 = 0$ represent two straight lines? Ans. $1\tfrac{5}{2}$.

14. For what values of H does the equation
$$x^2 + 2Hxy + y^2 - 5x - 7y + 6 = 0$$
represent straight lines? Ans. $\tfrac{5}{6}$ and $\tfrac{5}{4}$.

Find the lines
$$(x + 2y - 2)(2x + y - 6) = 0, \quad (x + 3y - 3)(3x + y - 6) = 0.$$

Problems on Loci. Construct a figure in accordance with the given conditions; choose a pair of coordinate axes, usually two principal lines in the figure; denote the coordinates of the moving point by (xy): then the required equation of the locus will be the algebraic relation between (xy) and the given data of the problem.

In a triangle ABC, given the base $AB = 2c$, to find the locus of the vertex C.

1. When $\overline{AC}^2 - \overline{BC}^2 = m$, a constant.

Solution. Take the base AB and a perpendicular through its middle point O, for the coordinate axes of X and Y; then the coordinates of C are (ON, CN) (xy), of A are $(-c, 0)$, of B are $(c, 0)$, and
$$\overline{AC}^2 = y^2 + (c - x)^2, \quad \overline{BC}^2 = y^2 + (c + x)^2.$$

$\therefore \overline{AC}^2 - \overline{BC}^2 = 4cx = m$ is the equation of the required locus, a straight line perpendicular to the axis of X.

2. When $\cot A + m \cot B = p$.

Solution. $\cot A = \dfrac{c+x}{y}, \quad \cot B = \dfrac{c-x}{y}.$

$\therefore \dfrac{c+x}{y} + \dfrac{m(c-x)}{y} = p,$ or $py + (m-1)x = c(m+1),$

and the locus is a straight line. If $m = 1$, $py = 2c$, and the locus is parallel to the base.

3. When $\overline{AC}^2 + \overline{BC}^2 = m^2$.   Ans. $x^2 + y^2 = \tfrac{1}{2}m^2 - c^2$.

4. When $AC : BC = k$.
  Ans. $(1 - k^2)(x^2 + y^2 + c^2) + 2c(1 + k^2)x = 0$.

5. When $\tan A \tan B = m$.   Ans. $y^2 + m^2 x^2 = m^2 c^2$.

6. When $A + B$ is given.   Ans. $x^2 + y^2 - 2cy \cot C = c^2$.

7. When $A - B = D$.   Ans. $x^2 - y^2 + 2xy \cot D = c^2$.

8. When $A = 2B$.   Ans. $3x^2 - y^2 + 2cx = c^2$.

9. When $\tan C = m \tan B$.   Ans. $m(x^2 + y^2 - c^2) = 2c(c - x)$.

10. When $AC + BC = m$. Produce the ordinate CN of the vertex C to a point P, such that $PN = AC$, or $BC = m - AC = m - y$; to find the locus of P.

Solution. By Geometry, $\overline{BC}^2 = \overline{AB}^2 + \overline{AC}^2 - 2 AB \cdot AN$.

$\therefore (m - y)^2 = 4c^2 + y^2 - 4c(c + x)$,   or $2my - 4cx = m^2$,

and the locus is a straight line.

11. Given the angle C, and $CA + CB = s$; to find the locus of the point P, which divides the side AB, such that $PA : PB = m : n$.

Solution. Take C as the origin, and CA, CB as the coordinate axes of X and Y; then (PN, PM) (xy) are the coordinates of P, and from the similar triangles CAB, NAP, MPB, we get

$$CA = \frac{(m+n)x}{n}, \quad CB = \frac{(m+n)y}{m}.$$

$$\therefore \frac{(m+n)x}{n} + \frac{(m+n)y}{m} = s, \quad \text{or} \quad \frac{x}{n} + \frac{y}{m} = \frac{s}{m+n},$$

and the locus is a straight line.

12. To find the locus of the points of intersection of the diagonals of the rectangles inscribed in a given triangle. Let b denote the base and h the altitude, which take for coordinate axes.

Ans. $\dfrac{2x}{b} + \dfrac{2y}{a} = 1$.

13. Any line is drawn parallel to the base of a given triangle; to find the locus of the intersections of the diagonals of the trapezoid thus formed. Take the vertex C as origin, and the sides $CA = a$, $CB = b$ as coordinate axes.   Ans. $bx - ay = 0$.

14. Let $AB = a$, $AC = b$ be the adjacent sides of a parallelogram; draw any line PP' parallel to AB, and any line QQ' parallel to AC; to find the locus of R, the intersection of PP' and QQ'.

Ans. $bx - ay = 0$.

15. Given an angle C and the area of the triangle; the opposite side AB is cut in a given ratio at P; to find the locus of P.

Solution. Let C be the origin, and $CA = x$, $CB = y$, the coordinate axes. The coordinates of $P(\alpha, \beta)$, which cut the line $A(x, 0)$ $B(0, y)$ in the ratio $n:m$, are $\alpha = \dfrac{mx}{m+n}$, $\beta = \dfrac{ny}{m+n}$ (Art. 5). The area is

$$\tfrac{1}{2} xy \sin C = \text{constant}, \text{ or } \tfrac{1}{2} \dfrac{(m+n)^2}{mn} \alpha\beta \sin C = \text{constant}.$$

$\therefore \alpha\beta =$ a constant, and $\alpha\beta$ can now be replaced by xy.

16. If the base instead of the area is given.

Solution. Then $c^2 = x^2 + y^2 - 2xy \cos C$. Substitute the values of x and y in terms of α and β, the coordinates of the required locus (Ex. 15).

$$\therefore \dfrac{\alpha^2}{m^2} + \dfrac{\beta^2}{n^2} - \dfrac{2\alpha\beta \cos C}{mn} = \dfrac{c^2}{(m+n)^2},$$

in which $\alpha\beta$ can be replaced by xy.

17. If the base AB pass through a fixed point (X', Y').

Solution. If the current coordinates of AB are (X, Y), its equation is

$$\dfrac{X}{x} + \dfrac{Y}{y} = 1, \text{ or } \dfrac{mX}{\alpha} + \dfrac{nY}{\beta} = m + n \text{ (Ex. 15)}.$$

$\therefore \dfrac{mX'}{\alpha} + \dfrac{nY'}{\beta} = m + n$ is the locus of $P(\alpha, \beta)$ when AB passes through a fixed point (X', Y').

18. Given the vertical angle C, and the sum of the reciprocals $\dfrac{1}{a} + \dfrac{1}{b} = \dfrac{1}{m}$ of the sides CA, CB; the base will pass through a fixed point.

Solution. The equation of AB is

$$\dfrac{x}{a} + \dfrac{y}{b} = 1, \text{ or } \dfrac{1}{a}(x - y) + \dfrac{y}{m} - 1 = 0,$$

since $\dfrac{1}{b} = \dfrac{1}{m} - \dfrac{1}{a}$. But $\dfrac{1}{a}$ is indeterminate; therefore, Art. 53, AB passes through the intersection of $x - y = 0$, $y - m = 0$.

CHAPTER IV.

THE CIRCLE.

The Equation of the Circle.

68. DEFINITION. — *A circle is the locus of a point which moves so that its distance from a fixed point, called the centre, is always equal to a constant length, called the radius.*

It follows from this definition that the position of the circle depends upon the position of its centre, and its size upon the length of its radius.

69. *To find the equation of the circle when the axes are rectangular.*

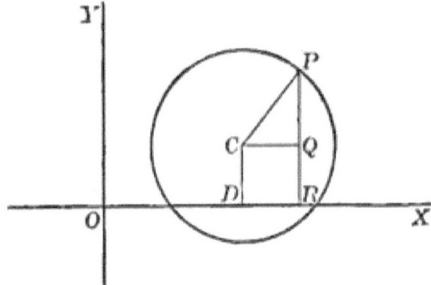

FIG. 32.

Let $P(xy)$ be any point on the locus, let $C(OD, CD)$ (d, e) be the coordinates of the centre, and let r be the constant length, or radius; then

$$CQ = x - d, \quad PQ = y - e, \quad CP = r,$$

and, by Geometry,

$$(x - d)^2 + (y - e)^2 = r^2, \qquad (a)$$

which is the required equation.

Special Cases. — I. *When the centre is at the origin.*

In this case $d = 0$, $e = 0$, and (a) becomes

$$x^2 + y^2 = r^2. \tag{b}$$

II. *When the axis of X is a diameter, and the axis of Y is a tangent.*

In this case $e = 0$, $d = r$, and (a) becomes

$$y^2 = 2rx - x^2 = (2r - x)x; \tag{c}$$

which shows that the ordinate y of any point on the curve is a mean proportional between the segments into which its foot divides the diameter.

70. The general equation of the second degree

$$Ax^2 + 2Hxy + By^2 + 2Gx + 2Fy + C = 0$$

can be reduced to the form of the equation of the circle,

$$(x - d)^2 + (y - e)^2 = r^2,$$

or $\qquad x^2 + y^2 - 2dx - 2ey + d^2 + e^2 - r^2 = 0,$

when $A = B$ and $H = 0$; for then it becomes

$$x^2 + y^2 + 2\frac{G}{A}x + 2\frac{F}{A}y + \frac{C}{A} = 0;$$

or, by completing the squares,

$$\left(x + \frac{G}{A}\right)^2 + \left(y + \frac{F}{A}\right)^2 = \frac{G^2 + F^2 - AC}{A^2}, \tag{d}$$

in which $\quad d = -\dfrac{G}{A}, \quad e = -\dfrac{F}{A}, \quad r = \dfrac{\sqrt{G^2 + F^2 - AC}}{A}.$

This radius is real when $G^2 + F^2 - AC$ is positive, in which case the circle is real; it is zero when $G^2 + F^2 - AC = 0$, in which case the circle is called the point-circle; and it is imaginary when $G^2 + F^2 - AC$ is negative, in which case the circle is imaginary, since no real values of x and y will satisfy its equation.

So long as A, G, F remain constant, the centre of the circle will remain fixed in position; but as the radius contains C in addition to these constants, it will vary when C varies, and equation (d) will then represent a series of *concentric* circles corresponding to the different values of C. It follows, therefore, *that circles whose equations differ only in their independent terms are concentric.*

Since the circle is always completely determined by three independent conditions, such as passing through three given points, or being tangent to three given straight lines, its equation will only need the three arbitrary and independent parameters G, F, C to satisfy these conditions. We may therefore make $A=1$, and then the coordinates of the centre are $(-G, -F)$, *the negatives of the halves of the coefficients of x and y, and the radius is* $\sqrt{G^2+F^2-C}$, *the square root of the sum of the squares of the coordinates of the centre diminished by the independent term C.*

71. *To find the equation of the circle in terms of its intercepts on the coordinate axes.*

For points on the axis of X, $y=0$, and the general equation of the circle

$$x^2 + y^2 + 2Gx + 2Fy + C = 0 \qquad (d)$$

becomes $x^2 + 2Gx + C = 0$, a quadratic whose roots x' and x'' are the intercepts on the axis of X. For $x=0$, (d) becomes $y^2 + 2Fy + C = 0$, a quadratic whose roots y' and y'' are the intercepts on the axis of Y. But, by Art. 20, the sums and products of these roots are

$$x' + x'' = -2G, \quad x'x'' = C; \quad y' + y'' = -2F, \quad y'y'' = C.$$

Therefore, by substitution, (d) becomes

$$x^2 + y^2 - (x'+x'')x - (y'+y'')y + \tfrac{1}{2}(x'x'' + y'y'') = 0,$$

or
$$\left(x - \frac{x'+x''}{2}\right)^2 + \left(y - \frac{y'+y''}{2}\right)^2 = \tfrac{1}{4}(x'^2 + x''^2 + y'^2 + y''^2),$$

which is the required equation.

If the axes are tangents to the circle, $x' = x''$ and $y' = y''$, for which, by Art. 20, $G^2 = C$, $F^2 = C$.

EXERCISES ON THE CIRCLE.

Find the equations of the circles corresponding to the following data:

1. Centre, $(2, 3)$; radius, 3.   *Ans.* $x^2 + y^2 - 4x - 6y + 4 = 0$.
2. Centre, $(-1, 2)$; radius, 5.   *Ans.* $x^2 + y^2 + 2x - 4y - 20 = 0$.
3. Centre, $(-5, -3)$; radius, 4.
                              Ans. $x^2 + y^2 + 10x + 6y + 18 = 0$.
4. Centre, $(0, -8)$; radius, 2.   *Ans.* $x^2 + y^2 + 16y + 60 = 0$.
5. Centre, (a, b); radius, $\sqrt{a^2 + b^2}$. *Ans.* $x^2 + y^2 - 2ax - 2by = 0$.
6. The centre in the first quadrant and both axes are tangents.
                              Ans. $x^2 + y^2 - 2rx - 2ry + r^2 = 0$.
7. The axis of Y is a diameter, and the axis of X a tangent.
                              Ans. $x^2 + y^2 \pm 2ry = 0$.
8. The line joining the origin and point (a, b) is a diameter.
                              Ans. $x^2 + y^2 - ax - by = 0$.
9. The line joining the points $(1, 2)$ and $(3, -4)$ is a diameter.
                              Ans. $x^2 + y^2 - 4x + 2y = 5$.
10. The line joining the points $(a, 0)$ and $(0, b)$ is a diameter.
                              Ans. $x^2 + y^2 - ax - by = 0$.
11. The line joining the points $(3, 4)$ and $(-3, -4)$ is a diameter.
                              Ans. $x^2 + y^2 = 25$.

EXERCISES ON THE CIRCLE.

Find the centres, radii, and intercepts of the following circles:

12. $x^2 + y^2 - 6x - 4y - 12 = 0$.

   Ans. $(3, 2)$, 5; on OX $(3 \pm \sqrt{21})$; on OY $(6$ and $-2)$.

13. $x^2 + y^2 - 6x - 4y = 0$.

   Ans. $(3, 2)$, $\sqrt{13}$; on OX $(0$ and $6)$; on OY $(0$ and $4)$.

14. $x^2 + y^2 + 10x - 2y + 17 = 0$.

   Ans. $(-5, 1)$, 3; on OX $(-5 \pm 2\sqrt{2})$; on OY $(1 \pm 4\sqrt{-1})$.

15. $x^2 + y^2 - x - y = 0$.

   Ans. $(\tfrac{1}{2}, \tfrac{1}{2})$, $\dfrac{1}{\sqrt{2}}$; on OX $(0$ and $1)$; on OY $(0$ and $1)$.

16. $4x^2 + 4y^2 + 4x - 8y + 3 = 0$.

   Ans. $(-\tfrac{1}{2}, 1)$, $\dfrac{1}{\sqrt{2}}$; on OX $(-\tfrac{1}{2} \pm \tfrac{1}{2}\sqrt{-2})$; on OY $(\tfrac{3}{2}$ and $\tfrac{1}{2})$.

17. $36x^2 + 36y^2 - 48x + 36y + 1 = 0$.

   Ans. $(\tfrac{2}{3}, -\tfrac{1}{2})$, $\sqrt{\tfrac{2}{3}}$; on OX $(\tfrac{2}{3} \pm \tfrac{1}{6}\sqrt{15})$; on OY $(-\tfrac{1}{2} \pm \tfrac{1}{3}\sqrt{2})$.

18. Given the circles
$$x^2 + y^2 + 2Gx + 2Fy + C = 0, \quad x^2 + y^2 + 2G'x + 2F'y + C' = 0.$$
(1) When are they concentric? (2) When tangents to the axes? (3) What the distance between centres? (4) When tangents internally? (5) When tangents externally? (6) When do they cut each other orthogonally, or cross at right angles?

   Ans. (1) $G' = G$, $F' = F$;

   (2) $G^2 = C$, $F^2 = C$, $G'^2 = C'$, $F'^2 = C'$;

   (3) $\sqrt{(G'-G)^2 + (F'-F)^2}$;

   (4) and (5) $\sqrt{(G'-G)^2 + (F'-F)^2} = \sqrt{G'^2 + F'^2 - C'} \mp \sqrt{G^2 + F^2 - C}$;

   (6) $G'^2 + F'^2 - C' + G^2 + F^2 - C = (G'-G)^2 + (F'-F)^2$;

   or $2GG' + 2FF' - C - C' = 0$.

19. Find the equation of the circle whose diameter is the line joining the points (a, b) and (a', b').

   Ans. $\left(x - \dfrac{a+a'}{2}\right)^2 + \left(y - \dfrac{b+b'}{2}\right)^2 = \dfrac{(a-a')^2 + (b-b')^2}{4}$.

The form of this answer suggests the method of solution. Since the lines through (a, b) and (a', b'), which meet in the point (xy) on the circle, are perpendicular to each other, we also have

$$\frac{y-b}{x-a} = -\frac{x-a'}{y-b'}, \text{ or } (y-b)(y-b') + (x-a)(x-a') = 0,$$

the required equation.

20. Show that the two circles $x^2 + y^2 - 2ax - 2by - 2ab = 0$ and $x^2 + y^2 + 2bx + 2ay - 2ab = 0$ cross at right angles.

21. Show that the circles $x^2 + y^2 + 2dx + k^2 = 0$ and $x^2 + y^2 + 2d'y - k^2 = 0$ intersect at right angles.

72. *To find the equation of the circle which passes through the three given points* $A(x'y')$, $B(x''y'')$, $C(x'''y''')$.

Since the required circle circumscribes the given triangle ABC, its centre (d, e) is equidistant from the vertices, and the equations

$$(d-x')^2 + (e-y')^2 = (d-x'')^2 + (e-y'')^2 = (d-x''')^2 + (e-y''')^2 = r^2$$

will give the required values of d and e and r.

The values of G, F, C, derived from the equations of condition,

$$x'^2 + y'^2 + 2Gx' + 2Fy' + C = 0,$$
$$x''^2 + y''^2 + 2Gx'' + 2Fy'' + C = 0,$$
$$x'''^2 + y'''^2 + 2Gx''' + 2Fy''' + C = 0,$$

and substituted in

$$x^2 + y^2 + 2Gx + 2Fy + C = 0$$

will also give the required equation.

In the same way we can find the values of G, F, C, which satisfy any other given conditions which the circle must fulfil.

73. *To find the equation of the circle when the axes are oblique.*

In this case, the triangle CPQ (Fig. 32) is oblique, and by (c), page 9, the required equation is

$$(x-d)^2 + (y-e)^2 + 2(x-d)(y-e)\cos\omega = r^2,$$

or
$$x^2 + 2xy\cos\omega + y^2 - 2(d+e\cos\omega)x - 2(e+d\cos\omega)y$$
$$+ d^2 + e^2 + 2de\cos\omega - r^2 = 0, \qquad (a)$$

which is of the form

$$x^2 + 2Hxy + y^2 + 2Gx + 2Fy + C = 0. \qquad (b)$$

By equating corresponding coefficients in (a) and (b),

$$d + e\cos\omega = -G, \qquad e + d\cos\omega = -F,$$
$$H = \cos\omega, \qquad d^2 + e^2 + 2de\cos\omega - r^2 = C,$$

from which we readily get

$$d\sin^2\omega = F\cos\omega - G,$$
$$e\sin^2\omega = G\cos\omega - F,$$
$$r^2\sin^2\omega = G^2 + F^2 - 2GF\cos\omega - C\sin^2\omega.$$

These equations will give the values of (d, e), r, and ω, which correspond to given values of H, G, F, C.

EXERCISES ON THE CIRCLE.

Find the equations of the circles which pass through the following given points:

1. Points $(0, 0)$, $(0, 5)$, $(3, 0)$. Ans. $x^2 + y^2 - 3x - 5y = 0$.
2. Points $(1, 2)$, $(-3, 0)$, $(0, -2)$.
    Ans. $14x^2 + 14y^2 + 18x - 8y - 72 = 0$.
3. Points $(0, 1)$, $(1, 0)$, $(2, 1)$. Ans. $x^2 + y^2 - 2x - 2y + 1 = 0$.
4. Points $(0, 0)$, $(a, 0)$, $(0, b)$. Ans. $x^2 + y^2 - ax - by = 0$.
5. Points $(a, 0)$, $(-a, 0)$, $(0, b)$.
    Ans. $b(x^2 + y^2) + (a^2 - b^2)y - a^2b = 0$.
6. Points $(2, 3)$, $(4, 5)$, $(6, 1)$.
    Ans. $3x^2 + 3y^2 - 26x - 16y + 61 = 0$.

7. Points $(4, 5)$, $(-2, 3)$; radius, 5.

Ans. $[x-(1\pm\tfrac{1}{2}\sqrt{6})]^2+[y+(-4\pm\tfrac{3}{2}\sqrt{6})]^2=25$.

8. Find the locus of the centres of all circles which pass through the points $(3, 4)$ and $(-5, 6)$. *Ans.* $y-4x=9$.

9. Find the locus of the centres of all the circles which pass through the points $(x'y')$ and $(x''y'')$.

Ans. $2(x'-x'')G+2(y'-y'')F+x'^2+y'^2-(x''^2+y''^2)=0$.

10. The centre of a circle is on the line $y-x+4=0$, and it passes through the points $(1,5)$ and $(4,6)$; its equation is
$$2x^2+2y^2-17x-y-30=0.$$

11. The equation of the circle $x^2+xy+y^2-4x+6y+1=0$ is referred to oblique axes; find ω, (d, e), and r.

Ans. $\omega=60°$, $(d,e),(\tfrac{14}{3},-\tfrac{16}{3})$, $r=\tfrac{1}{3}\sqrt{219}$.

12. The equation of the circle $x^2+y^2+xy+2x+2y=0$ is referred to oblique axes; find ω, (d,e), and r.

Ans. $\omega=60°$, (d,e), $(-\tfrac{2}{3},-\tfrac{2}{3})$; $r=\dfrac{2}{\sqrt{3}}$.

13. The diameter of a circle is the line joining the points $(x'y')$ and $(x''y'')$; show that its equation, when the axes are oblique, is
$$(x-x')(x-x'')+(y-y')(y-y'')$$
$$+[(y-y')(x-x'')+(y-y'')(x-x')]\cos\omega=0.$$

14. Show that the equation of the circle which passes through the three points $A(x'y')$, $B(x''y'')$, $C(x'''y''')$ is

$(x^2+y^2)[x'(y''-y''')+x''(y'''-y')+x'''(y'-y'')]$
$-(x'^2+y'^2)[x''(y'''-y)+x'''(y-y'')+x(y''-y''')]$
$+(x''^2+y''^2)[x'''(y-y')+x(y'-y''')+x'(y'''-y)]$
$-(x'''^2+y'''^2)[x(y'-y'')+x'(y''-y)+x''(y-y')]=0$.

Let $D(xy)$ be any fourth point on the circle; then, since the expressions in the brackets [ ] are the double areas of the triangles ABC, etc., this equation may be geometrically expressed by

$$\overline{OD}^2\cdot ABC+\overline{OB}^2\cdot CDA=\overline{OA}^2\cdot BCD+\overline{OC}^2\cdot DAB.$$

The Polar Equation of the Circle.

74. *To find the polar equation of the circle.*

Let O be the pole, and OX the initial line; also let $C(\rho', a)$ and $P(\rho, \theta)$ be the polar coordinates of the centre and of any point on the curve, and let r equal the radius CP.

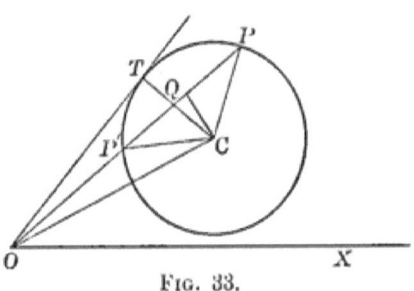

Fig. 33.

Then in the triangle COP we have, Trigonometry, Art. 146,

$$\overline{OP}^2 + \overline{OC}^2 - 2\,OP \cdot OC \cos COP = \overline{CP}^2,$$

or $\quad \rho^2 - 2\rho\rho' \cos(\theta - a) + \rho'^2 - r^2 = 0,$ $\quad\quad\quad (a)$

the required equation.

This quadratic shows that for a given value of θ there are two values of ρ, OP and OP'.

The sum of these roots is, Art. 20,

$$OP + OP' = 2\rho' \cos(\theta - a) = 2.OQ;$$

their product is

$$OP \cdot OP' = \rho'^2 - r^2 = \overline{OT}^2;$$

and they are equal, that is, $OP = OP' = OT$, when

$$\rho'^2 \cos^2(\theta - a) = \rho'^2 - r^2,$$

or $\quad \theta = a \pm \sin^{-1}\dfrac{r}{\rho'},$

are the limits within which θ must lie for real values of ρ.

These results are readily verified by the figure. Special cases of the polar equation of the circle are given in the exercises on page 23.

EXERCISES ON THE POLAR EQUATION OF THE CIRCLE.

Find the polar equations of the circle which correspond to the following rectangular equations, the pole at the origin O, and the axis of X the polar axis:

1. $x^2 + y^2 = r^2$.   Ans. $\rho = r$.

2. $(x-d)^2 + (y-e)^2 = r^2$.   Ans. $\rho^2 - 2\rho\rho' \cos(\theta - \alpha) + \rho'^2 - r^2 = 0$.

3. $y^2 + x^2 = 2rx$.   Ans. $\rho = 2r \cos \theta$.

4. $x^2 + y^2 + 2Gx + 2Fy + C = 0$.
    Ans. $\rho^2 + 2(G \cos \theta + F \sin \theta)\rho + C = 0$.

5. If through any fixed point O any secant be drawn cutting the circle in P and P', show that the product $OP \cdot OP'$ is constant and equal to the square of the tangent through the same point.

6. If through a fixed point O any secant be drawn to the circle, and OQ is the arithmetic mean between OP and OP', Fig. 33, to find the locus of Q.

    Ans. $\rho = \rho' \cos \theta$, a circle having OC as a diameter. (See Ex. 3.)

7. If OQ is a harmonic mean between OP and OP', that is, if $OQ = \dfrac{2\, OP \cdot OP'}{OP + OP'}$, then the locus of Q is $\rho \cos(\theta - \alpha) = \dfrac{\rho'^2 - r^2}{\rho'}$, the equation of a straight line perpendicular to OC, Art. 35, at a distance $\dfrac{\rho'^2 - r^2}{\rho'}$ from the pole.

8. Any straight line is drawn from a fixed point O, meeting a fixed circle in P, and a point Q is taken on this line, such that the rectangle $OQ \cdot OP = k$, a constant; show that the locus of Q is the circle

$$\rho^2(\rho'^2 - r^2) - 2k\rho'\rho \cos(\theta - \alpha) + k^2 = 0.$$

The Straight Line and the Circle.

75. *To find the coordinates of the points in which a chord intersects a circle.*

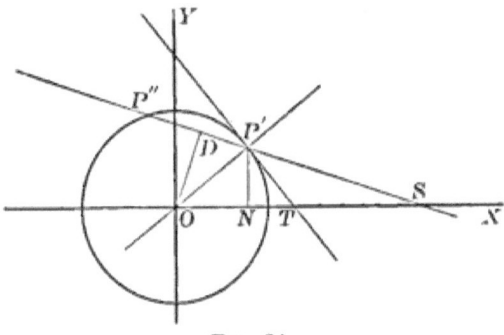

Fig. 34.

Let $x^2 + y^2 = r^2$ and $x\cos a + y\sin a = p$

be the equations of the circle and of the chord SP''. Since the required points P' and P'' are on both loci, their coordinates will be found by eliminating x and y between their equations. Equating values of y, we get

$$\frac{p - x\cos a}{\sin a} = \sqrt{r^2 - x^2},$$

or
$$x = p\cos a \pm \sin a \sqrt{r^2 - p^2}. \tag{a}$$

Eliminating x between this equation and that of the chord, we get

$$y = p\sin a \mp \cos a \sqrt{r^2 - p^2} \tag{b}$$

There are two different real points of intersection when $p^2 < r^2$; two real but coincident points when $p^2 = r^2$; and two imaginary points when $p^2 > r^2$.

If $y = mx + b$ is the chord, then the coordinates of P' and P'' are

$$x = \frac{-mb \pm \sqrt{(1+m^2)r^2 - b^2}}{1+m^2}, \qquad y = \frac{b \pm m\sqrt{(1+m^2)r^2 - b^2}}{1+m^2}. \tag{c}$$

These roots are real if $(1+m^2)r^2 > b^2$; equal if $(1+m^2)r^2 = b^2$; and imaginary if $(1+m^2)r^2 < b^2$.

76. *To find the equation of the locus of the middle points of a system of parallel chords of the circle* $x^2 + y^2 = r^2$.

Let $P'P''$, Fig. 34, be any one of a system of parallel chords. The values of the coordinates $P'(x'y')$ and $P''(x''y'')$ found in Art. 75, and (c), Art. 5, give

$$y = \frac{y'+y''}{2} = p \sin a, \quad x = \frac{x'+x''}{2} = p \cos a,$$

the coordinates of the middle point of the chord $P'P''$. By division,

$$\frac{y}{x} = \tan a, \quad \text{or} \quad y = -\frac{1}{m}x;$$

and since this equation is true for the middle point of any chord of the system, it is the equation of the required locus, and shows that the locus is a straight line perpendicular to the system of chords, that it passes through the centre of the circle, and is therefore a diameter of the circle.

77. *To find the equation of the chord which intersects a given circle in the given points* $(x'y')$, $(x''y'')$.

Let $x^2 + y^2 = r^2$ and $y - y' = \frac{y'-y''}{x'-x''}(x-x')$, Art. 39, be the equation of the circle and of a straight line passing through any two given points. Since these points are to be on the circle, these equations must be combined subject to this condition.

For the points $P'(x'y')$ and $P''(x''y'')$, Fig. 34, we have

$$y'^2 + x'^2 = r^2, \quad y''^2 + x''^2 = r^2.$$

By subtraction,

$$y'^2 - y''^2 = -(x'^2 - x''^2),$$

or
$$(y'-y'')(y'+y'') = -(x'-x'')(x'+x''),$$

or
$$\frac{y'-y''}{x'-x''} = -\frac{x'+x''}{y'+y''}.$$

EQUATIONS OF CHORDS OF THE CIRCLES. 149

Substituting this value of $\dfrac{y'-y''}{x'-x''}$ in the equation of the line, it becomes

$$y - y' = -\frac{x' + x''}{y' + y''}(x - x'),$$

or $\qquad y(y' + y'') + x(x' + x'') = r^2 + y'y'' + x'x'',\qquad (a)$

the required equation of the chord $SP'P''$.

78. *To find the equation of the chord which passes through the points of intersection of the two given circles*

$$x^2 + y^2 + 2Gx + 2Fy + C = S = 0, \quad x^2 + y^2 + 2G'x + 2F'y + C' = S' = 0.$$

We may find the coordinates of the points of intersection of these two circles, as in Art. 19, and then find the equation of the chord which passes through these points; or we may proceed as follows:

By Art. 53, the equation $L + kM = 0$ represents all straight lines which pass through the point of intersection ($L = 0, M = 0$) of the two given straight lines $L = 0$ and $M = 0$.

In the same way the equation

$$S + kS' = (1+k)(x^2+y^2) + 2(G+kG')x + 2(F+kF')y + C + kC' = 0 \;(a)$$

represents a system of circles all of which pass through the two points ($S = 0$, $S' = 0$), in which the given circles $S = 0$ and $S' = 0$ intersect. In the single case, however, in which the arbitrary constant $k = -1$, this equation reduces to

$$S - S' = 2(G - G')x + 2(F - F')y + C - C' = 0,\qquad (b)$$

which is of the first degree, and is therefore the required equation of the chord which passes through the intersections of the two given circles.

This common chord is also called the *radical axis* of the two circles.

To find the equation of the radical axis of two given circles, we have the following simple rule:

Make the coefficients of x^2 and y^2 the same in their equations, and then subtract one equation from the other.

The radical axis is always a real line whether the two circles meet in two real points or in two imaginary points; and becomes a common tangent when they meet in two coincident points.

EXERCISES ON THE CHORDS OF CIRCLES.

For examples on the intersections of loci, see Art. 20.

Find the locus of the middle points of a system of parallel chords for the circle:

1. $x^2 + y^2 = r^2$, for chords making 45° with the axis of X.

    Ans. $y = -x$.

2. $(x-d)^2 + (y-e)^2 = r^2$, for chords parallel to $\frac{x}{a} + \frac{y}{b} = 1$.

    Ans. $ax - by = ad - be$.

3. $x^2 + y^2 + 2Gx + 2Fy + C = 0$, for chords perpendicular to $Ax + By + C = 0$. *Ans.* $Ax + By + BF + AG = 0$.

Find the equations and lengths of the following chords for the circle $x^2 + y^2 = 25$:

4. Whose middle point is (1, 2). *Ans.* $2y + x = 5$; $4\sqrt{5}$.

5. Through $(-1, -3)$ parallel to $3x + 4y = 3$.

    Ans. $3x + 4y + 15 = 0$; 8.

6. Through (2, 1) perpendicular to $3x + 4y = 3$.

    Ans. $3y - 4x + 5 = 0$; $4\sqrt{6}$.

Find the equation of the radical axis and of the line through the centres for each of the following pairs of circles, and show that they are perpendicular to each other:

7. $x^2 + y^2 - 5x + 3y - 12 = 0$, $x^2 + y^2 + 3x - 4y + 5 = 0$.

    Ans. $7y - 8x = 17$; $16y + 14x = 11$.

8. $ax^2 + ay^2 - bx + cy + d = 0$, $x^2 + y^2 + cx - by + d' = 0$.

    Ans. $(ac + b)x - (ab + c)y = d - ad'$;
    $(ab + c)x + (ac + b)y = \frac{1}{2}(b^2 - c^2)$.

EXERCISES ON THE CHORDS OF CIRCLES. 151

9. $5x^2 + 5y^2 - 3x + 7y - 15 = 0$, $3x^2 + 3y^2 + 6x - 5y + 5 = 0$.

   Ans. $39x - 46y + 70 = 0$; $460x + 390y + 135 = 0$.

Find the length of the chords which each of the following circles cuts from the corresponding line, and also the equation of the circle described on the chord as a diameter:

10. $\begin{cases} x^2 + y^2 = r^2; & \text{chord, } 2\sqrt{r^2 - p^2}. \\ x\cos a + y\sin a = p; & \text{circle, } x^2 + y^2 - 2p(x\cos a + y\sin a) = r^2 - 2p^2. \end{cases}$

11. $\begin{cases} x^2 + y^2 = 25; & \text{chord, } 7\sqrt{2}. \\ x - y = 1; & \text{circle, } (x^2 + y^2) - (x - y) - 24 = 0. \end{cases}$

12. $\begin{cases} x^2 + y^2 = 65; & \text{chord, } \sqrt{10}. \\ 3x + y = 25; & \text{circle, } x^2 + y^2 - 5(3x + y) + 60 = 0. \end{cases}$

13. $\begin{cases} x^2 + y^2 - 6x + 2y - 15 = 0; & \text{chord, } 6. \\ 3y - 4x = 5; & \text{circle, } 5(x^2 + y^2) + 2(x - 7y) - 35 = 0. \end{cases}$

14. $\begin{cases} x^2 + y^2 = r^2; \text{ chord, } 2\left[\dfrac{(1 + m^2)r^2 - b^2}{1 + m^2}\right]^{\frac{1}{2}}. \\ y = mx + b; \text{ circle, } (1 + m^2)(x^2 + y^2) - 2b(y - mx) \\ \qquad\qquad\qquad\qquad\qquad\qquad\qquad = (1 + m^2)r^2 - 2b^2. \end{cases}$

15. $\begin{cases} x^2 + y^2 = r^2; \qquad \text{chord, } 2\left[\dfrac{(A^2 + B^2)r^2 - C^2}{A^2 + B^2}\right]^{\frac{1}{2}}. \\ Ax + By + C = 0; \text{ circle, } (A^2 + B^2)(x^2 + y^2) + 2C(Ax + By) \\ \qquad\qquad\qquad\qquad\qquad\qquad\qquad = (A^2 + B^2)r^2 - 2C^2. \end{cases}$

16. $\begin{cases} x^2 + y^2 = 2rx; \text{ chord, } \dfrac{2r}{\sqrt{1 + m^2}}. \\ y = mx; \qquad \text{circle, } (1 + m^2)(x^2 + y^2) - 2r(x + my) = 0. \end{cases}$

17. $\begin{cases} x^2 + y^2 = r^2; \qquad \text{chord, } \sqrt{4r^2 - (d^2 + e^2)}. \\ (x - d)^2 + (y - e)^2 = r^2; \text{ circle, } 2(x^2 + y^2) - 2(dx + ey) \\ \qquad\qquad\qquad\qquad\qquad\qquad\qquad = 2r^2 - (d^2 + e^2). \end{cases}$

Find the equations of the circles which are tangents

18. To the lines $x = 0$, $y = 0$, $x = c$.

    Ans. $4(x^2 + y^2) - 4c(x \mp y) + c^2 = 0$.

19. To the lines $x = 0$, $x = a$, and $3x + 4y + 5a = 0$.

    Ans. $x^2 + y^2 - ax + 2ay + a^2 = 0$, $x^2 + y^2 - ax + \tfrac{9}{2}ay + \tfrac{81}{16}a^2 = 0$.

20. To the lines $y=0$, $3y-4x=0$, $3y+4x=a$.

Ans. $9(x^2+y^2)- 3a(2x+y)+a^2=0$;
$64(x^2+y^2)-16a(x-2y)+a^2=0$;
$64(x^2+y^2)- 8a(2x+y)+a^2=0$;
$144(x^2+y^2)+24a(x-2y)+a^2=0$.

21. Circumscribing the triangle $y=0$, $3y-4x=0$, $3y+4x=a$.

Ans. $x^2+y^2-\dfrac{a}{4}x-\dfrac{7a}{86}y=0$.

79. *To find the equation of a straight line tangent to the circle* $x^2+y^2=r^2$.

DEFINITION.—*A straight line is tangent to a curve when it passes through any two coincident points on the curve.*

Since for the coincidence of the points P' and P'', or for tangency, $p=\pm r$, that is, $OD=\pm OP'$, Fig. 34, then the equation of the secant, Art. 75, becomes

$$x\cos a + y\sin a = \pm r, \tag{a}$$

the equation of two parallel lines both tangent to the circle $x^2+y^2=r^2$, the one at the point P', and the other at the end of the diameter through P', since the distance between the parallels is $2r$.

Equation (a) can be readily expressed in terms of the direction parameter m; for $\cot a = -\tan\theta = -m$, and (a) becomes

$$y = -\cot a \cdot x \pm r\csc a,$$

or $\qquad y = mx \pm r\sqrt{1+m^2},$ (b)

in which m is the tangent of the angle STP'.

Since m is arbitrary, (b) represents all possible pairs of parallel tangents to the circle $x^2+y^2=r^2$.

Instead of expressing the equation of the tangent in terms of the direction parameter m, we may express it in terms of the coordinates $(ON, P'N)$, $(x'y')$ of P', the point of tangency;

for, since $\cot TOP' = \dfrac{x'}{y'} = -\tan STP' = -m$, equation (b) becomes

$$xx' + yy' = \pm r^2, \qquad (c)$$

which represents the pair of parallel tangents at the point $P'(x'y')$, and the point $(-x', -y')$, the opposite end of the diameter from P'.

80. *To find the equation of the tangent to a given circle at the point $P'(x'y')$.*

The secant $SP'P''$, Fig. 34, will become a tangent when the points of intersection P' and P'' are made coincident. If the secant is supposed to revolve about the point P', it is plain that the point P'' may be made to approach P', and the two points will be coincident when $x'' = x'$ and $y'' = y'$, for which equation (a) of the secant, Art. 77, becomes

$$xx' + yy' = r^2,$$

the same equation of the tangent TP' as found in Art. 79. This equation of the tangent may also be found as follows:

The equation of the circle

$$x^2 + y^2 = r^2, \quad \text{or} \quad xx + yy = r^2,$$

shows that each point $P(xx, yy)$ on the curve is doubly represented, or may be regarded as two coincident points. If one of these points is $(x'y')$, then $xx + yy = r^2$ becomes $xx' + yy' = r^2$, an equation of the first degree, and therefore of a straight line, passing through the two coincident points (xy) and $(x'y')$ on the circle, which by definition is a tangent.

If the equation of the circle is

$$(x-d)^2 + (y-e)^2 = r^2, \quad \text{or} \quad (x-d)(x-d) + (y-e)(y-e) = r^2,$$

then $(x-d, y-e)$, $(x-d, y-e)$ are the coordinates of two coincident points on the circle, origin at the centre, and

$$(x-d)(x'-d)+(y-e)(y'-e)=r^2$$

is the equation of a straight line passing through the two coincident points $(x-d, y-e)$, $(x'-d, y'-e)$ on the circle, and is therefore a tangent at the point $(x'y')$, referred to the origin.

The coordinates (xy) in the general equation of the second degree,

$$Ax^2 + 2Hxy + By^2 + 2Gx + 2Fy + C = 0,$$

or $\quad Axx + H(xy+xy) + Byy + G(x+x) + F(y+y) + C = 0,$

doubly represent each point on the curve, and if one of them is $(x'y')$, then

$$Ax'x + H(x'y+xy') + By'y + G(x+x') + F(y+y') + C = 0$$

is an equation of the first degree, and therefore of a straight line, which passes through the two coincident points (xy) and $(x'y')$ on the curve, and is therefore a tangent at the point $(x'y')$.

81. *To find the equation of a normal to a given circle at the given point $P'(x'y')$.*

DEFINITION. — *A normal to a curve at any point on the curve is the straight line which is perpendicular to the tangent to the curve at this point.*

The equation of the tangent to the circle $x^2 + y^2 = r^2$ at the point $P'(x'y')$, Fig. 34, is

$$yy' + xx' = r^2, \quad \text{or} \quad y = -\frac{x'}{y'} x + \frac{r^2}{y'}.$$

The equation of a line through $P'(x'y')$ perpendicular to the tangent, is, Art. 46,

$$y - y' = \frac{y'}{x'}(x - x'), \quad \text{or} \quad yx' - xy' = 0,$$

which is the required equation, and shows that the normal to a circle always passes through the centre.

The equations of the normal and tangent to a circle at a given point $P'(x'y')$ may be found as follows:

LENGTHS OF TANGENT AND SUBTANGENT. 155

Assuming that the normal $P'O$ passes through the centre, as we may from Geometry, then its equation is $y = \frac{y'}{x'}x$, and the equation of the tangent TP' is

$$y - y' = -\frac{x'}{y'}(x - x'), \quad \text{or} \quad yy' + xx' = r^2,$$

as already found.

82. *To find the lengths of the tangent, subtangent, normal, and subnormal for a given point $P'(x'y')$ on the circle $x^2 + y^2 = r^2$.*

DEFINITION. — *The lengths of the tangent and normal to a curve are the distances measured on these lines from the point of tangency to the points in which they respectively cut the axis of X. The subtangent and subnormal are the respective projections of the tangent and normal on the axis of X.*

In the case of the circle $x^2 + y^2 = r^2$, $P'T$, $P'O$, NT, ON, Fig. 34, are the tangent, normal, subtangent, and subnormal for the point $P'(x'y')$.

We have at once the

$$\text{normal } OP' = r = \sqrt{x'^2 + y'^2}, \quad \text{subnormal } ON = x'.$$

The intercept OT of the tangent $yy' + xx' = r^2$ is, for $y = 0$,

$$OT = x = \frac{r^2}{x'} = \frac{x'^2 + y'^2}{x'},$$

and the subtangent

$$NT = OT - ON = \frac{x'^2 + y'^2}{x'} - x' = \frac{y'^2}{x'}.$$

The tangent is a mean proportional between its intercept on the axis of X and the subtangent;

$$\therefore \text{tangent} = PT = \sqrt{OT \cdot NT} = \sqrt{\frac{r^2}{x'} \cdot \frac{y'^2}{x'}} = \frac{ry'}{x'}.$$

83. *Tangents are drawn through a given point $T(h, k)$ to the circle $x^2 + y^2 = r^2$; to find the coordinates of the points of contact.*

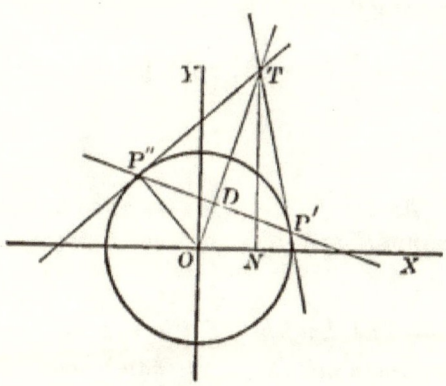

FIG. 35.

If the tangent $xx' + yy' = r^2$ to the circle $x^2 + y^2 = r^2$ at the point $P'(x'y')$ passes through the point $T(h, k)$, then

$$hx' + ky' = r^2, \text{ and } x'^2 + y'^2 = r^2,$$

both contain the required coordinates $(x'y')$ of the point of contact.

Solving these equations, we get

$$x' = \frac{r^2 h \pm rk\sqrt{h^2 + k^2 - r^2}}{h^2 + k^2}, \quad y' = \frac{r^2 k \mp rh\sqrt{h^2 + k^2 - r^2}}{h^2 + k^2},$$

the required coordinates of the points of contact; from which it follows that two tangents can be drawn to the circle through the given point $T(h, k)$.

These tangents are real for $h^2 + k^2 > r^2$; that is, when $T(h, k)$ is outside the circle; they are real and coincident when $h^2 + k^2 = r^2$, that is, when the point $T(h, k)$ is on the circle; and they are imaginary when $h^2 + k^2 < r^2$, that is, when the point $T(h, k)$ is within the circle.

84. *To find the length of the tangent drawn from a given point $T(h, k)$ to a given circle $x^2 + y^2 = r^2$.*

Let $T(ON, TN)$, (h, k), Fig. 35, be the given point; then $\overline{OT}^2 = h^2 + k^2$, $\overline{OP''}^2 = r^2$, and

$$TP'' = \sqrt{\overline{OT}^2 - \overline{OP''}^2} = \sqrt{h^2 + k^2 - r^2}$$

is the length required.

If the centre of the circle is at the point (d, e), then $\overline{OT}^2 = (h - d)^2 + (k - e)^2$, and the required length is

$$TP'' = \sqrt{(h - d)^2 + (k - e)^2 - r^2}.$$

These expressions for TP'' are obtained from the corresponding equations of the circle

$$x^2 + y^2 - r^2 = 0, \tag{a}$$

$$(x - d)^2 + (y - e)^2 - r^2 = 0, \tag{b}$$

by simply putting (h, k) in place of (x, y) and extracting the square root.

Since the general equation

$$x^2 + y^2 + 2Gx + 2Fy + C = 0$$

can be put in the form of (b), it follows that

$$TP'' = \sqrt{h^2 + k^2 + 2Gh + 2Fk + C}.$$

85. *To find the equation of the straight line passing through the points P' and P'' of tangency of the two tangents drawn through the given point $T(h, k)$. This line $P'P''$ is called the chord of contact.*

In Art. 83, the values of the coordinates $P'(x'y')$, $P''(x''y'')$ of the points of contact were found. Their differences are

$$x' - x'' = \frac{2rk\sqrt{h^2 + k^2 - r^2}}{h^2 + k^2}, \quad y' - y'' = -\frac{2rh\sqrt{h^2 + k^2 - r^2}}{h^2 + k^2}.$$

Therefore $\dfrac{y'-y''}{x'-x''} = -\dfrac{h}{k}$, which shows that the chord of contact $P'P''$ is perpendicular to OT. Fig. 35.

The general equation $y - y' = \dfrac{y'-y''}{x'-x''}(x-x')$ becomes

$$y - \frac{r^2k - rh\sqrt{h^2+k^2-r^2}}{h^2+k^2} = -\frac{h}{k}\left(x - \frac{r^2h + rk\sqrt{h^2+k^2-r^2}}{h^2+k^2}\right),$$

which readily reduces to

$$hx + ky = r^2, \tag{a}$$

the required equation of the chord of contact.

Second Solution. — The equations of the tangents at the points P' and P'' are

$$xx' + yy' = r^2, \quad xx'' + yy'' = r^2.$$

But since both of these tangents pass through the point $T(h, k)$, we have the two equations of condition,

$$hx' + ky' = r^2, \quad hx'' + ky'' = r^2,$$

which show that the points $(x'y')$ and $(x''y'')$ are both on the line whose equation is $hx + ky = r^2$, since they both satisfy it. Therefore $hx + ky = r^2$ is the equation of the chord of contact as before found.

This chord of contact will be a real line for all real values of (h, k) and r^2, the parameters of its equation $hx + ky = r^2$; that is, for any real circle and for any real point in the plane of this circle.

But when the point (h, k) is within the circle, the tangents TP' and TP'', as well as $(x'y')$ and $(x''y'')$ the coordinates of their points of contact are imaginary, and we have the remarkable proposition that a real line can be drawn through the imaginary points of contact of two imaginary tangents drawn from a real point within the circle.

EXERCISES ON TANGENTS, NORMALS, AND CHORDS.

To find the equations of tangents to given circles which make given angles with the axis of X:

CIRCLE.	ANGLE.	TANGENT.
1. $x^2 + y^2 = 25$,	$\frac{\pi}{4}$,	$y - x \mp 5\sqrt{2} = 0$.
2. $x^2 + y^2 = 100$,	$\frac{\pi}{3}$,	$y - x\sqrt{3} \pm 20 = 0$.
3. $(x-d)^2 + (y-e)^2 = r^2$,	$\theta = \tan^{-1} m$,	$y - e = m(x-d) \pm r\sqrt{1+m^2}$.
4. $x^2 + y^2 + 2Gx + 2Fy + C = 0$,	$\theta = \tan^{-1} m$,	$y + F = m(x+G)$ $\pm [(G^2 + F^2 - C)(1+m^2)]^{\frac{1}{2}}$.
5. $x^2 + y^2 = 2rx$,	$\theta = \tan^{-1} m$,	$y = m(x-r) \pm r\sqrt{1+m^2}$.
6. $x^2 + y^2 - 6x + 10y - 2 = 0$,	$\theta = \tan^{-1}(-\frac{3}{4})$,	$4y + 3x = 19$ or $= -41$.
7. $x^2 + y^2 + 8x - 6y - 24 = 0$,	$\theta = \tan^{-1}(-\frac{1}{2})$,	$2y + x = 2 \pm 7\sqrt{5}$.

To find the conditions that the following loci shall be tangents:

8. $(x-d)^2 + (y-e)^2 = r^2$,	$x \cos a + y \sin a = p$,	$r = d \cos a + e \sin a - p$.
9. $(x-d)^2 + (y-e)^2 = r^2$,	$y = mx + b$,	$r = \dfrac{e - md - b}{\sqrt{1+m^2}}$.
10. $(x-d)^2 + (y-e)^2 = r^2$,	$Ax + By + C = 0$,	$r = \dfrac{Ad + Be + C}{\sqrt{A^2 + B^2}}$.
11. $x^2 + y^2 = 2rx$,	$y = mx + b$,	$r^2 = 2rmb + b^2$.
12. $x^2 + y^2 + 2Gx + 2Fy + C = 0$,	$ax + by = 1$,	$G^2 + F^2 - C = \dfrac{(aG + bF + 1)^2}{a^2 + b^2}$.
13. $x^2 + y^2 = r^2$,	$Ax + By + C = 0$,	$r^2 = \dfrac{C^2}{A^2 + B^2}$.
14. $x^2 + y^2 - 4x + 6y + C = 0$,	$3x + 4y = 2$,	$C = \frac{261}{25}$.
15. $x^2 + y^2 + 2Gx + 2Fy + C = 0$,	$y = mx$,	$m = \dfrac{GF \pm \sqrt{C(G^2 + F^2 - C)}}{C - F^2}$.
16. $x^2 + y^2 + 2Gx + 2Fy + C = 0$,	$x^2 + y^2 + 2Fx + 2Gy + C = 0$,	$r = \dfrac{1}{\sqrt{2}}(F - G)$.

To find the equations of tangents and normals to the following circles, in terms of the coordinates of the points of tangency:

CIRCLE.	POINT.	TANGENT.	NORMAL.
17. $x^2 + y^2 = 25$,	$(4, 3)$,	$4x + 3y = 25$,	$4y - 3x = 0$.
18. $(x-2)^2 + (y-3)^2 = 10$,	$(5, 4)$,	$3x + y = 19$,	$3y - x = 7$.
19. $(x-c)^2 + (y-2c)^2 = 25c^2$,	$(5c, 5c)$,	$4x + 3y = 35c$,	$4y - 3x = 5c$.
20. $x^2 + y^2 - 14x - 4y - 5 = 0$,	$(10, 9)$,	$3x + 7y = 93$,	$3y - 7x + 43 = 0$.
21. $x^2 + y^2 + 2Gx + 2Fy + C = 0$,	$(x'y')$,		

$$x(x'+ G)+ y(y'+ F)+Gx'+ Fy'+ C = 0, \text{ tangent};$$
$$y(x'+ G)-x(y'+ F)+ Fx'- Gy'= 0, \text{ normal}.$$

To find the length of the tangent drawn from a given point to a given circle, and also the equation of the chord of contact:

CIRCLE.	POINT.	LENGTH.	EQUATION OF CHORD.
22. $x^2 + y^2 = 80$,	$(12, 6)$,	10,	$6x + 3y = 40$.
23. $(x-3)^2 + (y+6)^2 = 50$,	$(-5, 3)$,	$\sqrt{95}$.	$8x - 9y = 28$.
24. $x^2 + y^2 + 4x - 6y + 12 = 0$,	$(9, 10)$,	13,	$11x + 7y = 0$.
25. $4(x^2 + y^2) - 3x + 5y - 57 = 0$,	$(5, 8)$,	9,	$37x + 69y = 89$.
26. $x^2 + y^2 + 2Gx + 2Fy + C = 0$,	$(0, 0)$,	$\sqrt{C}$,	$Gx + Fy + C = 0$.

27. The length of the tangent from (G, F) to the circle $x^2 + y^2 - 6 = S = 0$ is twice the length of the tangent from (G, F) to the circle $x^2 + y^2 + 3(x + y) = S' = 0$. Show that $G^2 + F^2 + 4G + 4F + 2 = 0$.

28. The length of the tangent from the point (xy) to the circle $x^2 + y^2 + 2x = S = 0$ is three times its length from the circle $x^2 + y^2 - 4 = S' = 0$. Show that the point must be on the circle $S - 9S' = 0$, or $4x^2 + 4y^2 - x - 18 = 0$.

29. Find the equation of the circle through the intersections of the circles $x^2 + y^2 + 2x + 3y - 7 = S = 0$ and $x^2 + y^2 + 3x - 2y - 1 = S' = 0$, and through the point $(1, 2)$. *Ans.* $x^2 + y^2 + 4x - 7y + 5 = 0$.

30. Find the equation of a circle through the points of intersection of $x^2 + y^2 - 4 = S = 0$ and $x^2 + y^2 - 2x - 4y + 5 = S' = 0$ and tangent to the line $x + y - 3 = 0$. *Ans.* $x^2 + y^2 + 2x + 4y - 13 = 0$.

Poles and Polars with respect to a Circle.

86. Since the parameters (h, k) and r^2 of the equation $hx + ky = r^2$ of the chord of contact show that the position of this chord depends upon the point (h, k) and the radius r of the given circle $x^2 + y^2 = r^2$, it follows that there is a fixed relation between the point (h, k), the line $hx + ky = r^2$, and the circle $x^2 + y^2 = r^2$, which is called the *polar relation* of the point and line with respect to the circle. The point is called the *pole* of the line with respect to the circle, and the line is called the *polar* of the point with respect to the circle.

The principal properties of poles and polars with respect to the circle are contained in the following elementary propositions:

I. *The polar is perpendicular to the line which joins the pole and centre of the circle.*

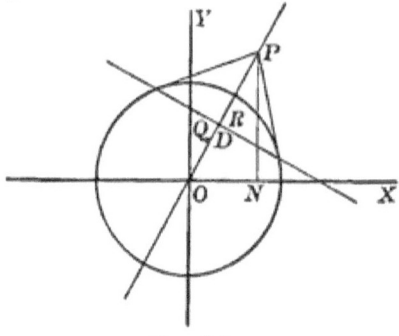

FIG. 36.

Let $P(h, k)$ be the pole; the equation of OP, the line joining the pole and centre of the circle $x^2 + y^2 = r^2$, is $hy - kx = 0$, to which the polar $hx + ky = r^2$ is perpendicular, by Art. 46.

II. *To construct the polar of the point $P(h, k)$ with respect to the circle $x^2 + y^2 = r^2$.*

The distance of the given pole $P(h, k)$ from the centre of the circle is $\sqrt{h^2 + k^2}$, and the distance of the polar from the centre of the circle is

$$OD = \frac{r^2}{\sqrt{h^2+k^2}} = \frac{r^2}{OP}.$$

Therefore, $\overline{OR}^2 = OD \cdot OP$; that is, the radius of the circle is a mean proportional between the distances of the pole and polar from the centre of the circle. Therefore, when the pole $P(h, k)$, or the distance $OP = \sqrt{h^2+k^2}$, is given, the distance OD is known; and conversely.

The relation $\overline{OR}^2 = OD \cdot OP$ involves the following properties:

III. *When the pole is outside the circle, the polar cuts the circle.*

IV. *When the pole is on the circle, the polar is the tangent at the pole.*

V. *When the pole is within the circle, the polar is without the circle.*

VI. *When the pole approaches indefinitely near the centre, the polar recedes indefinitely from the centre.*

VII. *The locus of the poles of a system of parallel polars is a straight line passing through the centre perpendicular to the polars.*

VIII. *If the polar of the point $P(h, k)$ passes through the point $Q(h', k')$, then will the polar of $Q(h', k')$ pass through $P(h, k)$.*

If $hx + ky = r^2$, the polar of the point $P(h, k)$, passes through the point $Q(h', k')$, then

$$hh' + kk' = r^2,$$

the condition that the polar of $P(h, k)$ passes through $Q(h', k')$ is also the condition that the polar $h'x + k'y = r^2$ of $Q(h', k')$ passes through $P(h, k)$, as was to be proved.

IX. *If the polar of $P(h, k)$ revolves about any fixed point $Q(h', k')$ in the plane of the circle, giving a system of polars,* then the locus of $P(h, k)$, the pole of this system, is a straight line which is the polar of $Q(h', k')$.

Since the polar $hx + ky = r^2$ of $P(h, k)$ passes through the fixed point $Q(h', k')$, the equation

$$hh' + kk' = r^2$$

is the locus of $P(h, k)$, which is a straight line and the polar of $Q(h', k')$.

X. *If a point $Q(h', k')$ moves along the polar of $P(h, k)$, then the polar of Q will revolve about P.*

Because the polar of Q in each new position will pass through P.

Propositions IX. and X. may be stated as follows:

If through any fixed point in the plane of a circle chords are drawn, and through the extremities of each tangents are drawn, then the intersection of each pair of tangents will lie on the same straight line.

If from any points in a given straight line pairs of tangents are drawn to a circle, the chords of contact will all pass through a fixed point.

XI. *If in any triangle $A(x'y')$, $B(x''y'')$, $C(x'''y''')$, any two vertices, as A and B, are the poles of their opposite sides, then the third vertex C is also the pole of its opposite side.*

Since the polars $xx' + yy' = r^2$, $xx'' + yy'' = r^2$ of $A(x'y')$ and $B(x''y'')$ both pass through $C(x'''y''')$, we have the equations of condition $x'''x' + y'''y' = r^2$, $x'''x'' + y'''y'' = r^2$, which show that $xx''' + yy''' = r^2$, the polar of $C(x'''y''')$ passes through A and B, and is therefore the equation of the side AB opposite C.

XII. Given a circle $x^2 + y^2 = r^2$, and the vertices $A(x'y')$, $B(x''y'')$, $C(x'''y''')$ of a triangle. The polars of A, B, C are the sides of a second triangle $A'B'C'$ called the *conjugate* of

164   PLANE ANALYTIC GEOMETRY.

ABC, the polars of A', B', C' being respectively the sides BC, AC, AB of the given triangle, by XI.

When the polars of A, B, C are respectively their opposite sides BC, AC, AB, the two triangles coincide, and the given triangle is called a *self-conjugate* triangle.

The lines AA', BB', CC', which join the corresponding vertices of a triangle and of its conjugate, meet in a point.

The equation of the line joining $A\,(x'y')$ to A', the intersection of the polars $xx''+yy''=r^2$, $xx'''+yy'''=r^2$, is, Art. 53,

$$(xx''+yy''-r^2)(x'x'''+y'y'''-r^2)$$
$$-(x'x''+y'y''-r^2)(xx'''+yy'''-r^2)=0.$$

In like manner, the equations of BB' and CC' are

$$(xx'''+yy'''-r^2)(x''x'+y''y'-r^2)$$
$$-(x''x'''+y''y'''-r^2)(xx'+yy'-r^2),$$

$$(xx'+yy'-r^2)(x'''x''+y'''y''-r^2)$$
$$-(x'''x'+y'''y'-r^2)(xx''+yy''-r^2);$$

and these lines pass through the same point, since the sum of their equations is zero.

EXERCISES ON POLES AND POLARS.

Given the following circles and poles, to find the corresponding polars:

	Circle.	Poles.	Polars.
1.	$x^2+y^2=9;$	$(8,3), (-5,4);$	$8x+3y=9,\quad 4y-5x=0.$
2.	$(x-1)^2+(y-2)^2=13;$	$(4,4), (8,-5);$	$3x+2y=20,\quad 7x-7y=6.$
3.	$x^2+y^2-3x-4y-8=0;$	$(-2,-3),(1,-4);$	$7x+10y=2,\quad x+12y+3=0.$
4.	$x^2+y^2+2Gx+2Fy+C=0;$	$(h,k), (0,0);$	$hx+ky+G(x+h)+F(y+k)+C=0,$ $Gx+Fy+C=0.$
5.	$4(x^2+y^2)-4x+6y-2=0;$	$(6,7), (-8,5);$	$22x+31y+7=0,\quad 23y-34x+29=0.$

EXERCISES ON POLES AND POLARS.

Given the following circles and polars, to find the corresponding poles:

CIRCLES.	POLARS.		POLES.
6. $x^2 + y^2 = 14$;	$3x + 4y = 7$,	$x - 2y = 2$;	$(6, 8), (7, -14)$.
7. $x^2 + y^2 = r^2$;	$ax + by = 1$,	$Ax + By + C = 0$;	(ar^2, br^2), $\left(-\dfrac{Ar^2}{C}, -\dfrac{Br^2}{C}\right)$.
8. $(x-1)^2 + (y-2)^2 = 12$;	$2x + 3y = 6$,	$3x - y = 2$;	$(-11, -16), (37, -10)$.
9. $(x+3)^2 + (y-5)^2 = 15$;	$2x - 3y + 18 = 0$,	$x + 2y = 2$;	$(7, -10), (-6, -1)$.

Given points of contact on a given circle, to find the equation of the chord of contact and the pole of this chord:

CIRCLES.	POINTS.	EQUATION OF CHORD.	POLE.
10. $x^2 + y^2 = 65$;	$(7, 4), (8, 1)$;	$y + 3x = 25$;	$(\tfrac{39}{5}, \tfrac{13}{5})$.
11. $x^2 + y^2 = 25$;	$(4, 3), (-3, -4)$;	$x - y = 1$;	$(25, -25)$.
12. $x^2 + y^2 - 5x - 3y + 6 = 0$;	$(1, 1), (2, 3)$;	$y - 2x + 1 = 0$;	$(\tfrac{1}{2}, \tfrac{5}{2})$.

13. Given the polars of $A(x'y')$ and $B(x''y'')$ with respect to the circle $x^2 + y^2 = r^2$. If AP is the perpendicular from A on the polar of B, and BQ the perpendicular from B on the polar of A, then
$$\frac{OA}{AP} = \frac{OB}{BQ}.$$

14. Show that the polar $hx + ky = r^2$ will touch the circle $x^2 + y^2 = 2rx$, if $k^2 + 2rh = r^2$.

15. The distances of two points from the centre of a circle are proportional to the distances of each from the polar of the other.

16. If a circle pass through the three given points $A(x'y')$, $B(x''y'')$, $C(x'''y''')$, then the polars of A, B, C with respect to this circle will form a second triangle $A'B'C'$. If A, A', B, B', C, C' are corresponding angles in the two triangles, show that the lines, AA', BB', CC' will meet in a point.

17. If any four points $A(x'y')$, $B(x''y'')$, $C(x'''y''')$, $D(x^{iv}y^{iv})$ lie on the circle $x^2 + y^2 = r^2$; if the tangents at A and B meet in P, and the tangents at C and D meet in Q, and the chords through A, B and C, D meet in R; then R is the pole of PQ.

18. Given the circle $x^2 + y^2 + 2Gx + 2Fy + C = 0$. Through the origin O draw two secants, one cutting the circle in the points A and A', and the other in B, B'. Suppose that the chords AB and $A'B'$ meet in P, and the chords AB' and BA' meet in Q; show that the origin is the pole of PQ.

19. If the four points in Ex. 17 are so taken that the tangents at A and B and the secant CD meet in a point, then will the tangents at C and D and the secant AB also meet in a point.

Elementary Propositions on Systems of Circles.*

87. *To find the equation of a system of circles concentric with the circle $S = 0$.*

In Art. 30 we found that the equation

$$Ax + By + C = P$$

represents a system of lines parallel to the line $P = 0$ so long as the parameters A, B, C are constant, P being a variable function of x and y for different lines of the system, but a constant function of x and y for each separate line of the system.

In the same way, we shall find that the equation

$$x^2 + y^2 + 2Gx + 2Fy + C = S \qquad (a)$$

represents a system of circles concentric with the given circle $S = 0$, whose fixed centre is $C(-G, -F)$ and radius $CT = \sqrt{G^2 + F^2 - C}$. Fig. 37.

For S is such a function of x and y that it will be a constant for the locus of any point $P(xy)$ in the plane of the circle $S = 0$ which moves so that its distance from the fixed point $C(-G, -F)$ remains constant; but will be a variable function of x and y

---

* The student who wishes a fuller treatment of this subject will find it in Salmon's "Conic Sections," or in Plücker's "Analytisch-Geometrische Entwicklungen."

when the point $P(xy)$ moves from one concentric circle to another of the system.

By writing (a) in the form

$$(x + G)^2 + (y + F)^2 = G^2 + F^2 - C + S,$$

we see at once that it represents a series of circles having the same centre $C(-G, -F)$ as the given circle $S = 0$, and radii depending upon the varying values of S.

S is positive for all concentrics external to $S = 0$, and negative for all concentrics internal to $S = 0$.

88. *To show that S is the square of the tangent to the given circle $S = 0$ drawn from any point $P(xy)$ in any one of its concentric circles.*

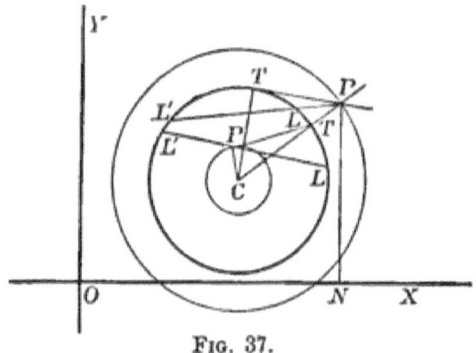

Fig. 37.

If $C(-G, -F)$ is the centre of $S = 0$, CT its radius, and $P(xy)$ any point on any circle concentric with $S = 0$, then $\overline{PT}^2 = \overline{CP}^2 - \overline{CT}^2$. But

$$\overline{CP}^2 = (x+G)^2 + (y+F)^2 \text{ and } \overline{CT}^2 = G^2 + F^2 - C.$$

$$\therefore \overline{PT}^2 = (x+G)^2 + (y+F)^2 - (G^2 + F^2 - C) = S = PL \cdot PL',$$

(Art. 74); and S is the square of the required tangent.

For all points $P(xy)$, on external concentrics, S is positive, and the tangents to $S = 0$ are real; and for all points $P(xy)$, on internal concentrics, S is negative, and the tangents to $S=0$ are imaginary. But in this case $-S = PL \cdot PL'$, the segments PL and PL' being measured in opposite directions from P.

89. *If $S=0$ and $S'=0$ are the equations of two given circles, centres at C and C', then $kS \pm lS' = 0$ is the equation of a system of circles all passing through the two points of intersection of the two given circles, k and l being arbitrary constants.*

First, the equation

$$kS \pm lS' = (k \pm l)(x^2 + y^2) + 2(kG \pm lG')x \\ + 2(kF \pm lF')y + kC \pm lC' = 0 \qquad (a)$$

represents circles referred to rectangular axes, since (Art. 70) it does not contain the product xy of the coordinates, and the coefficients of x^2 and y^2 are equal; and second, they all pass through the points of intersection of $S=0$ and $S'=0$, since the coordinates $(S=0, S'=0)$ of these points satisfy the equation $kS \pm lS' = 0$.

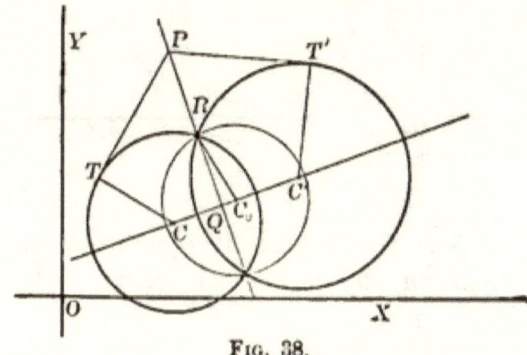

Fig. 38.

The coordinates of the centres of the two systems of circles, $kS + lS' = 0$ and $kS - lS' = 0$, are

$$\left(-\frac{kG+lG'}{k+l}, -\frac{kF+lF'}{k+l}\right), \quad \left(-\frac{kG-lG'}{k-l}, -\frac{kF-lF'}{k-l}\right),$$

which, by Art. 5, are the points found by dividing the distance between the centres $C(-G, -F)$ and $C''(-G', -F'')$, called the *central line*, internally and externally in the ratio $k : l$. It follows then that all the centres of the system $kS + lS' = 0$ are on the central line, and all the centres of the system $kS - lS' = 0$ are on the central line produced.

If we denote the radii of the circles of these two systems by $R_\pm$, and also put

$$r^2 = G^2 + F^2 - C, \quad r'^2 = G'^2 + F'^2 - C', \quad D^2 = (G - G')^2 + (F - F')^2,$$

then
$$R_\pm = \frac{(k \pm l)(kr^2 \pm lr'^2) \mp klD^2}{(k \pm l)^2}.$$

The circle of each of these systems for which $k = l$ is the *limiting circle* of the system.

As k and l approach equality, the centre of $kS + lS = 0$ approaches $C_0\left(-\dfrac{G + G'}{2}, -\dfrac{F + F'}{2}\right)$, the middle point of the central line, as the limit, its radius becomes

$$R_+ = \frac{2(r^2 + r'^2) - D^2}{4},$$

and its equation is

$$S + S' = 2(x^2 + y^2) + 2(G + G')x + 2(F + F')y + C + C' = 0.$$

As k and l approach equality, the centre of $kS - lS = 0$ recedes from C_0, its radius increases indefinitely; at the limit the circle becomes the straight line PQ, the common chord, or radical axis of all the circles of the system, and its equation is

$$S - S' = 2(G - G')x + 2(F - F')y + C - C' = 0.$$

The equation (a) may be written in a simple form by taking the central line CC' and the radical axis PQ as the axes of coordinates.

Let the constant length $RQ = d$ or $d\sqrt{-1}$ according as $S = 0$, $S' = 0$ intersect in two real or in two imaginary points; let the coordinates of the centre of the system of circles be $(\pm h, 0)$, h being a variable distance from the origin Q. Then

$$(x \pm h)^2 + y^2 = h^2 \pm d^2,$$

or $$x^2 + y^2 \pm 2hx \mp d^2 = 0,$$

is the equation of the system of circles all passing through the two points of intersection of $S=0$ and $S'=0$.

90. *To find the locus of a point $P(xy)$ which moves so that the tangents drawn from it to the given circles $S=0$ and $S'=0$ are always equal.*

By Art. 88, S and S' are the squares of the tangents drawn from the point $P(xy)$ to the circles $S=0$ and $S'=0$. Since these tangents are equal, we have at once

$$S = S' \quad \text{or} \quad S - S' = 0,$$

the equation of the required locus, which, by Art. 89, is the radical axis of the two given circles.

It is obvious also that $kS - lS' = 0$ is the equation of the locus of a point which moves so that the lengths of the tangents drawn from the point $P(xy)$ to the circles $S=0$ and $S'=0$ are in the ratio of $\sqrt{l} : \sqrt{k}$; and that the locus is a circle passing through the intersections of $S=0$ and $S'=0$.

91. *To show that the radical axes of the three given circles $S=0$, $S'=0$, $S''=0$, taken two and two meet in a point.*

The equations of the three radical axes are

$$S - S' = 0, \quad S' - S'' = 0, \quad S'' - S = 0,$$

and their sum is zero. Therefore the radical axes meet in a point. This point is called the *radical centre* of the three circles.

SYSTEMS OF CIRCLES.

92. *To find the coordinates of the internal and external points of section of the central line of the two given circles $S = 0$ and $S' = 0$ in which the two pairs of common tangents intersect.*

DEFINITIONS. — These points are called the *centres of similitude* of the two circles.

A *direct* tangent touches the two circles on the same side of the central line, and meets the central line produced in the *direct* centre of similitude.

An *inverse* tangent touches the two circles on opposite sides of the central line which it meets in the *inverse* centre of similitude.

The centres of similitude $P_i(xy)$ and $P_e(xy)$, together with the centres C, C' and the points of tangency T, T' are the vertices of similar right triangles whose homologous sides are in the ratio $r : r'$ of the radii of the two circles. Therefore, by Art. 5,

$$\left(-\frac{r'G + rG'}{r + r'},\ -\frac{r'F + rF''}{r + r'}\right),\ \left(-\frac{r'G - rG'}{r - r'},\ -\frac{r'F - rF''}{r - r'}\right)$$

are the required coordinates of the centres of similitude.

It is also obvious that lines drawn through the ends of any two parallel radii both lying on the same side of the central line, or lying on opposite sides of the central line, will pass through a centre of similitude.

Or, if these parallel radii are divided both internally or externally in the ratio of $r : r'$, the lines drawn through these points will also pass through a centre of similitude.

93. *To find the equations of the two circles which have their centres at the centres of similitude of the circles $S = 0$ and $S' = 0$ and pass through the intersections of these circles.*

If r and r' are the radii of $S = 0$ and $S' = 0$, then

$$r'S - rS' = 0, \qquad r'S + rS' = 0$$

are the required equations, since their centres are the same as the centres of similitude, and they pass through the intersections of $S=0$ and $S'=0$, by Art. 89.

94. *To find the four common tangents to two given circles $S=0$ and $S'=0$.*

The tangent to $S=0$ in terms of m is of the form

$$y = mx \pm r\sqrt{1+m^2}, \quad \text{or} \quad (y-e) = m(x-d) \pm r\sqrt{1+m^2},$$

according as the centre is at the origin $(0,0)$ or at the point (d, e). Since the required tangents must pass through the centres of similitude, put the coordinates of these centres for x and y in the equations of the tangents, and then find the values of m. Since the equations of the tangents to $S=0$ and $S'=0$ are quadratics in m, we shall find four values of m, which will give the four common tangents required.

95. *To find the condition that the two circles $S=0$ and $S'=0$, shall cut each other at right angles, or orthogonally.*

DEFINITION. — *The direction of a curve at any point is the same as the direction of the tangent at this point.*

It follows then that the angle made by two curves at a point of intersection is the same as the angle made by their tangents at this point. If two circles cross at right angles, their tangents, and therefore their radii at the point of crossing, are perpendicular to each other. Therefore

$$(G-G')^2 + (F-F')^2 = G^2 + F^2 - C + G'^2 + F'^2 - C',$$

or $\qquad 2GG' + 2FF' = C + C',$

is the required condition.

96. *To find the equation of the circle having its centre at the radical centre of three given circles, $S=0$, $S'=0$, $S''=0$, and cutting them orthogonally.*

Let $(x_c y_c)$ be the coordinates of the radical centre, and (xy) any point on the required circle; then

$$(x-x_c)^2 + (y-y_c)^2 = S_{cc} = S'_{cc} = S''_{cc}$$

is its equation in which S_{cc} denotes the square of the tangent from the radical centre to the circle $S=0$.

97. *Let $P(h, k)$, Fig. 37, be any point in the plane of the circle $S = 0$; to find the the equation of the circle orthogonal to $S = 0$, having $P(h, k)$ as its centre.*

Let (xy) be any point on the orthogonal circle; then

$$(x-h)^2 + (y-k)^2 = \overline{CT}^2 = S = h^2 + k^2 + 2hG + 2kF + C,$$

or $\quad x^2 + y^2 - 2h(x+G) - 2k(y+F) - C = 0,$

is the required equation.

98. *To find the equation of the system of circles which cut the two given circles $S' = 0$ and $S'' = 0$ orthogonally.*

If the circle $S=0$ cuts both $S'=0$ and $S''=0$ orthogonally, then the two equations of condition

$$2GG' + 2FF' - C - C' = 0,$$
$$2GG'' + 2FF'' - C - C'' = 0,$$

will give the values of any two of the three parameters G, F, C in terms of the third; and by substituting these values in $S=0$, we shall have the required equation. Since this equation still contains one arbitrary parameter, the circle can be made to fulfil one additional condition, such as passing through a given point or cutting a third circle $S''' = 0$ orthogonally.

The values of G, F, C found by solving the equations

$$2GG' + 2FF' - C - C' = 0,$$
$$2GG'' + 2FF'' - C - C'' = 0,$$
$$2GG''' + 2FF''' - C - C''' = 0,$$

will make the circle $S=0$ orthogonal to the three given circles.

EXERCISES ON SYSTEMS OF CIRCLES.

Given the circle $x^2 + y^2 - 2x + 4y + 1 = 0$, to find the equation of the concentric circle —

1. Through $(3, 2)$.   Ans. $x^2 + y^2 - 2x + 4y - 15 = 0$.
2. Through $(-5, 7)$.   Ans. $x^2 + y^2 - 2x + 4y - 112 = 0$.
3. Through $(-3, -5)$.   Ans. $x^2 + y^2 - 2x + 4y - 20 = 0$.

To find the length of the tangent to the circle $x^2 + y^2 + 6x - 4y + 4 = 0$ from —

4. The point $(3, 4)$.   Ans. $\sqrt{31}$.
5. The point $(-5, 6)$.   Ans. $\sqrt{11}$.
6. The point $(3, -7)$.   Ans. $6\sqrt{3}$.
7. The point $(-3, -5)$.   Ans. $2\sqrt{10}$.

To find the equation of the circle orthogonal to $x^2 + y^2 - 6x + 4y - 12 = 0$ having its centre at —

8. The point $(5, 3)$.   Ans. $x^2 + y^2 - 10x - 6y + 30 = 0$.
9. The point $(-3, 7)$.   Ans. $x^2 + y^2 + 6x - 14y - 34 = 0$.
10. The point $(-2, -5)$.   Ans. $x^2 + y^2 + 4x + 10y + 20 = 0$.

Given the circles
$$S = x^2 + y^2 - 2x - 4y - 2 = 0,$$
$$S' = x^2 + y^2 - 6x - 4 = 0,$$
$$S'' = x^2 + y^2 + 8x + 2y + 8 = 0,$$
to find their —

11. Radical axes.
    Ans. $2x - 2y + 1 = 0$, $7x + y + 6 = 0$, $5x + 3y + 5 = 0$.

12. Central lines.
    Ans. $y + x - 3 = 0$, $7y - x + 3 = 0$, $5y - 3x - 7 = 0$.

13. Radical centre and common orthogonal circle.
    Ans. $(-\frac{13}{16}, -\frac{5}{16})$, $(x + \frac{13}{16})^2 + (y + \frac{5}{16})^2 = \frac{299}{128}$.

EXERCISES ON SYSTEMS OF CIRCLES.

14. Given $x^2+y^2-4x-2y+4=0$, $x^2+y^2+4x+2y-4=0$, to find the centres of similitude and the four common tangents.

Ans. Direct centre, $(4, 2)$; tangents, $3y-4x+10=0$, $y=2$.

Inverse centre, $(1, \frac{1}{2})$; tangents, $4y+3x-5=0$, $x=1$.

15. Given $x^2+y^2-10x-4y+20=0$, $x^2+y^2-2y-3=0$, to find the centres of similitude and the four common tangents.

Ans. Direct centre, $(-10, -1)$; tangents, $12y-5x=38$, $y+1=0$.

Inverse centre, $(2, \frac{7}{5})$; tangents, $y+2.4x=6.2$, $x=2$.

16. Determine the circle $S = x^2+y^2+2Gx+2Fy+C=0$ so that it shall cut the circles
$$S' = x^2+y^2-4x-2y+4=0, \quad S'' = x^2+y^2+4x+2y-1=0$$
orthogonally, and show that the resulting equation represents a system of circles having a common radical axis.

17. If a circle cuts $S'=0$, $S''=0$, $S'''=0$ orthogonally, it will cut any circle $kS'+lS''+mS'''=0$ orthogonally.

18. If AB be the diameter of a given circle, the polar of A with respect to any circle which cuts the given circle orthogonally will pass through B.

19. The square of the tangent from any point of one circle to another is proportional to the perpendicular from that point upon their radical axis.

20. Find the equation of the circle which passes through the intersections of the circles $x^2+y^2=25$, $x^2+y^2-2x+4y=31$, and the point $(-8, 9)$.

21. Show that the length of the common chord of the two circles
$$(x-a)^2+(y-b)^2=c^2, \quad (x-b)^2+(y-a)^2=c^2,$$
is $[4c^2-2(a-b)^2]^{\frac{1}{2}}$.

22. If $\dfrac{G}{G'} = \dfrac{F}{F'}$, show that the circles
$$x^2+y^2+2Gx+2Fy=0, \quad x^2+y^2+2G'x+2F'y=0,$$
touch at the origin.

PROBLEMS ON THE CIRCLE.

1. A point so moves that the square of its distance from a fixed point varies as its perpendicular distance from a fixed straight line; show that the locus is a circle.

2. A point so moves that the sum of the squares of its distances from the four sides of a given square is constant; show that the locus is a circle.

3. The locus of a point, the sum of the squares of whose distances from n fixed points is constant, is a circle.

4. Find the locus of a point whose distances from two fixed points have a given ratio.

5. Show that the circle whose equation is
$$x^2 + y^2 + (2G + AC)x + (2F + BC)y = 0$$
passes through the origin and the points of intersection of
$$Ax + By = 1 \text{ and } x^2 + y^2 + 2Gx + 2Fy + C = 0.$$

6. What is the locus of the middle point of a straight line whose length is l, and whose ends move on a pair of rectangular coordinate axes?

7. Find the equation of the circle which passes through the origin and cuts off the positive intercepts a and b from the coordinate axes X and Y, respectively.

8. Find the equation of a straight line passing through the point $(x'y')$ from which the circle $x^2 + y^2 = r^2$ cuts a chord whose length is d.

9. Find the equation of the chord of the circle $x^2 + y^2 = r^2$ whose middle point is $(x'y')$.

10. A and B are two fixed points, and P so moves that $PA = k \cdot PB$. Show that the locus of P is a circle, and that all the circles for different values of k have a common radical axis.

11. Find the locus of a point which so moves that the square of its distance from the base of an isosceles triangle is equal to the product of its distances from the other sides.

12. Find the locus of a point whose polars with respect to two given circles make a given angle with one another.

13. Show that the radical axis of two circles bisects their four common tangents.

14. If a circle be described on the line joining the centres of similitude of two given circles as a diameter, show that the tangents drawn from any point on this circle to the two given circles are in the ratio of the corresponding radii.

15. Find the locus of a point such that tangents from it to two concentric circles are inversely as their radii.

16. A, B, C, D, are four fixed points on a straight line, and P a moving point such that the angles APB and CPD are equal; show that the locus of P is a circle.

17. Find the equation for determining the values of r for the points of intersection of the circle and the straight line whose equations are
$$r = 2a\cos\theta \text{ and } r\cos(\theta - a) = p.$$
Show that $p = 2a\cos^2\frac{a}{2}$ when the straight line becomes a tangent.

18. The polar equation of the circle on (a, a) and (b, β) as a diameter is
$$r^2 - [a\cos(\theta - a) + b\cos(\theta - \beta)] + ab\cos(a - \beta) = 0.$$

19. A circle passes through the origin and cuts the rectangular coordinate axes X and Y in points P and Q, such that the line PQ always passes through a fixed point (h, k); show that the locus of the centre of the circle is $\frac{x}{h} + \frac{y}{k} = 2.$

CHAPTER V.

THE CONIC SECTIONS.

The Equations of the Conic Sections.

99. The sections of a right cone made by a plane may be designated by the general term "the conic," and are included in the following simple definition.

A conic is the locus of a point which moves so that its distance from a fixed point is in a constant ratio to its distance from a fixed straight line.

The fixed point is called the *focus* of the conic; the fixed straight line is called its *directrix;* and the constant ratio, which is denoted by e, is called the *eccentricity* of the conic.

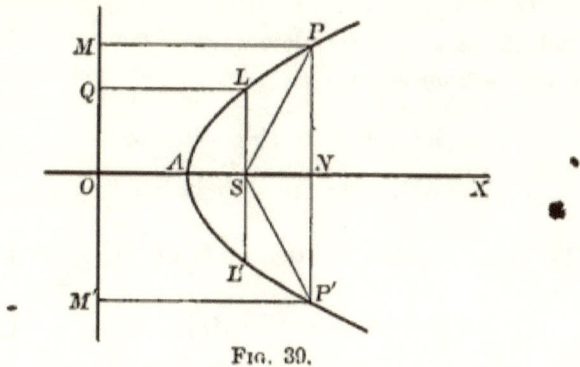

Fig. 30.

Let S be the focus, and the line OM the directrix. Through the focus draw SO perpendicular to OM, and take OX, OM as the rectangular axes of the conic. Put the given distance $OS = 2p$. Let $P(ON, PN)$, (xy) be any point on the conic, and let $PS = r$; then, by definition, $\dfrac{PS}{PM} = e$, or $r = ex$.

EQUATIONS OF THE CONIC SECTIONS. 179

When $e = 1$, the conic is called the *parabola*.
When $e < 1$, the conic is called the *ellipse*.
When $e > 1$, the conic is called the *hyperbola*.

100. *To find the equation of the conic referred to the rectangular axes OX, OM.*

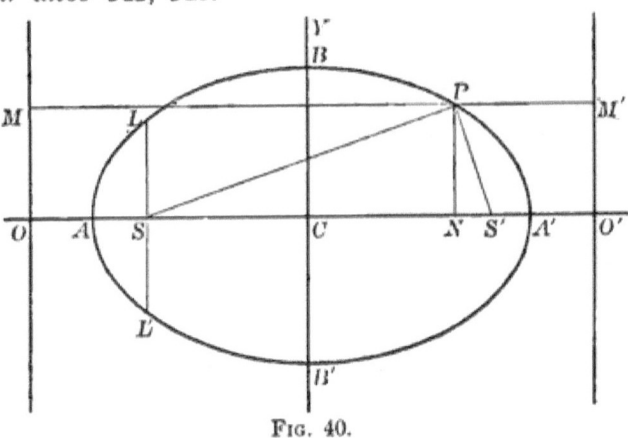

FIG. 40.

From the definition we get

$$\overline{PS}^2 = e^2 \overline{PM}^2.$$

But $\quad \overline{PS}^2 = \overline{PN}^2 + \overline{SN}^2 = y^2 + (x - 2p)^2, \quad \overline{PM}^2 = x^2.$

$$\therefore y^2 + (x - 2p)^2 = e^2 x^2 \tag{a}$$

is the required equation.

To find the points in which the conic cuts the axis of X, which are called its vertices.

For points on the axis of X, $y = 0$, for which (a) becomes

$$(x - 2p)^2 = e^2 x^2.$$

$$\therefore x_1 = \frac{2p}{1+e} = OA, \quad x_2 = \frac{2p}{1-e} = OA',$$

are the required distances from the directrix to the vertices A and A' of the conic.

For points on the axis OM, or directrix, $x = 0$, for which (a) becomes $y^2 + 4p^2 = 0$, which shows that y is imaginary, or that the conic does not cut the directrix.

For the parabola (Fig. 39),

$$e = 1, \quad x_1 = \frac{2p}{1+e} = p = OA, \quad x_2 = \frac{2p}{1-e} = \infty.$$

For the ellipse (Fig. 40),

$$e < 1, \quad x_1 = \frac{2p}{1+e} > p = OA, \quad x_2 = \frac{2p}{1-e} > 2p = OA'.$$

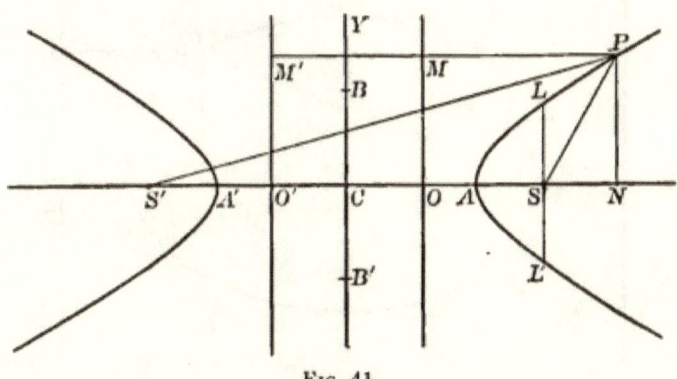

Fig. 41.

For the hyperbola (Fig. 41),

$$e > 1, \quad x_1 = \frac{2p}{1+e} < p = OA, \quad x_2 = \frac{2p}{1-e} < 0 = -OA'.$$

101. *The following properties of the conic are already apparent or are readily deduced.*

I. The conic is symmetrical with reference to the axis of x, since the definition $PS = ePM$ for any point P above the axis of x holds equally for the corresponding point P' below this axis (Fig. 39).

II. The conic does not cut the directrix.

III. The point C midway between the vertices A and A' is called the *centre* of the conic.

IV. The parabola is said to have but one vertex, since the other lies at an infinite distance from the directrix. The parabola is a non-central curve.

EQUATIONS OF THE CONIC SECTIONS.

Since, from the definition, the fixed point S might have been taken to the left of the directrix, it follows:

V. That if p is minus, the parabola lies to the left of the directrix.

VI. That the ellipse has two vertices A, A', and two foci S, S', lying between the two directrices OM and OM'.

VII. That the hyperbola has two vertices A, A' and two foci S, S' lying outside of the directrices OM and OM'.

DEFINITIONS. — The distance AA' between the vertices is called the *major* or *transverse axis* of the conic.

The middle point C of the major axis is called the *centre* of the conic.

The distance BB' which the conic cuts from the perpendicular to the major axis drawn through the centre C is called the *minor* or *conjugate axis*.

The distance LL' which the conic cuts from the perpendicular to the major axis drawn through the focus S is called the *latus-rectum*. It is the double ordinate through the focus.

VIII. *To find the length AA' of the major axis of the conic, which denote by $2a$, and the distance of its centre C from the directrix.* (Figs. 40, 41.)

Length $\quad AA' = OA' - OA = \dfrac{2p}{1-e} - \dfrac{2p}{1+e} = \dfrac{4pe}{1-e^2} = \pm 2a.$

Distance $\quad OC = OA + AC = \dfrac{2p}{1+e} + \dfrac{2pe}{1-e^2} = \dfrac{2p}{1-e^2} = \pm \dfrac{a}{e};$

in which the signs simply indicate the direction of the vertex A' from the vertex A, $+$ for the ellipse and $-$ for the hyperbola.

IX. *To find the length BB' of the minor axis of the conic, denoted by $2b$ for the ellipse, and by $2b\sqrt{-1}$ for the hyperbola.*

The abscissa of the centre C is $\dfrac{a}{e}$, and $2p = \dfrac{a(1-e^2)}{e}$, by VIII., which substituted for x and $2p$ in equation (a), Art. 100, give

$$y^2 = a^2 - \left(\dfrac{a}{e} - \dfrac{a(1-e^2)}{e}\right)^2 = a^2(1-e^2) = \pm \overline{CB}^2 = \pm b^2,$$

the plus sign corresponding to the ellipse and the minus sign to the hyperbola;

$\therefore BB' = 2b$ for the ellipse, and $2b\sqrt{-1}$ for the hyperbola.

X. *To find the length LL' of the latus-rectum of the conic.*

The abscissa of the focus S is $2p$, which substituted for x in equation (a) gives $y^2 = 4p^2 e^2$. The double ordinate

$$LL' = 2y = 4pe = 2a(1-e^2), \text{ by VIII.}, = \dfrac{2b^2}{a}, \text{ by IX.}$$

XI. *To find the eccentricity e in terms of a and b, the semi-axes.*

From IX.

$$a^2(1-e^2) = \pm b^2. \quad \therefore e = \dfrac{\sqrt{a^2 \mp b^2}}{a}.$$

XII. *To find the distance SC between the focus and the centre, which denote by c. This distance is called the linear eccentricity of the conic.*

$$SC = OC - OS = \dfrac{a}{e} - 2p = ae, \text{ by VIII.} \quad \therefore ae = c, \text{ or } \dfrac{c}{a} = e;$$

that is, the ratio of the linear eccentricity to the semi-major axis is the eccentricity e.

XIII. For any point $P(xy)$ on the ellipse or hyperbola the lines PS and PS' are called the focal distances of this point. Figs. 40, 41.

For the ellipse the sum, and for the hyperbola the difference, of the focal distances of any point $P(xy)$ on the curve is equal to the major axis $AA' = 2a$.

By definition
$$PS = ePM, \quad PS' = ePM', \quad \text{and} \quad OC = \frac{a}{e}, \text{ by VIII.}$$

For the ellipse
$$PS + PS' = e(PM + PM') = e \cdot 2 OC = 2a.$$

For the hyperbola
$$PS' - PS = e(PM' - PM) = e \cdot 2 OC = 2a.$$

XIV. *To express the focal distances of any point $P(xy)$ on the ellipse or hyperbola in terms of the abscissa of this point.*

$$\left. \begin{aligned} PS &= ePM = e(OC + CN) = e\left(\frac{a}{e} + x\right) = a + ex \\ PS' &= ePM' = e(O'C - CN) = e\left(\frac{a}{e} - x\right) = a - ex \end{aligned} \right\} \text{ellipse.}$$

$$\left. \begin{aligned} PS &= ePM = e(CN - OC) = e\left(x - \frac{a}{e}\right) = ex - a \\ PS' &= ePM' = e(CN + O'C) = e\left(x + \frac{a}{e}\right) = ex + a \end{aligned} \right\} \text{hyperbola.}$$

These equations are called the *linear equations* of the ellipse and hyperbola, the origin being at the centre C.

102. *To find the equation of the conic when the origin is at the vertex A. (Fig. 39.)*

Since the new origin A is on the axis of X, the ordinate y will remain the same, but we must put $x = x' + \frac{2p}{1+e}$. Art. 21, Case I. Substituting this value of x in equation (a) and dropping the prime, it becomes

$$y^2 + (1 - e^2)x^2 = 4pex, \qquad (b)$$

the required equation of the conic having its origin at the vertex A.

For $e=1$, (b) reduces to $y^2 = 4px$, which is the simplest form of the equation of the parabola.

For the ellipse and hyperbola, (b) may be written

$$y^2 = (1-e^2)\left(\frac{4pe}{1-e^2}x - x^2\right) = \pm\frac{b^2}{a^2}(2ax - x^2), \tag{b'}$$

since $1-e^2 = \pm\frac{b^2}{a^2}$, by IX., and $\frac{4pe}{1-e^2} = 2a$, by VIII.,

the origin for both curves being at the left-hand vertex, A for the ellipse and A' for the hyperbola (Figs. 40, 41).

If we put the semi-latus-rectum, Art. 101, X., $2pe = P$, then by VIII. (b) may be written

$$y^2 = 4pex - (1-e^2)x^2 = 4pex \pm \frac{2pe}{a}x^2 = 4pex \mp \frac{P}{a}x^2,$$

which shows that when $e=1$, the semi-transverse axis becomes infinite, and both the ellipse and hyperbola become the parabola $y^2 = 4px$.

103. *To find the equations of the conic when the origin is at the focus S. (Fig. 38.)*

In this case $y = y'$ and $x = x' + 2p$, since the origin is changed from O to S, a distance equal to $2p$. Substituting this value of x in equation (a), Art. 100, we get at once

$$y^2 + x^2 = e^2(x + 2p)^2 = (ex + 2pe)^2,$$

or $\quad y^2 + (1-e^2)x^2 - 2Pex - P^2 = 0, \tag{c}$

in which the eccentricity and semi-latus-rectum are the given parameters. For the parabola $e=1$, and its equation becomes $y^2 = 4p(x+p)$.

104. *To find the equations of the ellipse and hyperbola when the origin is at the centre C. (Figs. 40, 41.)*

In this case, $y = y'$ and $x = x' + a$ if we remove the origin from the left-hand vertex to the centre C.

Substituting this value of x in (b'), Art. 102, we get

$$y^2 = \pm\frac{b^2}{a^2}[2a(x+a) - (x+a)^2] = \frac{\pm b^2}{a^2}(a^2 - x^2).$$

$$\therefore \frac{y^2}{\pm b^2} = 1 - \frac{x^2}{a^2} \text{ or } \frac{x^2}{a^2} + \frac{y^2}{\pm b^2} = 1 \qquad (c)$$

are the required equations of the ellipse and hyperbola when the origin is at the centre C.

We may also find these equations as follows:

$$\overline{PN}^2 + \overline{SN}^2 = \overline{PS}^2, \text{ or } \overline{PN}^2 + (SC + CN)^2 = \overline{PS}^2, \text{ Fig. 40.}$$

But by XII. and XIV., Art. 101,

$$PN = y, \quad SC + CN = ae + x, \quad PS = a + ex.$$
$$\therefore y^2 + (ae + x)^2 = (a + ex)^2,$$

or $\quad \dfrac{x^2}{a^2} + \dfrac{y^2}{a^2(1-e^2)} = 1$, or $\dfrac{x^2}{a^2} + \dfrac{y^2}{\pm b^2} = 1$, by IX., Art. 101.

These equations may be written

$$\frac{(1-e^2)^2}{4p^2 e^2} x^2 + \frac{1-e^2}{4p^2 e^2} y^2 = 1, \qquad (c')$$

since from VIII. and IX., Art. 101, we readily get

$$a^2 = \frac{4p^2 e^2}{(1-e^2)^2}, \quad \pm b^2 = \frac{4p^2 e^2}{1 - e^2}.$$

If the coordinates of the vertex are (h, k), and the new axes are parallel to the old, the equation $y^2 = 4px$ of the parabola referred to this new origin and axes is

$$(y - k)^2 = 4p(x - h);$$

and when (h, k) are the coordinates of the centre of the ellipse and hyperbola, then their equations become

$$\frac{(x-h)^2}{a^2} + \frac{(y-k)^2}{\pm b^2} = 1.$$

The Ellipse. — When $b > a$, BB' becomes the major axis on which the foci lie.

When $b = a$, the ellipse becomes the circle $x^2 + y^2 = a^2$, having the major axis as a diameter, and is called the *major auxiliary circle*.

When $a = b$, the ellipse becomes the circle $x^2 + y^2 = b^2$, having the minor axis as a diameter, and is called the *minor auxiliary circle*.

The Hyperbola. — When the axis AA' becomes imaginary, and BB' real, the hyperbola $\dfrac{x^2}{-a^2} + \dfrac{y^2}{b^2} = 1$ is called the *conjugate* of $\dfrac{x^2}{a^2} + \dfrac{y^2}{-b^2} = 1$.

When $b = a$, the conjugate hyperbolas become $y^2 - x^2 = a^2$ and $x^2 - y^2 = a^2$, and are called *equilateral conjugates*.

105. *To find the equation of the conic when the focus S is on the directrix.*

In this case $2p = 0$, and equation (b), Art. 102, becomes
$$y^2 + (1 - e^2)x^2 = 0 \text{ or } y = \pm x\sqrt{e^2 - 1},$$
two straight lines which are real for $e > 1$, coincident for $e = 1$, and imaginary for $e < 1$, and must be classed among the conic sections.

106. DEFINITIONS. — A *diameter* of an ellipse or hyperbola is any line drawn through the centre C and terminating in the curve.

The axes are called the *principal* diameters. In the parabola a diameter is any line *parallel* to the axis of the curve.

107. *A diameter of the ellipse or hyperbola is bisected at the centre C.*

If $P(xy)$ is any point on the curve,
$$\frac{x^2}{a^2} + \frac{y^2}{\pm b^2} = 1,$$
then the point $(-x, -y)$ is also on the curve, for its coordinates satisfy the equations of the curve.

But the line joining the points (xy) and $(-x, -y)$ is bisected at the centre C.

108. *To trace the general form of the conic from its equation.*

For this purpose let us take $y^2 = 4px$, the equation of the parabola, and $\dfrac{x^2}{a^2} + \dfrac{y^2}{\pm b^2} = 1$, the equations of the ellipse and hyperbola, as the simplest forms, and for this reason the ones to be mainly used hereafter in deducing the more important properties of these curves.

First, let us examine the parabola (Fig. 39):

I. The equation $y = \pm \sqrt{4px}$ shows that for every value of x there are two equal values of y with opposite signs. The curve is therefore symmetrical with respect to the axis of X.

II. As there is no limit to the positive values of x, there is no limit to the corresponding plus and minus values of y. Therefore the halves of the curve above and below the axis of X extend to infinity.

III. For any negative value of x, y is imaginary. Therefore no part of the curve lies to the left of the axis of Y through the vertex A.

Second, the ellipse (Fig. 40):

IV. The equation $y = \pm \dfrac{b}{a}\sqrt{a^2 - x^2}$ shows that y is imaginary for values of x numerically greater than $+a$ or $-a$. Therefore the curve lies between the parallels $x = \pm a$, through A and A', and is symmetrical with respect to the major axis AA'.

V. The equation of the ellipse can also be written

$$x = \pm \dfrac{a}{b}\sqrt{b^2 - y^2},$$

which shows that the curve lies between the parallels $y = \pm b$, through B and B', and that the curve is symmetrical with respect to the minor axis BB'.

VI. The curve lies wholly within the rectangle whose area is $4ab$ to which it is tangent at the points A, A', B, B'.

Third, the hyperbola (Fig. 41);

VII. The equation $y = \pm \dfrac{b}{a}\sqrt{x^2 - a^2}$ show that y is imaginary for all values of x numerically less than $+a$ or $-a$. Therefore the curve lies outside of the parallels $x = \pm a$, through the vertices A and A'.

VIII. The curve is symmetrical with respect to the axis of x, and as y is real for all values of x numerically greater than $\pm a$, both branches of the curve extend to infinity.

IX. The curve is also symmetrical with reference to the minor axis BB'.

109. *To construct a conic, having given its equation.*

This method has been given in Art. 18 on the construction of loci.

110. *To trace a conic by points determined from known properties of the curve.*

The Parabola (Fig. 39). — Having given OS the distance between the focus and directrix, which lay down on any assumed line OX. The point A midway between O and S is the vertex of the curve. At any point N on the axis draw a line perpendicular to the axis. With S as a centre and ON as a radius, find the points P and P' on this perpendicular, which are points on the curve, since $SP = ON = PM$, by definition. In this way find as many points as are needed, through which to trace a sufficiently near approximation to the true curve.

The Ellipse (Fig. 40). — Having given $SC = c$, the linear eccentricity, and $AA' = 2a$, the major axis. On any assumed line OO' lay off $2SC = SS'$, and from C, the middle of SS', lay off $CA = \pm a$, and find the major axis AA'. Now since for any point P on the locus, $PS + PS' = 2a$, by Art. 101, XIII., we have this simple rule for finding points on the curve. Divide AA' into any two parts. With these parts as radii and the foci as centres, describe arcs of circles. These circles will meet on the ellipse since the sum of their radii equals AA'. In this way find as many points as are needed to trace the curve with sufficient accuracy.

The Hyperbola (Fig. 41). — Having given $SC = c$ the linear eccentricity, and $AA' = 2a$ the major axis. As in the ellipse find the points S, S', A, A' on any assumed line. Since for any point P on the locus $PS' - PS = 2a$, we have the following simple rule for finding points on the curve. On $A'A$ produced take any point X, then $A'X - AX = A'A$. With $A'X$ and AX as radii, describe circles with the foci S, S' as centres; these circles will intersect on the curve since the difference of their radii equals $A'A$. In this way find as many points as are needed to trace the curve.

111. *To trace a conic by motion with the data given in the last article.*

The Parabola (Fig. 39). — Let the right-hand edge of a ruler coincide with the directrix OM. Place the altitude of a triangular ruler against the directrix ruler, and let its base coincide with MP. Take a string equal to the base MB of the triangular ruler, fasten one end of it at the focus S, and the other end at the end B of the base of the triangular ruler. A pencil tightly pressed against the string will touch the base at P, a point on the curve, since $SP = PM$. As the triangular ruler slides along the directrix ruler, the point P of the pencil will move along and trace the curve.

The Ellipse (Fig. 40). — Find the points S, S', as in the last article. Take a string equal in length to $AA' = 2a$; fasten an end at each focus, press a pencil tightly against the string;

then its point P is on the curve, since $PS + PS' = 2a =$ the length of the string. The point of the pencil will move along and trace the curve, since $PS + PS'$ is equal to $2a$ for all points on the curve.

The Hyperbola (Fig. 41). — Find the points S, S', as in the last article. Let the length r of a ruler exceed $AA' = 2a$ by d, and take a string equal to d in length; then $r - d = 2a$. Fasten one end of the ruler at the focus S' about which it can turn; at the other or movable end of the ruler fasten one end of the string, and the other end of the string at the focus S'. Now press the string tightly with a pencil against the ruler at the point P, which is also on the curve, since $PS' - PS = 2a$. The point P of the pencil will move along and trace the right-hand branch of the hyperbola. The left-hand branch can be traced in the same way by fastening the ruler at S and the string at S'.

112. *To find the equation of the conic referred to a new pair of axes respectively parallel to OX and OM (Fig. 39), when the coordinates of the old origin are (h, k).*

The equations for transformation of coordinates are

$$x = h + x', \quad y = k + y'. \tag{Art. 21}$$

The Parabola. — If we take the equation $y^2 = 4px$, the coordinates of the vertex are (h, k), and we get

$$(y - k)^2 = 4p(x - h),$$

or $\quad y^2 - 2ky - 4px + k^2 + 4hp = 0,$

for the transformed equation.

This equation is of the general form

$$y^2 + 2Fy + 2Gx + C = 0;$$

and if we suppose that the two are identical, then we find by comparing coefficients that

EQUATIONS OF THE CONIC SECTIONS.

$$h = \frac{F^2 - C}{2G}, \quad k = -F, \quad 2p = -G.$$

Therefore, the latus-rectum $= -2G$; the vertex is $\left(\frac{F^2 - C}{2G}, -F\right)$; the focus is $\left(\frac{F^2 - G^2 - C}{2G}, -F\right)$; the equation of the axis is $y = -F$; the equation of the directrix is
$$x = \frac{G^2 + F^2 - C}{2G}.$$

The Ellipse and Hyperbola. — Since the equations of the ellipse and hyperbola, when the centre is the origin, are of the simple forms

$$\frac{x^2}{a^2} + \frac{y^2}{b^2} = 1, \text{ and } \frac{x^2}{a^2} + \frac{y^2}{-b^2} = 1,$$

it will be easier to proceed, as in Art. 70, to determine under what conditions the equation

$$Ax^2 + By^2 + 2Gx + 2Fy + C = 0 \tag{a}$$

will represent the one or the other of these curves, for which neither A nor B is zero.

By completing the squares (a) becomes

$$A\left(x + \frac{G}{A}\right)^2 + B\left(y + \frac{F}{B}\right)^2 = \frac{AF^2 + BG^2 - ABC}{AB} = K,$$

in which $\left(-\frac{G}{A}, -\frac{F}{B}\right)$ are the coordinates of the centre.

If we take the centre as the origin, then (a) reduces to the simple form

$$Ax^2 + By^2 = K, \tag{b}$$

and the axes of the conic coincide with the coordinate axes. Now let us find the major and minor axes. By making y and x equal zero in succession, we get the intercepts on the coordinate axes, which are

$$x^2 = \frac{K}{A} = a^2, \quad y^2 = \frac{K}{B} = b^2,$$

or
$$2a = 2\sqrt{\frac{K}{A}}, \quad 2b = 2\sqrt{\frac{K}{B}},$$

the required axes of the conic.

If K is minus, we can make it plus by changing all the signs in equation (b).

It follows then, assuming that K is always plus, that

I. If A and B are plus, (b) represents an ellipse.

II. If $A < B$, then $2a$ is the major axis.

III. If A and B are both minus, (b) represents an imaginary ellipse.

IV. If A and B have opposite signs, (b) represents an hyperbola.

The following results are readily found:

V. The semi-axes are $a = \sqrt{\frac{K}{A}}, \quad b = \sqrt{\frac{K}{B}}.$

VI. The eccentricity is $e = \sqrt{\frac{(B-A)}{B}}.$

VII. The linear eccentricity $SC = ae = \sqrt{\frac{K(B-A)}{AB}}.$

VIII. The distance $OS = 2p = \sqrt{\frac{KA}{B(B-A)}}.$

IX. The latus-rectum is $4pe = \frac{2}{B}\sqrt{KA}.$

X. The distance $OC = \frac{a}{e} = \sqrt{\frac{KB}{A(B-A)}}.$

XI. The coordinates of the centre are $\left(-\frac{G}{A}, -\frac{F}{B}\right).$

113. *To find the polar equation of the conic, the pole being at the focus S, and the axis of X the polar axis.* (Fig. 39.)

The rectangular equation of the conic, when the origin is at the focus S, is

$$y^2 + x^2 = e^2(x + 2p)^2 \quad \text{(Art. 103)}. \tag{c}$$

If (r, θ) are the polar coordinates of any point $P(xy)$ on the curve, then $x = r\cos\theta$, $y = r\sin\theta$, and (c) becomes

$$r^2 = e^2(r\cos\theta + 2p)^2,$$

or $\quad r = \pm e(r\cos\theta + 2p),$

or $\quad r = \dfrac{2pe}{1 - e\cos\theta}, \quad r' = -\dfrac{2pe}{1 + e\cos\theta},$

in which the positive root $r = SP$ is one segment of the focal chord PSP'', and the negative root $r' = SP''$, the other segment measured in the opposite direction from the focus S.

If the vectorial angle is measured from the line SA, that is, from the vertex A to the right, then these values of r become

$$\left.\begin{array}{l} r = \dfrac{2pe}{1 - e\cos(\pi - \theta)} = \dfrac{2pe}{1 + e\cos\theta}, \\[2mm] r' = -\dfrac{2pe}{1 + e\cos(\pi - \theta)} = -\dfrac{2pe}{1 - e\cos\theta}, \end{array}\right\} \tag{a}$$

the best forms for use.

If the polar axis makes an angle a with SA, then

$$r = \dfrac{2pe}{1 + e\cos(\theta - a)}, \quad r' = -\dfrac{2pe}{1 - e\cos(\theta - a)}.$$

If the semi-latus-rectum is denoted by P, as in Art. 102, then by X., Art. 101,

$$2pe = a(1 - e^2) = \dfrac{b^2}{a} = P,$$

and equations (a) become

$$r = \dfrac{P}{1 + e\cos\theta}, \quad r' = -\dfrac{P}{1 - e\cos\theta}.$$

PLANE ANALYTIC GEOMETRY.

We can deduce a positive value of r' from the value of r by increasing θ by π; then

$$r' = \frac{P}{1 + e\cos(\theta + \pi)} = \frac{P}{1 - e\cos\theta}.$$

114. *To find the polar equation of the ellipse and hyperbola when the pole is at the centre.*

The equation $\dfrac{x^2}{a^2} + \dfrac{y^2}{\pm b^2} = 1$ becomes

$$\frac{r^2 \cos^2\theta}{a^2} + \frac{r^2 \sin^2\theta}{\pm b^2} = 1, \text{ or } r^2 = \frac{\pm a^2 b^2}{a^2 \sin^2\theta \pm b^2 \cos^2\theta},$$

which may readily be reduced to

$$r = \frac{\pm b^2}{1 - e^2 \cos^2\theta},$$

the required equation.

EXERCISES ON EQUATIONS OF THE CONIC.

Given the values of (p, e), the axis of the curve, and the origin; to find the lengths of the semi-axes a, b; the linear eccentricity c; the latus-rectum LL'; and the equation of the curve. (See Figs. 39, 40, 41.)

Given $p = 2$, $e = \frac{1}{2}$, for the ellipse:

1. $a = \frac{8}{3}$, $b = \frac{4}{3}\sqrt{3}$, $c = \frac{4}{3}$, $LL' = 4$.
2. $4y^2 + 3x^2 - 32x + 64 = 0$, origin on axis at O.
3. $4y^2 + 3x^2 - 16x = 0$, origin on axis at A.
4. $4y^2 + 3x^2 - 8x - 16 = 0$, origin on axis at S.
5. $12y^2 + 9x^2 = 64$, origin on axis at C.
6. $12y^2 + 9x^2 - 96y - 54x + 209 = 0$, coordinates of C (3, 4).

Given $p = 5$, $e = \frac{3}{2}$, for the hyperbola:

7. $a = 12$, $b = 6\sqrt{-5}$, $c = 18$, $LL' = 30$.
8. $5x^2 - 4y^2 + 80x - 400 = 0$, origin on axis at O.

9. $5x^2 - 4y^2 + 120x = 0$, origin on axis at A.
10. $5x^2 - 4y^2 + 180x + 900 = 0$, origin on axis at S.
11. $5x^2 - 4y^2 = 720$, origin on axis at C.
12. $5x^2 - 4y^2 + 200x - 8y + 1276 = 0$, coordinates of S $(-2,-1)$.

Given $p = 3$, $e = 1$, for the parabola:
13. $LL' = 12$, $y^2 = 12x$, origin on axis at A.
14. $y^2 - 12x + 36 = 0$, origin on axis at O.
15. $y^2 - 12x - 36 = 0$, origin on axis at S.
16. $y^2 - 6y - 12x - 57 = 0$, coordinates of A $(-5, 3)$.
17. $y^2 + 2y - 12x - 11 = 0$, coordinates of S $(2, -1)$.

To find the equations of the following lines when the axis of X is the axis of the conic, and the origin is (Fig. 39):

18. At O. The equation of LL' is $x = 2p$; of OL is $y = ex$; of SQ is $y + ex = 2pe$; of AQ is $y + e(1+e)x = 2pe$; of AL is $y - (1+e)x + 2p = 0$.

19. At A. Of OM is $x = -\dfrac{2p}{1+e}$; of LL' is $x = \dfrac{2pe}{1+e}$; of OL is $y - ex = \dfrac{2pe}{1+e}$; of SQ is $y + ex = \dfrac{2pe^2}{1+e}$; of AL is $y = (1+e)x$; of AQ is $y + e(1+e)x = 0$.

20. At S. Of OM is $x = -2p$; of AY is $x = -\dfrac{2pe}{1+e}$; of OL is $y - ex = 2pe$.

21. At C. Of AB is $ay - bx = ab$ (Fig. 40); of LB' is $aey + (2pe + b)x + abe = 0$.

Take a new pair of axes (Fig. 40) respectively parallel to the axis of the conic and its directrix. Let (h, k) be the coordinates of the vertex A; then the following lines and points are expressed by:

22. $O\left(h - \dfrac{2p}{1+e}, k\right)$; $S\left(h + \dfrac{2pe}{1+e}, k\right)$; $C\left(h + \dfrac{2pe}{1-e^2}, k\right)$; $S'\left(h + \dfrac{2pe}{1-e}, k\right)$; $L\left(h + \dfrac{2pe}{1+e}, k + 2pe\right)$;

$$B\left(h+\frac{2pe}{1-e^2},\ k+\frac{2pe}{\sqrt{1-e^2}}\right);$$

the equation of OM is $x=h-\frac{2p}{1+e}$; of LL' is $x=h+\frac{2pe}{1+e}$;

of OO' is $y=k$; of BB' is $x=h+\frac{2pe}{1-e^2}$.

Find the latus-rectum, vertex, focus, axis, and directrix of each of the following parabolas:

23. $y^2-6y-8x-7=0.$
24. $x^2+10x-2y+4=0.$
25. $5y^2-10y-9x-10=0.$
26. $3x^2+12x-8y=0.$
27. $y^2-4y-5x-16=0.$
28. $5x^2-3x+2y-4=0.$
29. $3y^2+9y-8x-2=0.$
30. $x^2-4x-2y+3=0.$
31. $y^2+8y-12x+2=0.$
32. $x^2+2Gx+2Fy+C=0.$

	Latus-Rectum.	Vertex.	Focus.	Equation of Axis.	Equation of Directrix.
(23)	8;	$(-2, 3)$;	$(0, 3)$;	$y=3$;	$x=-4.$
(24)	2;	$(-5, -\frac{21}{2})$;	$(-5, -10)$;	$x=-5$;	$y=-11.$
(25)	$\frac{9}{5}$;	$(-\frac{5}{3}, 1)$;	$(-\frac{75}{60}, 1)$;	$y=1$;	$x=-\frac{127}{60}.$
(26)	$\frac{8}{3}$;	$(-2, -\frac{3}{2})$;	$(-2, -\frac{5}{6})$;	$x=-2$;	$y=-\frac{13}{6}.$
(27)	5;	$(-4, 2)$;	$(-\frac{11}{4}, 2)$;	$y=2$;	$x=-\frac{21}{4}.$
(28)	$-\frac{2}{5}$;	$(\frac{3}{10}, \frac{83}{40})$;	$(\frac{3}{10}, \frac{17}{8})$;	$x=\frac{3}{10}$;	$y=\frac{83}{40}.$
(29)	$\frac{8}{3}$;	$(-\frac{35}{32}, -\frac{3}{2})$;	$(-\frac{41}{96}, -\frac{3}{2})$;	$y=-\frac{3}{2}$;	$x=-\frac{149}{96}.$
(30)	2;	$(2, -\frac{1}{2})$;	$(2, 0)$;	$x=2$;	$y=-1.$
(31)	12;	$(-\frac{7}{6}, -4)$;	$(+\frac{11}{6}, -4)$;	$y=-4$;	$x=-\frac{25}{6}.$
(32)	$-2F$;	$\left(-G, \frac{G^2-C}{2F}\right)$;	$\left(-G, \frac{G^2-F^2-C}{2F}\right)$;	$x=-G$;	$y=\frac{G^2+F^2-C}{2F}.$

Find the semi-axes, eccentricity, latus-rectum, centre, and foci of each of the following ellipses and hyperbolas:

33. $9x^2+25y^2-36x-150y+36=0.$
34. $16x^2+25y^2+32x-100y-284=0.$
35. $4x^2+9y^2-8x+18y-23=0.$

36. $20x^2 + 24y^2 + 80x + 144y + 281 = 0$.

37. $3x^2 + 4y^2 - 6x + 8y + 1 = 0$.

38. $3x^2 - 2y^2 + 6x - 4y + 2 = 0$.

39. $5x^2 - 8y^2 + 3x - 2y + 1 = 0$.

40. $4x^2 - 12y^2 + 6x + 8y - 2 = 0$.

	Semi-Axes.	Eccentricity.	LL'.	Centres.	Foci.
(33)	$5, 3$;	$\frac{4}{5}$;	$\frac{18}{5}$;	$(2, 3)$;	$(2 \pm 4, 3)$.
(34)	$5, 4$;	$\frac{3}{5}$;	$\frac{32}{5}$;	$(-1, 2)$;	$(-1 \pm 3, 2)$.
(35)	$3, 2$;	$\frac{1}{3}\sqrt{5}$;	$\frac{8}{3}$;	$(1, -1)$;	$(1 \pm \sqrt{5}, -1)$.
(36)	$\frac{1}{2}\sqrt{3}, \frac{1}{2}\sqrt{10}$;	$\frac{1}{2}\sqrt{6}$;	$\frac{2}{3}\sqrt{3}$;	$(-2, -3)$;	$(-2 \pm \frac{1}{2}\sqrt{2}, -3)$.
(37)	$\sqrt{2}, \frac{1}{2}\sqrt{6}$;	$\frac{1}{2}$;	$\frac{3}{2}\sqrt{2}$;	$(1, -1)$;	$(1 \pm \frac{1}{2}\sqrt{2}, -1)$.
(38)	$\frac{1}{2}\sqrt{-3}, \frac{1}{2}\sqrt{2}$;	$\frac{1}{2}\sqrt{15}$;	$\frac{5}{3}\sqrt{2}$;	$(-1, -1)$;	$(-1, -1 \pm \frac{1}{6}\sqrt{30})$.
(39)	$\frac{3}{20}\sqrt{-6}, \frac{3}{20}\sqrt{15}$;	$\frac{1}{5}\sqrt{65}$;	$\frac{4}{25}\sqrt{15}$;	$(-\frac{3}{10}, -\frac{1}{8})$;	$(-\frac{3}{10}, -\frac{1}{8} \pm \frac{3}{40}\sqrt{39})$.
(40)	$\frac{1}{12}\sqrt{105}, \frac{1}{12}\sqrt{-35}$;	$\frac{1}{3}\sqrt{12}$;	$\frac{7}{5}$;	$(-\frac{3}{4}, \frac{1}{3})$;	$(-\frac{3}{4} \pm \frac{1}{6}\sqrt{35}, \frac{1}{3})$.

41. For what value of p will the parabola $y^2 = 4px$ pass through the point $(x'y')$? *Ans.* $p = \dfrac{y'^2}{4x'}$.

42. The focal distance of any point (xy) on the parabola $y^2 = 4px$ is $p + x$.

43. When the axis of the parabola and its directrix are coordinate axes, its equation is $y^2 - 4px + 4p^2 = 0$.

44. With the same axis of X, and the focus at the origin, the equation is $y^2 - 4px - 4p^2 = 0$.

45. In the parabola $y^2 = 4px$ the squares of the ordinates of any two points on the curve have the same ratio as the abscissæ of these points.

46. A parabola has the point A as its vertex, and passes through the points B, B' (Fig. 40); find its equation. *Ans.* $y^2 = 2pex$.

47. If S is the vertex, and the parabola passes through the points B, B', what is its equation? *Ans.* $y^2 = 2px$.

48. If B' is the vertex of a parabola and it passes through the foci S, S', what is its equation?
Ans. $x^2 = \dfrac{a^2 - b^2}{b} y$.

49. If the parabola has its vertex at B' and passes through A, A', its equation is $x^2 = \dfrac{a^2}{b} y$.

50. How are the four parabolas $y^2 = \pm 4px$, $x^2 = \pm 4py$ situated? They meet in the four points $(\pm 4p, \pm 4p)$. The circle through these points is $x^2 + y^2 = 32p^2$ (1), and the circle through the four foci is $x^2 + y^2 = p^2$ (2). The length of the tangent from any point in (1) to (2) is $p\sqrt{31}$.

51. For what point on the parabola is the ordinate n times the abscissa? *Ans.* $\left(\dfrac{4p}{n^2}, \dfrac{4p}{n}\right)$.

52. If the focal distance of a point on a parabola is n times the latus-rectum, what is the abscissa of the point? *Ans.* $x = (4n - 1)p$.

53. What are equations of the focal chords for $x = 4(n - 1)p$?
Ans. $2(n - 1)y \mp \sqrt{4n - 1}\,(x - p) = 0$.

54. In the ellipse or hyperbola the semi-conjugate axis is a mean proportional between the focal segments SA and SA' of the transverse axis. (Figs. 40, 41.)

Note.— In the ellipse the focus is a point of internal section; while in the hyperbola the focus is a point of external section.

55. In the ellipse and hyperbola the semi-latus-rectum is a harmonic mean between the focal segments of the transverse axis.

56. In any conic the semi-latus-rectum is a harmonic mean between the segments r and r' of any focal chord.

57. In any conic the sum of the reciprocals of any two perpendicular focal chords is $\dfrac{2 - e^2}{4pe}$.

EXERCISES ON EQUATIONS OF THE CONIC.

58. If the segments of any two perpendicular focal chords of a conic are SP, SP', and SQ, SQ', show that

$$\frac{1}{SP \cdot SP'} + \frac{1}{SQ \cdot SQ'} = \frac{2-e^2}{4p^2e^2}.$$

59. If in any conic r', r, r'' are the radius vectors which correspond to $\theta - 60°$, θ, $\theta + 60°$, then $\dfrac{1}{r'} + \dfrac{1}{r''} - \dfrac{1}{r} = \dfrac{1}{2pe}$.

60. If r and r' are the lengths of any two perpendicular radius vectors drawn from the vertex of the parabola $y^2 = 4px$, show that $(rr')^{\frac{4}{3}} = 16p^2(r^{\frac{2}{3}} + r'^{\frac{2}{3}})$.

61. In the ellipse or hyperbola the squares of the ordinates of any two points on the curve are in the same ratio as the products of the segments of the transverse axis made by the feet of these ordinates.

62. If through a fixed point O on the axis of a parabola any chord POP' be drawn, show that the product of the ordinates of P and P' will be constant; as will also the product of the abscissae of these points.

63. Two straight lines are drawn through the vertex of a parabola, at right angles to each other, meeting the curve in P and Q; show that the line PQ cuts the axis in a fixed point.

64. If the circle $x^2 + y^2 + 2Gx + 2Fy + C = 0$ cuts the parabola $y^2 - 4px = 0$ in four points, the algebraic sum of the ordinates of these points will be zero.

65. Show that the area of the triangle inscribed in the parabola $y^2 = 4px$ is $\dfrac{1}{8p}(y' - y'')(y'' - y''')(y''' - y')$ in which y', y'', y''' are the ordinates of the angular points.

66. Find the equation of the circle which passes through the vertex and the ends of the latus-rectum of the parabola $y^2 = 4px$.

Ans. $y^2 = (5p - x)x$.

67. If PSQ be a focal chord of a parabola, and PA meets the directrix in M, then MQ is parallel to the axis of the parabola.

68. Circles are described on any two perpendicular focal chords of a parabola as diameters; show that their common chord passes through the vertex of the parabola.

69. The equation of the ellipse which has the point $(-1, 1)$ for its focus, the line $4x - 3y = 0$ for its directrix, and $\frac{5}{8}$ for its eccentricity is $20x^2 + 24xy + 27y^2 + 72(x - y + 1) = 0$.

70. Show that the circle described on SP as a diameter (Fig. 40) touches the circle described on the transverse axis as a diameter.

71. If the ordinate MP of an hyperbola be produced to Q, so that $MQ = SP$, find the locus of Q.

72. A chord PQ of an ellipse is perpendicular to the major axis AA'; PA, QA are produced to meet at R; show that the locus of R is an hyperbola having the same axis as the ellipse.

73. In an ellipse, if PP' and QQ' be focal chords at right angles to each other, then

$$\frac{1-e^2}{SP \cdot SP'} + \frac{1-e^2}{SQ \cdot SQ'} = \frac{1}{AC^2} + \frac{1}{BC^2}.$$

74. Find the equation of the ellipse referred to coordinate axes passing through the ends of the minor axis, and meeting at one end of the major axis.

CHAPTER VI.

THE PARABOLA $y^2 = 4px$.

The Parabola is Concave towards the Axis.

115. *Prove that the parabola $y^2 = 4px$ is concave towards its axis AX. (Fig. 42.)*

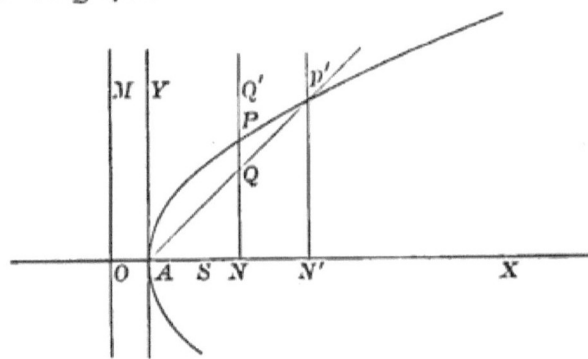

FIG. 42.

Let $P'(x'y')$ be any point on the curve; draw AP', and at any point N between A and N' erect a perpendicular to the axis AX. Let $P(xy)$ and $Q(xk)$ be corresponding points on the curve and line.

For the point $Q(xk)$, on the line AP', $k = \dfrac{y'}{x'}x$, and

$$k^2 = \frac{y'^2}{x'^2}x^2 = \frac{4p}{x'}x^2, \text{ since } y'^2 = 4px'.$$

For the point $P(xy)$, on the curve, $y^2 = 4px$; and by division $\dfrac{y^2}{k^2} = \dfrac{x'}{x}$. But $x' > x$, and therefore $y > k$; that is, the curve is concave towards AX, as was to be proved.

The ellipse and hyperbola, origin at left-hand vertex,

$$y^2 = \frac{\pm b^2}{a^2}(2ax - x^2) = \frac{\pm b^2}{a^2}\left(\frac{2a}{x} - 1\right)x^2, \qquad (b') \text{ Art. 102}$$

are also concave towards the axis of x.

For the point $Q(xk)$, on the line AP',

$$k = \frac{y'}{x'}x, \quad \text{and} \quad k^2 = \frac{y'^2}{x'^2}x^2 = \frac{\pm b^2}{a^2}\left(\frac{2a}{x'} - 1\right)x^2, \qquad (c)$$

and dividing (b') by (c),

$$\frac{y^2}{k^2} = \frac{\dfrac{2a}{x} - 1}{\dfrac{2a}{x'} - 1}.$$

But for $x' > x$, $\dfrac{2a}{x} - 1 > \dfrac{2a}{x'} - 1$, and $\therefore y > k$; that is, the curve is concave towards the axis of X.

116. *To find when the point (h, k) is without, on, or within the parabola $y^2 = 4px$. (Fig. 42.)*

The three points Q', P, Q have the same abscissa $AN = x = h$, and $4ph$ is the same for each; but for the ordinates, $Q'N > PN > QN$ (Art. 115). Therefore

For the point $Q'(hk)$ without the curve, $k^2 - 4ph > 0$.
For the point $P(hk)$ on the curve, $k^2 - 4ph = 0$.
For the point $Q(hk)$ within the curve, $k^2 - 4ph < 0$.

In the same way it can be shown that for the ellipse and hyperbola, $\dfrac{x^2}{a^2} + \dfrac{y^2}{\pm b^2} = 1$, we have

$$ak^2 \pm b^2h^2 \mp a^2b^2 > 0, \ = 0, \ < 0,$$

for points without, on, or within the curves.

The Parabola and the Secant Line.

117. *To find the equation of the secant which intersects the parabola* $y^2 = 4px$ *in the two given points* $P'(x'y')$ *and* $P''(x''y'')$. *(Fig. 43.)*

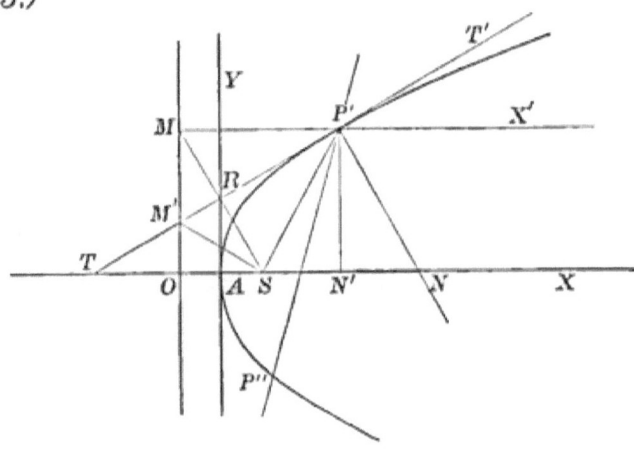

Fig. 43.

For these given points $y^2 = 4px$ becomes

$$y'^2 = 4px', \quad y''^2 = 4px'',$$

from which, by subtraction and division, we get

$$\frac{y' - y''}{x' - x''} = \frac{4p}{y' + y''} = m, \qquad (a)$$

which is the tangent of the angle θ, which the secant $P'P''$ makes with the axis of X (Art. 4).

This value of m, substituted in the equation of the straight line through two given points,

$$y - y' = \frac{y' - y''}{x' - x''}(x - x'),$$

gives
$$y - y' = \frac{4p}{y' + y''}(x - x'),$$

or
$$(y' + y'') y - 4px - y'y'' = 0, \qquad (b)$$

the required equation of the secant $P'P''$.

The Tangent and Normal.

118. *To find the equations of the tangent and normal to the parabola $y^2 = 4px$ at the point $P'(x'y')$.* *(Fig. 43.)*

I. *The Tangent.* — The secant $P'P''$ will become a tangent at the point $P'(x'y')$ by revolving it about P' until P'' and P' coincide, when $x'' = x'$ and $y'' = y'$. Introducing these conditions into (b), the equation of the secant (Art. 117), it becomes

$$2y'y - 4px - y'^2 = 0,$$

or $\qquad yy' = 2p(x + x'),\qquad\qquad$ (a)

the required equation of the tangent $P'T$.

II. *The Normal.* — The equation of a line through the point $P'(x'y')$, perpendicular to the tangent (a), is (Art. 50)

$$y - y' = -\frac{y'}{2p}(x - x'),$$

or $\qquad (y - y')2p + (x - x')y' = 0,\qquad$ (b)

the required equation of the normal $P'N$.

119. *To find the equations of the tangent and normal in terms of the direction parameter m.*

I. *The Tangent.* — The equation $yy' = 2p(x + x)$ of the tangent may be written

$$y = \frac{2px}{y'} + \frac{2px'}{y'} = \frac{2px}{y'} + \frac{y'}{2}.$$

But $\dfrac{2p}{y'} = m$, and $\dfrac{y'}{2} = \dfrac{p}{m}$, and therefore

$$y = mx + \frac{p}{m} \qquad\qquad (a)$$

is the required equation.

Since m is arbitrary, this is the equation of any tangent to the parabola $y^2 = 4px$.

If the coordinates of the vertex A are (h, k), then the equation of the tangent is

$$y - k = m(x - h) + \frac{p}{m}. \tag{b}$$

II. *The Normal.* — The equation of the normal may be written

$$y = -\frac{y'}{2p}x + y' + \frac{x'y'}{2p} = -\frac{y'}{2p}x + y' + \frac{y'^3}{8p^2}.$$

But $-\dfrac{y'}{2p} = m$, from which we get

$$y' = -2pm, \text{ and } \frac{y'^3}{8p^2} = -pm^3.$$

$$\therefore y = mx - 2pm - pm^3 \tag{c}$$

is the required equation.

III. *To find the subtangent TN' and the subnormal $N'N$. (Fig. 43.)*

(1) *The Subtangent.* — For $y = 0$, the tangent $yy' = 2p(x + x')$ gives

$$2p(x + x') = 0, \text{ or } x = -x' = -AT;$$

that is, $TA = AN'$, or the subtangent TN' is bisected by the vertex A. Therefore $TN' = 2x'$.

(2) *The Subnormal.* — For $y = 0$, the normal

$$(y - y')2p + (x - x')y' = 0$$

gives $\quad x = 2p + x' = AN;$

or $\quad x - x' = 2p = AN - AN' = N'N;$

which shows that the subnormal $N'N$ is constant and equal to OS, the distance from the focus to the directrix.

IV. *To find the value of m for the tangent to the parabola $y^2 = 4px$, which passes through the point (h, k).*

If the tangent $y = mx + \dfrac{p}{m}$ passes through (h, k), then

$$k = mh + \frac{p}{m},$$

from which we readily get

$$m = \frac{k \pm \sqrt{k^2 - 4ph}}{2h},$$

which shows that two real tangents can be drawn to the parabola when the point (h, k) is without the curve; two real and coincident ones when (h, k) is on the curve; and two imaginary ones when (h, k) is within the curve (Art. 116).

120. *A pair of perpendicular tangents to the parabola intersect on the directrix.*

If the equation of one tangent is $y = mx + \dfrac{p}{m}$, the perpendicular one is $y = -\dfrac{x}{m} - mp$. Eliminating y, by subtraction, we get

$$\left(m + \frac{1}{m}\right)x + \left(m + \frac{1}{m}\right)p = 0,$$

or $\qquad x = -p$,

which is the equation of the directrix, the locus of the intersections of all perpendicular tangents.

121. *The tangent $P'T$ makes equal angles with the focal distance $P'S$ and the axis of X; the normal $P'N$ makes equal angles with the focal distance $P'S$ and the diameter $P'X'$. (Fig. 43.)*

I. *The Tangent.*—Since $TA = AN'$ and $AS = OA$, we get by addition

$$TA + AS = AN' + OA,$$

or $\qquad TS = ON' = SP'$;

or the triangle $P'ST$ is isosceles, and $TP'S = P'TS = TP'M$.

II. *The Normal.* — The angles $NP'S$ and $NP'X'$ each equals a right angle diminished by the equal angles

$$TP'S = T'P'X'. \quad \therefore NP'S = NP'X'.$$

This shows that a ray of light, or a wave of sound, which is parallel to the axis AX, and strikes the curve at any internal point P', will be reflected to the focus; or if originating at the focus, will be reflected in lines parallel to the axis.

122. *A line drawn through the focus $S(p, 0)$ perpendicular to the tangent at $P'(x'y')$ meets it on AY, the tangent at the vertex A; and meets the diameter $P'X'$ corresponding to this tangent on the directrix at M. (Fig. 43.)*

The line through $S(p, 0)$ perpendicular to the tangent $yy' = 2p(x + x')$ is (Art. 50)

$$y = -\frac{y'}{2p}(x - p).$$

The point of intersection of these two lines is $R\left(0, \frac{y'}{2}\right)$, which is on the line $x = 0$, or AY.

Since $OS = 2AS$, $OM = 2RA = y'$, the ordinate of any point on $P'X'$, and M is on the diameter $P'X'$.

Also, since the tangent $P'T$ is perpendicular to SM at its middle point, *all points on this tangent are equidistant from the focus S, and the point M in which its diameter $P'X'$ meets the directrix.*

123. *A line drawn through the focus $S(p, 0)$ perpendicular to the focal distance SP' meets the tangent $P'T$ on the directrix OM. (Fig. 43.)*

The equation of the line SP', through $S(p, 0)$ and $P'(x'y')$, is (Art. 39)

$$y = \frac{y'}{x' - p}(x - p);$$

and the equation of the line SM' perpendicular to SP' is (Art. 46)

$$y = -\frac{x'-p}{y'}(x-p).$$

The point of intersection of SM' and the tangent

$$yy' = 2p(x+x') \text{ is } M'\left(-p, \frac{2p(x'-p)}{y'}\right),$$

a point on the directrix OM whose equation is $x = -p$.

124. *To draw a tangent to the parabola $y^2 = 4px$ at the point $P'(x'y')$. (Fig. 43.)*

Draw a perpendicular to the focal distance SP', and the line drawn through the points M' and P' will be the required tangent.

Or, take $N'N = 2p$, and draw the normal $P'N$. Through P', perpendicular to the normal, draw the required tangent.

125. *Through any point P without the parabola to draw two tangents.*

With P as a centre, and PS as a radius, describe a circle cutting the directrix in the points M and M'. The diameters of the parabola drawn through these points will intersect the curve in the required points of tangency.

For the point P on each tangent will be equidistant from S and the point in which its diameter meets the directrix (Art. 122).

126. *A tangent is drawn to the parabola $y^2 = 4px$ through a given point $P(h, k)$; to find the coordinates of the point of tangency. (Fig. 44.)*

The tangent to the parabola at the point $P'(x'y')$ is

$$yy' = 2p(x+x');$$

and since the tangent is drawn through the point $P(h, k)$, we have
$$ky' = 2p(h + x').$$
But for the point of contact $y'^2 = 4px'$, and by combining these equations, we readily get

$$y' = k \pm \sqrt{k^2 - 4ph}, \qquad (a)$$

$$x' = \frac{k^2 - 2ph \pm k\sqrt{k^2 - 4ph}}{2p}, \qquad (b)$$

the required coordinates of the point of tangency.

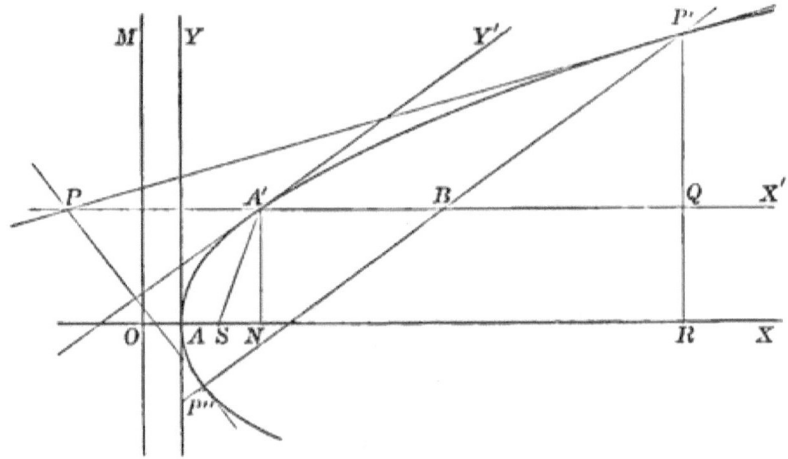

Fig. 44.

These values show that there are two real tangents for $k^2 - 4ph > 0$, a point without the curve; two real and coincident tangents for $k^2 - 4ph = 0$, a point on the curve; and two imaginary tangents for $k^2 - 4ph < 0$, a point within the curve.

127. *To find the equation of $P'P''$, the chord of contact, of two tangents drawn to the parabola from any point $P(h, k)$.*

The differences of the roots in (*a*) and (*b*) (Art. **126**) are

$$y' - y'' = 2\sqrt{k^2 - 4ph} \text{ and } x' - x'' = \frac{k\sqrt{k^2 - 4ph}}{p};$$

and the equation of $P'P''$ is therefore
$$y - (k + \sqrt{k^2 - 4ph}) = \frac{2p}{k}\left(x - \frac{k^2 - 2ph + k\sqrt{k^2 - 4ph}}{2p}\right),$$
which readily reduces to
$$ky = 2p(x + h),$$
the required equation.

Second Solution. — Since the tangents at $P'(x'y')$ and $P''(x''y'')$ both pass through the point $P(h, k)$, we have the two equations of condition
$$ky' = 2p(x' + h), \quad ky'' = 2p(x'' + h),$$
which show that $ky = 2p(x + h)$ is the required equation of the chord $P'P''$, since it is satisfied by the coordinates $(x'y')$ and $(x''y'')$ of the points of tangency.

As in the circle (Art. 86), this chord is also called the polar of the point $P(h, k)$ with respect to the parabola $y^2 = 4px$.

EXERCISES AND PROBLEMS ON CHORDS, TANGENTS, AND NORMALS.

1. Are the points $(3, 6)$, $(2, 4)$, $(4, 5)$, without, on, or within the parabola $y^2 = 8x$?

2. What is the equation of the secant which intersects the parabola $y^2 = 4x$ in the points $(1, 2)$ and $(2, 4)$? *Ans.* $3y - 2x = 4$.

Find the points in which the following parabolas and straight lines meet:

3. $y^2 = 8x$, $\quad 3y - 9x = 2$. $\quad$ Touch at $(\frac{2}{9}, \frac{4}{3})$.

4. $y^2 = 4px$, $\quad y = 3x - p$. $\quad$ Intersect in $(p, 2p)$, $\left(\frac{p}{9}, -\frac{2}{3}p\right)$.

5. $y^2 = 4px$, $\quad y = 2x + \frac{p}{2}$. $\quad$ Touch at $\left(\frac{p}{4}, p\right)$.

6. $y^2 - 7x - 8y + 14 = 0$, $\quad 7x + 6y = 13$. $\quad$ Touch at $(1, 1)$.

7. For what value of p will the parabola $y^2 = 4px$ pass through the point $(4, -8)$? $\quad$ *Ans.* $p = 4$.

EXERCISES AND PROBLEMS ON CHORDS, ETC.

8. Find the equations of the tangent and normal to the parabola $y^2 = 9x$ at the point $(1, 3)$.

Ans. Tangent, $2y - 3x = 3$; normal, $3y + 2x = 11$.

9. Find the lengths of the tangent, normal, subtangent, and subnormal of the last example. *Ans.* $\sqrt{13}$; $\frac{3}{2}\sqrt{13}$; 2; $4\frac{1}{2}$.

10. For what value of p will the parabola $y^2 = 4px$ touch the line $2y - 3x = 1$? *Ans.* $p = \frac{3}{4}$.

11. Show that $y - x - p = 0$, $y + x + p = 0$ are the tangents at the ends of the latus-rectum of the parabola $y^2 = 4px$, and that $y + x - 3p = 0$, $y - x + 3p = 0$ are the normals at the same points.

12. The equation of the line through the vertex A and the upper end L of the latus-rectum of the parabola $y^2 = 4px$ is $y = 2x$. (Fig. 39, Art. 99.)

13. What angle does the line AL make with the tangent at L?
Ans. $\tan^{-1}\frac{1}{3}$.

14. Find the point in which the normal at L again meets the parabola $y^2 = 4px$, and also the length of the intercepted chord.
Ans. $(9p, -6p)$; $8p\sqrt{2}$.

15. The tangents at the ends L and L' of the latus-rectum are perpendicular to each other.

16. The tangents to the parabola $y^2 = 4px$ at the points $(x'y')$ and $\left(\dfrac{p^2}{x'}, -\dfrac{4p^2}{y'}\right)$ are perpendicular to each other.

17. Show that for all values of m the line $y = m(x+p) + \dfrac{p}{m}$ is tangent to the parabola $y^2 = 4p(x+p)$.

18. Tangents to the parabolas $y^2 = 4p(x+p)$ and $y^2 = 4p'(x+p')$, which are perpendicular to each other, meet on the straight line $x + p + p' = 0$.

19. What angles do the tangents to the parabola $y^2 = 8x$, which pass through the point $(3, 7)$, make with the axis of x?
Ans. $\theta = \tan^{-1} 2$, or $= \tan^{-1}\frac{1}{3}$.

20. Tangents to the parabola $y^2 = 9x$ pass through the point $(4, 10)$; find the coordinates of the points of tangency.

Ans. $(36, 18)$; $(\frac{4}{9}, 2)$.

21. Find the coordinates of the point of tangency when the tangent makes an angle of $30°$ with the axis of x. Ans. $(3p, 2p\sqrt{3})$.

22. The length of the perpendicular from O (Fig. 43) on the tangent at $P'(x'y')$ is $\dfrac{p(x'-p)}{\sqrt{p(x'+p)}}$.

23. The perpendicular from the focus S on the tangent at $P'(x'y')$ is a mean proportional between p and $P'S$, the focal distance of the point of tangency.

24. Show that the length of a focal chord of the parabola $y^2 = 4px$, which makes an angle θ with the axis of x, is $\dfrac{4p}{\sin^2\theta}$.

25. Show that the product of the segments of any focal chord of the parabola $y^2 = 4px$ equals p times the length of this chord.

26. Show that the sum of the subtangent and subnormal for any point $P'(x'y')$ on the parabola $y^2 = 4px$ equals one-half of the focal chord parallel to the tangent.

27. Find the equations of the tangents to the parabola $y^2 = 4px$, which are respectively parallel and perpendicular to the line $\dfrac{x}{a} + \dfrac{y}{b} = 1$.

Ans. $aby + b^2x + a^2p = 0$; $aby - a^2x - b^2p = 0$.

28. The locus of the intersection of the tangents $y = mx + \dfrac{p}{m}$ and $y = m'x + \dfrac{p}{m'}$ is a straight line when mm' is constant; and when $mm' = -1$, this line is the directrix.

29. Find the points of contact of the tangents the perpendiculars on which from O (Fig. 43) are equal to one-fourth of the latus-rectum. Ans. $(0, 0)$; $(3p, 2p\sqrt{3})$.

30. A circle has its centre at A (Fig. 43), and its diameter is $3AS$; show that the common chord of the circle and parabola bisects AS.

31. The tangent at any point $P'(x'y')$ of the parabola $y^2 = 4px$ will meet the directrix and latus-rectum produced in two points equidistant from the focus.

Poles and Polars of the Parabola.

128. *If the polar of the point $P(h,k)$ passes through the point $Q(h',k')$, then will the polar of $Q(h',k')$ pass through $P(h,k)$.*

If $ky = 2p(x+h)$, the polar of $P(h,k)$, passes through $Q(h',k')$, then
$$kk' = 2p(h'+h),$$
the condition that the polar of $P(h,k)$ passes through $Q(h',k')$ is also the condition that the polar $k'y = 2p(x+h')$ of $Q(h',k')$ passes through $P(h,k)$, as was to be proved.

129. *The polar of the focus $S(p,0)$ is the directrix.*

For the polar of $S(p,0)$ is $x = -p$, the equation of the directrix.

130. *The polar of any point Q, on the directrix, passes through the focus $S(p,0)$.*

This follows from Arts. 128, 129. It follows, then, that if tangents be drawn to the parabola from any point on the directrix, the chords of contact will all pass through the focus.

Diameters of the Parabola.

131. *The locus of the middle points of a system of parallel chords of a parabola is a straight line parallel to the axis of the parabola. (Fig. 44.)*

In Art. 117 we found that

$$\frac{4p}{y'+y''} = m \qquad (a)$$

is the tangent of the angle θ, which the chord $P'P''$ makes with the axis of x. But for B, the middle point of this chord, $2y = y' + y''$, by (c), Art. 5, and (a) becomes

$$y = \frac{2p}{m} = 2p \cot \theta. \qquad (b)$$

But since $P'P''$ is any chord of the system, (b) is the equation of the required locus, which is a straight line $A'X'$, parallel to AX, the axis of the parabola, and is therefore a diameter of the curve.

132. *To find the equation of the locus of the intersections of the pairs of tangents to the parabola drawn through the ends of a system of parallel chords. (Fig. 44.)*

The equations of the tangents through the ends $P'(x'y')$, $P''(x''y'')$, of any chord $P'P''$ of the system are

$$yy' = 2p(x + x'),$$
$$yy'' = 2p(x + x''),$$

which by subtraction give

$$y(y' - y'') = 2p(x' - x'') = \tfrac{1}{2}(y'^2 - y''^2).$$

$$\therefore y = \frac{y' + y''}{2} = 2p \cot \theta$$

is the required equation of the locus, which by Art. 131 is also the equation of the locus of the middle points of the system of parallel chords.

133. *The tangent to the parabola at the vertex A' of the diameter $A'X'$ is parallel to the chords bisected by this diameter. (Fig. 44.)*

The ordinate $A'N$ of A' is $y = \frac{2p}{m}$ (Art. 131); therefore $\frac{4p^2}{m^2} = 4px$, and $x = \frac{p}{m^2} = AN$ is the abscissa of A'. The line through $A'\left(\frac{p}{m^2}, \frac{2p}{m}\right)$, parallel to the chord $P'P''$, is

$$y - \frac{2p}{m} = m\left(x - \frac{p}{m^2}\right) = mx - \frac{p}{m},$$

or
$$y = mx + \frac{p}{m},$$

which is the equation of a tangent (Art. 119).

The Parabola referred to Oblique Axes.

134. *To find the equation of the parabola referred to any diameter $A'X'$ and the tangent $A'Y'$ at the vertex A' of this diameter as coordinate axes. (Fig. 44.)*

Let the rectangular coordinates of P' be $(AR, P'R)$, (xy), and its oblique coordinates be $(A'B, P'B)$, $(x'y')$.

Then $AR = AN + A'B + BQ$, $P'R = A'N + P'Q$.

But $A'N = 2p \cot \theta$, $AN = p \cot^2 \theta$, (Arts. 131, 133)

and $BQ = y' \cos \theta$, $P'Q = y' \sin \theta$.

$\therefore x = p \cot^2 \theta + x' + y' \cos \theta$, $y = 2p \cot \theta + y' \sin \theta$.

Substituting these values of x and y in $y^2 = 4px$, we have

$$(2p \cot \theta + y' \sin \theta)^2 = 4p(p \cot^2 \theta + x' + y' \cos \theta),$$

which readily reduces to

$$y'^2 = \frac{4p}{\sin^2 \theta} x'.$$

But $\frac{p}{\sin^2 \theta} = SA' = p'$ = the focal distance of the vertex A'; for

$$SA' = ON = AS + AN = p + p \cot^2 \theta = \frac{p}{\sin^2 \theta} = p'.$$

Therefore, omitting primes on x and y, we have

$$y^2 = 4p'x,$$

the required equation of the parabola referred to a diameter and to the tangent at its vertex as coordinate axes.

135. As the equations of the parabola $y^2 = 4px$ and $y^2 = 4p'x$, referred to rectangular and oblique coordinates, do not differ in form, but only in the value of the parameters p and p', it follows that $y = mx + \dfrac{p'}{m}$ is a tangent for all values of m; the tangent at $P'(x'y')$ is $yy' = 2p'(x + x')$; for $y = 0$, $x = -x'$, that is, the subtangent is bisected by the vertex A'; the polar of the point $P(h, k)$ is $ky = 2p'(x + h)$; and the locus of the middle points of a system of chords, whose direction parameter is m, is
$$y = \frac{2p'}{m}.$$

PROBLEMS AND EXERCISES ON THE PARABOLA.

1. When the axis of the parabola and its directrix are coordinate axes, the tangent at the point $P'(x'y')$ is $yy' = 2p(x + x' - 2p)$.

2. When the origin is at the focus S, the tangent at $P'(x'y')$ is $yy' = 2p(x + x' + 2p)$.

The general equation of the parabola, whose axis is parallel to the axis of x, is $y^2 + 2Fy + 2Gx + C = 0$ (Art. 112). Find the equation of the parabola whose axis is parallel to the axis of x, when it passes through three given points.

3. The points $(3, 1)$, $(2, -2)$, $(-1, 5)$.
   Ans. $4y^2 - 3y + 21x - 64 = 0$.

4. The points $(2, -1)$, $(1, 0)$, $(3, 2)$.
   Ans. $2y^2 - y - 3x + 3 = 0$.

5. The points $(1, 1)$, $(0, 0)$, $(-1, 5)$.
   Ans. $3y^2 - 13y + 10x = 0$.

6. What is the polar of the point $(1, 2)$ with respect to the parabola $y^2 + 2x - 6y + 15 = 0$?   *Ans.* $x - y + 10 = 0$.

7. In what points does this polar intersect the parabola?
   Ans. $(-5, 5)$, $(-11, -1)$.

8. What are the tangents at these points?
   Ans. $2y + x = 5$, $4y - x = 7$.

9. Show that these tangents intersect in the point $(1, 2)$.

10. The polars of all points on the latus-rectum, or latus-rectum produced, pass through the point 0 (Fig. 39).

11. Tangents are drawn at the ends of two focal radii of the parabola $y^2 = 4px$; show that the angle between these radii is double the angle between the tangents.

12. What is the equation of the diameter of the parabola $y^2 = 4px$ which corresponds to the system of chords parallel to $x \cos a + y \sin a = p$? $\qquad$ Ans. $y = -2p \tan a$.

13. In the parabola $y^2 = 4px$ the chord bisected at the point (h, k) is $k(y - k) = 2p(x - h)$.

14. Show that the parameter of any diameter of the parabola $y^2 = 4p'x$ is the double ordinate through the focus.

15. The equation of the parabola referred to the tangents at the ends of the latus-rectum as coordinate axes is $\sqrt{x} + \sqrt{y} = \sqrt{2p\sqrt{2}}$.

16. The equation of the parabola referred to the tangent and normal at the upper end of the latus-rectum is $(y - x)^2 = 8px\sqrt{2}$.

17. Find the equation of the circle whose diameter is SP'', the focal distance of $P''(x'y')$ on the parabola $y^2 = 4px$.
$\qquad$ Ans. $x^2 + y^2 - x(p + x') - yy' + px' = 0$.

18. Find the equation of the tangent which is parallel to the polar of $(-1, 2)$ with respect to the parabola $y^2 = 12x$; also find the coordinates of the points of tangency. $\qquad$ Ans. $y - 3x = 1$, $(\frac{1}{3}, 2)$.

19. Find the equation of the locus of the middle points of a system of chords parallel to the polar of the point $(5, 3)$ with respect to the parabola $y^2 = 8x$. $\qquad$ Ans. $y = 3$.

20. Show that the tangents through the ends of any one of these chords intersect on the diameter $y = 3$.

21. The line $y = mx + b$ intersects the parabola $y^2 = 4px$ in two points; find the ordinate of the point midway between them.

22. If the normal at any point $P'(x'y')$ on the parabola $y^2 = 4px$ again meets the curve in Q, and $P'S = r$, and l is the length of the perpendicular from S to the tangent at P'; then $P'Q = \dfrac{4lr}{r - p}$.

23. Show that the circle described on SP' as a diameter touches the tangent at the vertex A of $y^2 = 4px$.

24. If the straight line $y = m(x - p)$ meets the parabola in $(x'y')$, $(x''y'')$, show that

$$x' + x'' = 2p + \frac{4p}{m^2}; \quad x'x'' = p^2; \quad y' + y'' = \frac{4p}{m}; \quad y'y'' = -4p^2.$$

25. A circle is described on a focal chord of $y^2 = 4px$ as a diameter; if m is the tangent of the angle which this chord makes with the axis of x, the equation of the circle is

$$x^2 + y^2 - 2px\left(1 + \frac{2}{m^2}\right) - \frac{4py}{m} - 3p^2 = 0.$$

26. Any circle described on a focal chord as diameter touches the directrix of the parabola.

27. If a chord of the parabola $y^2 = 4px$ is a tangent to the parabola $y^2 = 8p(x - c)$, show that the straight line $x = c$ bisects that chord.

28. Find the length of the perpendicular from an external point (h, k) on the chord of contact of the tangents to $y^2 = 4px$ drawn through (h, k).

$$\text{Ans.} \quad \frac{k^2 - 4ph}{\sqrt{k^2 + 4p^2}}.$$

29. From an external point (h, k) tangents are drawn to the parabola $y^2 = 4px$; the length of the chord of contact is

$$\frac{(k^2 + 4p^2)^{\frac{1}{2}}(k^2 - 4ph)^{\frac{1}{2}}}{p}.$$

30. From an external point (h, k) two tangents are drawn to the parabola $y^2 = 4px$; the area of the triangle formed by the tangents and chord of contact is $\dfrac{(k^2 - 4ph)^{\frac{3}{2}}}{2p}$.

PROBLEMS AND EXERCISES ON THE PARABOLA.

31. The normal at L, the upper end of the latus-rectum, meets $y^2 = 4px$ again in a point P. Show that the diameter on which the tangents at L and P intersect passes through the other end L' of the latus-rectum.

32. If the chord PQ is a normal at P, and the tangents at P and Q meet in T, show that PT is bisected by the directrix.

33. Find the ordinate of the point on $y^2 = 4px$ at which the tangent makes equal angles with the coordinate axes. *Ans.* $y = 2p$.

34. Find the locus of the intersection of perpendiculars from the focus S on the normal to $y^2 = 4px$. *Ans.* $y^2 = p(x-p)$.

35. Two normals to the parabola $y^2 = 4px$ are always at right angles to each other; find the locus of their intersections.
Ans. $y^2 = p(x - 3p)$.

36. Find the condition that the line $\dfrac{x}{a} + \dfrac{y}{b} = 1$ shall be tangent to the parabola $y^2 = 4px$. *Ans.* $b^2 = -ap$.

37. If two tangents be drawn to $y^2 = 4px$, and a third be drawn parallel to their chord of contact, show that the third will bisect the parts of the other two included between their point of intersection and points of contact.

38. If θ, θ' be the inclinations to the axis of two tangents to $y^2 = 4px$ drawn through the point (h, k), then
$$\tan\theta + \tan\theta' = \frac{k}{h}; \quad \tan\theta \tan\theta' = \frac{p}{h}.$$

39. If the line $y = mx$ is the locus of the pole (h, k) with respect to the parabola $y^2 = 4px$, then the polars of (h, k) all pass through the point $\left(0, \dfrac{2p}{m}\right)$.

CHAPTER VII.

THE ELLIPSE $\frac{x^2}{a^2} + \frac{y^2}{b^2} = 1$.

The Ellipse and the Secant Line.

136. *To find the equation of the secant which intersects the ellipse* $\frac{x^2}{a^2} + \frac{y^2}{b^2} = 1$ *in the two given points* $P'(x'y')$ *and* $P''(x''y'')$. (*Fig. 45.*)

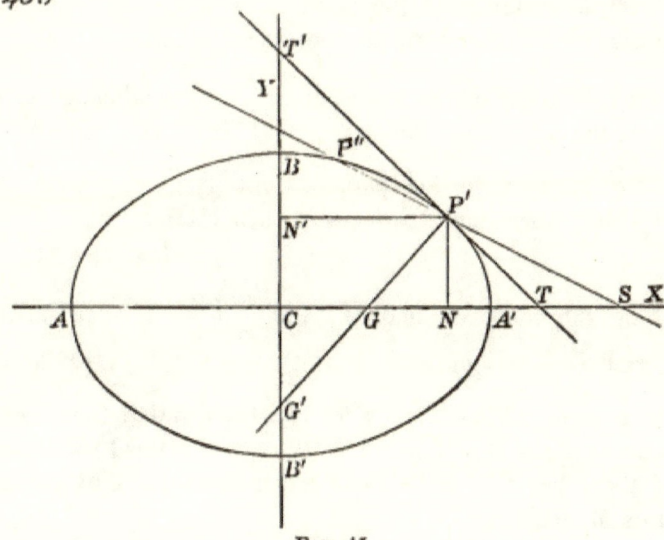

Fig. 45.

For these given points the equations of condition are

$$\frac{x'^2}{a^2} + \frac{y'^2}{b^2} = 1, \quad \frac{x''^2}{a^2} + \frac{y''^2}{b^2} = 1;$$

from which, by subtraction and division, we get

$$\frac{y' - y''}{x' - x''} = -\frac{b^2(x' + x'')}{a^2(y' + y'')} = m = \tan XSP'' \quad (a) \text{ (Art. 4)}$$

THE TANGENT AND NORMAL.

This value of m, substituted in

$$y - y' = \frac{y' - y''}{x' - x''}(x - x'),$$

the equation of the straight line through two given points, gives

$$y - y' = -\frac{b^2(x' + x'')}{a^2(y' + y'')}(x - x'), \tag{b}$$

the required equation of the secant $P'P''$, which may be reduced to the form

$$\frac{x(x' + x'')}{a^2} + \frac{y(y' + y'')}{b^2} = 1 + \frac{x'x''}{a^2} + \frac{y'y''}{b^2}, \tag{c}$$

since $\quad \dfrac{x'^2}{a^2} + \dfrac{y'^2}{b^2} = 1.$

The Tangent and Normal.

137. *To find the equations of the tangent and normal to the ellipse at the given point $P'(x'y')$. (Fig. 45.)*

I. *The Tangent.* — The secant $P'P''$ will become a tangent at the point $P'(x'y')$ by revolving it about P' until P'' and P' coincide, when $x'' = x'$ and $y'' = y'$.

Introducing these conditions into equations (*a*) and (*c*) (Art. 136), they become

$$m = -\frac{b^2 x'}{a^2 y'} = \tan XTP',$$

and $\quad \dfrac{yy'}{b^2} + \dfrac{xx'}{a^2} = 1. \tag{a}$

Therefore (*a*) is the equation of the tangent $P'T$.

II. *The Normal.* — The equation of the normal is (Art. 50)

$$y - y' = -\frac{1}{m}(x - x') = \frac{a^2 y'}{b^2 x'}(x - x'),$$

or $\quad \dfrac{a^2(x - x')}{x'} = \dfrac{b^2(y - y')}{y'}. \tag{b}$

138. *To find the length of the subtangent NT, and of the subnormal GN. (Fig. 45.)*

I. *The Subtangent NT.* — For $y = 0$, the equation (a) of the tangent (Art. 137) gives

$$x = \frac{a^2}{x'} = CT, \text{ or } CT \cdot CN = \overline{CA}'^2.$$

$$\therefore NT = CT - CN = \frac{a^2}{x'} - x' = \frac{a^2 - x'^2}{x'}.$$

Also, for $x = 0$,

$$y = \frac{b^2}{y'} = CT', \text{ or } CT' \cdot CN' = \overline{CB}^2.$$

$$\therefore N'T' = CT' - CN' = \frac{b^2}{y'} - y' = \frac{b^2 - y'^2}{y'}$$

= the subtangent on the axis of Y.

II. *The Subnormal GN.* — For $y = 0$, the equation (b) of the normal (Art. 137) gives

$$x - x' = -\frac{b^2 x'}{a^2}, \text{ or } x = \left(1 - \frac{b^2}{a^2}\right)x' = e^2 x' = CG.$$

$$\therefore GN = CN - CG = x' - e^2 x' = (1 - e^2)x' = \frac{b^2}{a^2}x'.$$

$$\therefore e^2 = \frac{CG}{CN}. \quad \therefore \frac{GN}{CN} = \frac{\overline{CB}^2}{\overline{CA}'^2}.$$

Also, for $x = 0$,

$$y - y' = -\frac{a^2}{b^2}y', \text{ or } y = \left(1 - \frac{a^2}{b^2}\right)y' = -\frac{a^2 e^2}{b^2}y' = CG'.$$

Also,

$$G'N' = CG' + CN' = \frac{a^2 e^2}{b^2}y' + y' = \frac{a^2}{b^2}y',$$

the subnormal on the axis of Y.

Also,

$$CT' \cdot CG' = \frac{b^2}{y'} \cdot \frac{a^2 e^2}{b^2}y' = a^2 e^2 = SC \cdot CS'.$$

THE TANGENT AND NORMAL.

139. *To find the equations of the tangent and normal in terms of the direction parameter m.*

I. *The Tangent*. — The equation (a) of the tangent (Art. 137) may be written

$$y = -\frac{b^2 x'}{a^2 y'} x + \frac{b^2}{y'}.$$

But $-\dfrac{b^2 x'}{a^2 y'} = m$; and since $\dfrac{x'^2}{a^2} + \dfrac{y'^2}{b^2} = 1$,

$$\frac{b^2}{y'} = \frac{b^2}{y'} \sqrt{\frac{x'^2}{a^2} + \frac{y'^2}{b^2}} = \pm \sqrt{a^2 \frac{b^4 x'^2}{a^4 y'^2} + b^2} = \pm \sqrt{a^2 m^2 + b^2};$$

and therefore

$$y = mx \pm \sqrt{a^2 m^2 + b^2} \qquad (a)$$

is the required equation in terms of m.

Since m is arbitrary, this is the equation of any tangent to the ellipse $\dfrac{x^2}{a^2} + \dfrac{y^2}{b^2} = 1$.

If the coordinates of the centre C of the ellipse are (h, k), then the equation of the tangent is

$$y - k = m(x - h) \pm \sqrt{a^2 m^2 + b^2}.$$

Equation (a) can be put in the form $x \cos a + y \sin a = p$; for $m = \tan \theta = -\cot a$, and it becomes

$$x \cos a + y \sin a = \sqrt{a^2 \cos^2 a + b^2 \sin^2 a} = a\sqrt{1 - e^2 \sin^2 a}, \quad (b)$$

the normal equation of the tangent.

II. *The Normal*. — The equation (b) of the normal (Art. 137) may be written

$$y = \frac{a^2 y'}{b^2 x'} x - \frac{a^2 y'}{b^2} + y' = \frac{a^2 y'}{b^2 x'} x - (a^2 - b^2) \frac{y'}{b^2}.$$

But $\dfrac{a^2 y'}{b^2 x'} = m$; and since $\dfrac{x'^2}{a^2} + \dfrac{y'^2}{b^2} = 1$,

$$\frac{y'}{b^2} = \frac{\dfrac{a^2}{x'} \times \dfrac{y'}{b^2}}{\dfrac{a^2}{x'}\sqrt{\dfrac{x'^2}{a^2}+\dfrac{y'^2}{b^2}}} = \pm \frac{\dfrac{a^2 y'}{b^2 x'}}{\sqrt{a^2 + b^2 \dfrac{a^4 y'^2}{b^4 x'^2}}} = \pm \frac{m}{\sqrt{a^2 + b^2 m^2}};$$

therefore

$$y = mx \pm \frac{(a^2 - b^2)m}{\sqrt{a^2 + b^2 m^2}} \tag{b}$$

is the required equation of the normal in terms of m.

This equation can be put in the normal form

$$x \cos a + y \sin a = p;$$

for $m = -\cot a$, and (b) readily reduces to

$$x \cos a + y \sin a = \frac{ae^2 \sin a \cos a}{\sqrt{1 - e^2 \cos^2 a}}. \tag{b}$$

140. *To find the value of m for the tangent to the ellipse $\dfrac{x^2}{a^2}+\dfrac{y^2}{b^2}=1$ which passes through the point $P(h, k)$.*

If the tangent $y = mx + \sqrt{a^2 m^2 + b^2}$ passes through the point $P(h, k)$, then

$$k = mh + \sqrt{a^2 m^2 + b^2};$$

from which we readily get

$$m = \frac{kh \pm \sqrt{a^2 k^2 + b^2 h^2 - a^2 b^2}}{h^2 - a^2},$$

which shows that two real tangents can be drawn to the ellipse when the point (h, k) is without the curve, two real and coincident ones when (h, k) is on the curve, and two imaginary ones when (h, k) is within the curve (Art. 116).

141. *To find the coordinates $(x'y')$ of the point of tangency of the tangent to the ellipse which passes through the point $P(h, k)$.*

THE CHORD OF CONTACT.

If the tangent $\dfrac{xx'}{a^2} + \dfrac{yy'}{b^2} = 1$ passes through $P(h, k)$, then the two equations

$$\frac{hx'}{a^2} + \frac{ky'}{b^2} = 1, \text{ and } \frac{x'^2}{a^2} + \frac{y'^2}{b^2} = 1,$$

give by solution the required coordinates

$$x' = \frac{a^2b^2h \pm a^2k\sqrt{a^2k^2 + b^2h^2 - a^2b^2}}{a^2k^2 + b^2h^2} = \frac{a^2b^2h \pm a^2kR}{a^2k^2 + b^2h^2}, \tag{a}$$

$$y' = \frac{a^2b^2k \mp b^2h\sqrt{a^2k^2 + b^2h^2 - a^2b^2}}{a^2k^2 + b^2h^2} = \frac{a^2b^2k \mp b^2hR}{a^2k^2 + b^2h^2}, \tag{b}$$

by putting $R = \sqrt{a^2k^2 + b^2h^2 - a^2b^2}$ for brevity; which show that two real tangents can be drawn to the ellipse when the point (h, k) is without the curve, two real and coincident ones when (h, k) is on the curve, and two imaginary ones when (h, k) is within the curve (Art. 116).

142. *To find the equation of the chord of contact of the two tangents drawn to the ellipse through the point $P(h, k)$.*

If $(x'y')$ and $(x''y'')$ are the coordinates of the points of tangency, then from (a) and (b) (Art. 141),

$$x' - x'' = \frac{2a^2kR}{a^2k^2 + b^2h^2}, \quad y' - y'' = -\frac{2b^2hR}{a^2k^2 + b^2h^2},$$

and $\quad \dfrac{y' - y''}{x' - x''} = -\dfrac{b^2h}{a^2k} = m \hfill \text{(Art. 4)}$

is the tangent of the angle which the chord of contact makes with the axis of X.

The equation of this chord is

$$y - \frac{a^2b^2k - b^2hR}{a^2k^2 + b^2h^2} = -\frac{b^2h}{a^2k}\left(x - \frac{a^2b^2h + a^2kR}{a^2k^2 + b^2h^2}\right),$$

which readily reduces to

$$\frac{ky}{b^2} + \frac{hx}{a^2} = 1. \tag{a}$$

Second Solution. — Since the tangents at the points $P'(x'y')$ and $P''(x''y'')$ both pass through the point $P(h, k)$, the equations of condition

$$\frac{ky'}{b^2} + \frac{hx'}{a^2} = 1, \quad \frac{ky''}{b^2} + \frac{hx''}{a^2} = 1,$$

show that $\frac{ky}{b^2} + \frac{hx}{a^2} = 1$ is the equation of the chord of contact, for it is satisfied by the coordinates $(x'y')$ and $(x''y'')$ of the points of tangency. As in the circle (Art. 86), this chord is also called the polar of the point $P(h, k)$ with respect to the ellipse $\frac{x^2}{a^2} + \frac{y^2}{b^2} = 1$.

143. *If the polar of the point $P(h, k)$ with respect to the ellipse $\frac{x^2}{a^2} + \frac{y^2}{b^2} = 1$ passes through the point $Q(h', k')$, then will the polar of $Q(h', k')$ pass through the point $P(h, k)$.*

This proposition is proved in the same way as the corresponding one relating to the parabola, in Art. 128. The polar of (h, k) is

$$\frac{ky}{b^2} + \frac{hx}{a^2} = 1;$$

and if (h, k) is the focus $S(-ae, 0)$, its polar is

$$-\frac{aex}{a^2} = 1, \quad \text{or} \quad x = -\frac{a}{e},$$

which is the equation of the corresponding directrix.

EXERCISES ON TANGENTS, NORMALS, AND POLARS.

1. Are the points $(3, 4)$, $(3, 1)$, $(2, 1)$ without, on, or within the ellipse $x^2 + 3y^2 = 12$?

2. Find the equations of the tangent and normal to the ellipse $2x^2 + 3y^2 = 11$ at the point $(2, 1)$.

Ans. $4x + 3y = 11$, $3x - 4y - 2 = 0$.

Given the ellipse $x^2 + 4y^2 = 13$, and the pole $(\tfrac{13}{5}, \tfrac{13}{10})$:

3. Find the equation of the polar. *Ans.* $x + 2y = 5$.

4. Find the points in which the polar intersects the ellipse.
Ans. $(2, \tfrac{3}{2})$, $(3, 1)$.

5. Find the equations of the tangents which pass through the pole $(\tfrac{13}{5}, \tfrac{13}{10})$. *Ans.* $2x + 6y = 13$, $3x + 4y = 13$.

6. Find the equations of the corresponding normals.
Ans. $2y - 6x + 9 = 0$, $3y - 4x + 9 = 0$.

7. In what points does the line passing through the pole perpendicular to the polar intersect the coordinate axes?
Ans. $(\tfrac{39}{20}, 0)$ on the axis of X, and $(0, -\tfrac{39}{10})$ on the axis of Y.

8. Is the line $y = x + \sqrt{\tfrac{5}{6}}$ tangent to $2x^2 + 3y^2 = 1$?

For the ellipse $a^2y^2 + b^2x^2 = a^2b^2$ find

9. The two tangents which make an angle of 60° with the axis of x. *Ans.* $y = x\sqrt{3} \pm \sqrt{3a^2 + b^2}$.

10. The two normals which make an angle of 45° with the axis of x.
Ans. $y = x \pm \dfrac{a^2 e^2}{\sqrt{a^2 + b^2}}$.

11. The ratio of the axes when the major axis equals n times the distance between the foci. *Ans.* $n : \sqrt{n^2 - 1}$.

12. The ratio of the axes when the centre and foci divide the major axis into four equal parts. *Ans.* $2 : \sqrt{3}$.

13. A tangent and normal each parallel to $3x - 4y = 5$.

14. A tangent and normal at L (Fig. 40, p. 179).
Ans. $y - ex = a$; $ey + x + ae^3 = 0$.

15. The equation of the line AL (Fig. 40), and the angle which AL makes with the tangent at L.
Ans. $y = (1 + e)(x + a)$, $\theta = \tan^{-1} \dfrac{1}{1 + e + e^2}$.

16. The equations of the four tangents which make equal intercepts on the axes. *Ans.* $y = \pm x \pm \sqrt{a^2 + b^2}$.

17. The eccentricity when the latus-rectum equals $\frac{1}{n}$ th of the minor axis. *Ans.* $e^2 = \dfrac{n^2 - 1}{n^2}$.

18. The eccentricity when the normal at L (Fig. 40) passes through B'. *Ans.* $e^4 + e^2 - 1 = 0$.

19. The equations of the lines $A'B$ and CL (Fig. 40), and the eccentricity when these lines are parallel.
Ans. $ay + bx = ab$; $a^2 ey + b^2 x = 0$; parallel if $2e^2 = 1$.

20. The coordinates of the point $(x'y')$ such that the intercepts on the coordinate axes by the tangent at this point are in the ratio $a : b$.
Ans. $x' = \dfrac{a}{\sqrt{2}}$, $y' = \dfrac{b}{\sqrt{2}}$.

21. The coordinates of a point $(x'y')$ such that the tangent at this point makes equal angles with the coordinate axes.
Ans. $x' = \dfrac{a^2}{\sqrt{a^2 + b^2}}$, $y' = \dfrac{b^2}{\sqrt{a^2 + b^2}}$.

22. The coordinates of the point $(x'y')$ for which the subtangent equals the subnormal. *Ans.* Same point as in last example.

23. The length of the perpendicular from the point $T(h, k)$ to its polar is $\dfrac{a^2 k^2 + b^2 h^2 - a^2 b^2}{\sqrt{a^4 k^2 + b^4 h^2}}$.

24. The area of the triangle formed by the tangents through the point $T(h, k)$ and the chord of contact is
$$\dfrac{(a^2 k^2 + b^2 h^2 - a^2 b^2)^{\frac{3}{2}}}{a^2 k^2 + b^2 h^2}.$$

25. If P' and P'' are the points of tangency, and C the centre of the ellipse, the area of the triangle $P''CP''$ is
$$\dfrac{a^2 b^2 \sqrt{a^2 k^2 + b^2 h^2 - a^2 b^2}}{a^2 k^2 + b^2 h^2}.$$

26. The area of the quadrilateral $TP'CP''$ is $\sqrt{a^2 k^2 + b^2 h^2 - a^2 b^2}$.

27. If any ordinate MP be produced to meet the tangent at the end of the latus-rectum, through the focus S in Q, then $QM = SP$.

28. The length of the chord cut from the line $y = mx + c$ is
$$\frac{2ab\left[(1+m^2)(a^2m^2+b^2-c^2)\right]^{\frac{1}{2}}}{a^2m^2+b^2}.$$

29. The coordinates of the middle point of this chord are
$$x = -\frac{a^2mc}{a^2m^2+b^2}, \quad y = \frac{b^2c}{a^2m^2+b^2}.$$

30. The coordinates of the points of tangency of the tangents parallel to $y = mx + c$ are
$$x' = \mp \frac{a^2m}{\sqrt{a^2m^2+b^2}}, \quad y' = \pm \frac{b^2}{\sqrt{a^2m^2+b^2}}.$$

31. The equations of the tangents at these points are
$$y = mx \pm \sqrt{a^2m^2+b^2}.$$

32. The circle described on the focal distance $P(x'y')$ $S(-ae, 0)$ is $y^2 + x^2 - yy' - x(x' - ae) - aex' = 0$.

33. If m and m' are the tangents of the angles which the two tangents through the point (h, k) make with the axis of X, then
$$m + m' = \frac{2kh}{h^2 - a^2}, \quad mm' = -\frac{b^2 - k^2}{h^2 - a^2}.$$

34. From any external point (h, k) two tangents are drawn; if x', x'' are the abscissas of the points of tangency, then
$$x' + x'' = \frac{2ha^2b^2}{a^2k^2 + b^2h^2}, \quad x'x'' = \frac{a^4(b^2 - k^2)}{a^2k^2 + b^2h^2}.$$

Elementary Propositions relating to Tangents and Normals.

144. *The locus of the intersections of all pairs of perpendicular tangents to the ellipse is the circle $y^2 + x^2 = a^2 + b^2$, which is called the director circle of the ellipse.* (Fig. 46.)

The tangent $P''I$ perpendicular to the tangent $P'T'$,

is
$$y - mx = \sqrt{a^2m^2 + b^2}, \qquad (a) \text{ (Art. 136)}$$
$$my + x = \sqrt{a^2 + b^2m^2}. \qquad \text{(Art. 46)}$$

230   PLANE ANALYTIC GEOMETRY.

Squaring both sides of these equations, adding, and dividing the sum by $(1 + m^2)$, gives

$$y^2 + x^2 = a^2 + b^2,$$

the equation of the director circle. If, then, with C as a centre and $\sqrt{a^2 + b^2}$ as a radius, a circle be described around the ellipse, it will pass through I, and also through the intersections of all other perpendicular tangents. It appears also that $CI = AB$.

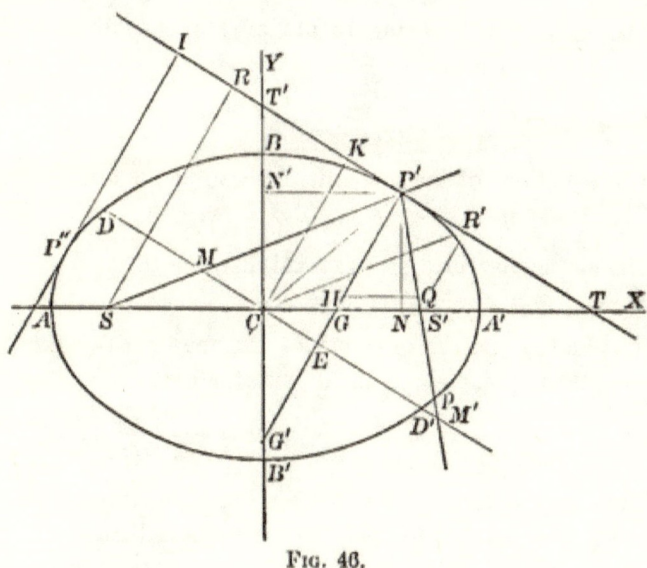

FIG. 46.

145. *If a line be drawn through the focus $S'(ae, 0)$ perpendicular to any tangent $P'T$, the locus of R', the point of intersection will be the major auxiliary circle $x^2 + y^2 = a^2$ of the ellipse. (Fig. 46.)*

The equation of a line through $S'(ae, 0)$, perpendicular to the tangent

$$y - mx = \sqrt{a^2m^2 + b^2}, \qquad (a) \text{ (Art. 136)}$$

is

$$y = -\frac{1}{m}(x - ae), \text{ or } my + x = ae. \quad \text{(Art. 46)}$$

Squaring both sides of these equations, adding, and dividing the sum by $(1 + m^2)$, gives

$$y^2 + x^2 = a^2,$$

the major auxiliary circle. We shall get the same result by taking the focus $S(-ae, 0)$. If, then, with C as a centre and a as a radius, a circle be described about the ellipse, it will pass through R and R', and also through the intersections of all other lines through the foci, which are perpendicular to the corresponding tangents. It also follows that $CR = CR' = CA'$.

146. *The normal bisects the angle $SP'S'$ between the focal distances $P'S = a + ex' = r$ and $P'S' = a - ex' = r'$.*

By Art. 101, XII., and Art. 138, II.,

$$SG = SC + CG = ae + e^2x', \quad GS' = CS' - CG = ae - e^2x'.$$

$$\therefore \frac{SG}{GS'} = \frac{ae + e^2x'}{ae - e^2x'} = \frac{a + ex'}{a - ex'} = \frac{P'S}{P'S'}. \qquad \text{(Art. 101, XIV.)}$$

Therefore, by Geometry, $SP'G = GP'S'$.

It follows that $SP'R = S'P'R'$; that is, *the tangent at any point P' makes equal angles with the focal distances of this point.*

147. *The product of the perpendiculars from the foci upon any tangent equals the square of the semi-minor axis;* or

$$SR \cdot S'R' = \overline{CB}^2 = b^2. \quad (Fig.\ 46.)$$

If we take the normal equation of the tangent at $P'(x'y')$,

$$x \cos \alpha + y \sin \alpha = a\sqrt{1 - e^2 \sin^2 \alpha}, \qquad \text{(Art. 139)}$$

then the perpendiculars from $S(-ae, 0)$ and $S'(ae, 0)$ on this tangent are (Art. 42)

$$SR = a\sqrt{1 - e^2 \sin^2 \alpha} + ae \cos \alpha,$$
$$S'R' = a\sqrt{1 - e^2 \sin^2 \alpha} - ae \cos \alpha.$$

$$\therefore SR \cdot S'R' = a^2(1 - e^2 \sin^2 a) - a^2 e^2 \cos^2 a = b^2,$$
since $a^2(1 - e^2) = b^2$ (Art. 101, IX.).

148. *To find the lengths of the perpendiculars from*
$S(-ae, 0)$, $C(0, 0)$, $S'(ae, 0)$, $G(e^2 x', 0)$, $G'\left(0, -\dfrac{a^2 e^2}{b^2} y'\right)$,
on the tangent TT' in terms of the focal distances of the point of tangency. (*Fig. 46.*)

The equation of the tangent TT' is
$$a^2 y'y + b^2 x'x = a^2 b^2. \qquad (a) \text{ (Art. 136)}$$

Then by Art. 42,
$$\overline{SR}^2 = \frac{(a^2 b^2 + b^2 aex')^2}{a^4 y'^2 + b^4 x'^2} = \frac{a^2 b^4 (a + ex')^2}{a^2 b^2 (a^2 - e^2 x'^2)} = \frac{b^2 r}{r'}, \qquad \text{(Art. 146)}$$
since from the equation of the ellipse $a^2 y^2 + b^2 x^2 = a^2 b^2$,
$$a^4 y'^2 + b^4 x'^2 = a^2 b^2 (a^2 - x'^2) + b^4 x'^2 = a^2 b^2 (a^2 - e^2 x'^2) = a^2 b^2 rr'.$$

In the same way find
$$\overline{S'R'}^2 = \frac{b^2 r'}{r}, \text{ and } \overline{CK}^2 = \frac{a^2 b^2}{rr'} = a^2(1 - e^2 \sin^2 a), \quad (b) \text{ (Art. 139)}$$
$$\overline{P'G}^2 = \frac{b^2}{a^2} rr', \text{ and } \overline{P'G'}^2 = \frac{a^2}{b^2} rr'. \quad \therefore \frac{P'G}{P'G'} = \frac{b^2}{a^2}.$$
$$\therefore SR \cdot S'R' = CK \cdot P'G = b^2, \quad CK \cdot P'G' = a^2,$$
$$P'G \cdot P'G' = rr', \quad P'G = \frac{b^2}{CK} = \frac{b^2}{a\sqrt{1 - e^2 \sin^2 a}} = \frac{a(1 - e^2)}{\sqrt{1 - e^2 \sin^2 a}}.$$

149. *A diameter DD', parallel to the tangent TT', cuts from the focal distance of the point of tangency a constant $P'M = a$, the semi-major axis.*

The similar triangles $MP'E$ and $SP'R$ give
$$\frac{P'M}{P'S} = \frac{P'E}{SR}, \text{ or } P'M = \frac{CK}{SR} \cdot P'S = a, \qquad \text{(Art. 148)}$$

DIAMETERS OF THE ELLIPSE. 233

150. If Q is the middle point of the focal chord $P'p$, and QH is parallel to GS', then the similar triangles $HP'Q$ and $GP'S'$ give

$$\frac{HQ}{P'Q} = \frac{GS'}{P'S'} = \frac{ae - e^2 x'}{a - ex'} = e; \text{ or } HQ = e \cdot \tfrac{1}{2} P'p.$$

Diameters of the Ellipse.

151. *To find the equation of the locus of the middle points of a system of parallel chords of an ellipse.* (Fig. 47.)

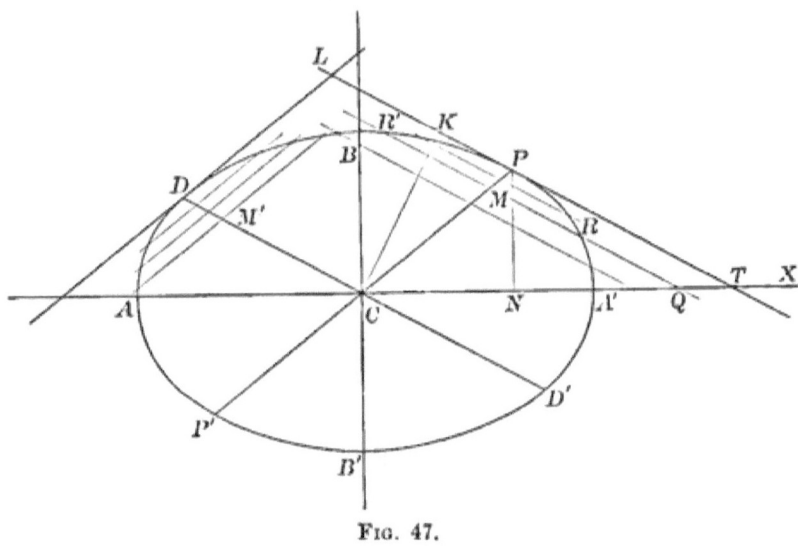

Fig. 47.

Let $R(x'y')$ and $R'(x''y'')$ be the two points in which any one of the chords of the parallel system intersects the ellipse. The tangent of the angle XQR', which this chord makes with the axis of X, is (a) (Art. 136)

$$m = -\frac{b^2(x' + x'')}{a^2(y' + y'')}.$$

If the coordinates of M, the middle point of RR', are (xy), then, $2x = x' + x''$, and $2y = y' + y''$.   (c) (Art. 5)

Making these substitutions in the value of m, it becomes

$$m = -\frac{b^2 x}{a^2 y}, \text{ or } y = -\frac{b^2}{a^2 m}x, \qquad (a)$$

the required equation of the locus, since it is true for the middle points M of all the chords of the system.

I. The equation (a) shows that the locus PC passes through C, the centre of the ellipse; it is therefore a diameter of the ellipse.

II. The tangent $XCP = -\dfrac{b^2}{a^2 m} = m'$, and the equation (a) of PC may be written $y = m'x$.

III. Therefore $mm' = -\dfrac{b^2}{a^2}$; that is, the product of the tangents of the angles which the parallel chords and the locus of their middle points make with the axis of X equals minus the square of the ratio of the semi-axes $b^2 : a^2$.

IV. If $a = b$, then $mm' = -1$; that is, in the circle the locus of any system of parallel chords is perpendicular to these chords.

V. In the ellipse, if $m = 0$, then $m' = \infty$; that is, the axes are the only diameters of the ellipse which are perpendicular to the systems of chords they bisect.

VI. The equation of the diameter DD' is $y = mx$, since it is one of the chords parallel to RR'; and it bisects a system of chords parallel to PP', whose equation is $y = m'x$, since the relation $mm' = -\dfrac{b^2}{a^2}$ remains unchanged.

VII. *Conjugate Diameters.* — Any two diameters, such as PP' and DD', each of which is parallel to the system of chords which the other bisects, are called *conjugate diameters;* and the points P and D are *conjugate points.*

DIAMETERS OF THE ELLIPSE.

Since the product mm' of the tangents of the angles $A'CP$ and $A'CD$ is minus, it follows that if the angle $A'CP$ is acute, the angle $A'CD$ must be obtuse, and that conjugate diameters always lie on opposite sides of BB', the minor axis.

VIII. The axes of the ellipse are the only pair of conjugate diameters which are perpendicular to each other.

IX. The tangent at P, the end of a diameter, is the *limiting chord* of the system which the diameter PC bisects, since the equal ordinates MR and MR' approach zero, as the points R and R' approach coincidence at P.

X. *Supplemental Chords.* — The two straight lines drawn from any point on the ellipse to the ends of any diameter are called *supplemental chords*.

XI. *Diameters which are respectively parallel to a pair of supplemental chords are conjugates.*

Let $P(x'y')$ and $P'(-x', -y')$ be the ends of any diameter; then
$$y - y' = m(x - x') \quad \text{and} \quad y + y' = m'(x + x') \qquad (a)$$
are the equations of the two chords. If these chords meet at any point D, on the ellipse $\dfrac{x^2}{a^2} + \dfrac{y^2}{b^2} = 1$, then for this point
$$y^2 - y'^2 = m'm(x^2 - x'^2).$$

But from the equation of the ellipse we have
$$y^2 - y'^2 = -\frac{b^2}{a^2}(x^2 - x'^2);$$

therefore $m'm = -\dfrac{b^2}{a^2}$, which shows that diameters parallel to these supplemental chords are conjugate, by VI.

Auxiliary Circles and the Eccentric Angle.

152. On the major and minor axes of the ellipse $\dfrac{x^2}{a^2}+\dfrac{y^2}{b^2}=1$ as diameters describe the auxiliary circles as in Fig. 48. From the point p on the major auxiliary circle, and the point p' on the minor auxiliary circle, drop the perpendiculars pN and $p'M$ on the axes AA' and BB', respectively. These perpendiculars meet the ellipse in the points P and P', called the *points* on the ellipse *corresponding* to the points p and p' on the auxiliary circles.

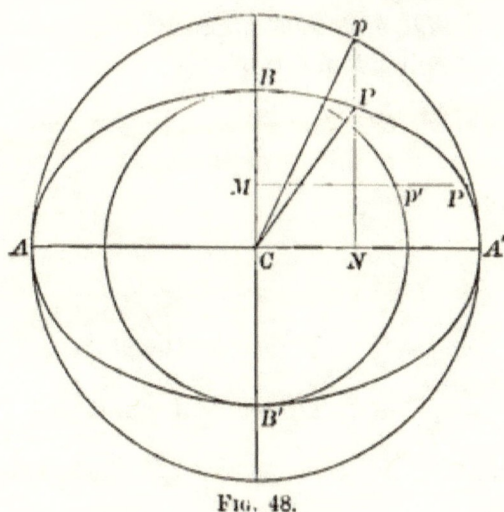

Fig. 48.

I. *The corresponding ordinates pN and PN are as $a:b$.*

Let $p(CN, pN)$, (x, Y) and $P(CN, PN)$, (x, y) be the coordinates of the corresponding points p and P. The equations of the ellipse and auxiliary circle for these points are

$$\frac{y^2}{b^2}+\frac{x^2}{a^2}=1, \text{ and } \frac{Y^2}{a^2}+\frac{x^2}{a^2}=1,$$

which, by subtraction, give

$$\frac{Y^2}{a^2}-\frac{y^2}{b^2}=0, \text{ or } \frac{Y}{y}=\frac{a}{b}, \text{ or } y=\frac{b}{a}Y,$$

which shows that the major auxiliary circle will become the ellipse by reducing all its ordinates in the ratio of $b:a$.

II. *The corresponding abscissas $p'M$ and PM on the minor auxiliary circle and the ellipse are in the ratio $b:a$.*

For the points P' and p', the equations of the ellipse and of the minor auxiliary circle are

$$\frac{y^2}{b^2} + \frac{x^2}{a^2} = 1, \text{ and } \frac{y^2}{b^2} + \frac{X^2}{b^2} = 1,$$

which give, by subtraction,

$$\frac{x^2}{a^2} - \frac{X^2}{b^2} = 0, \text{ or } \frac{X}{x} = \frac{b}{a}, \text{ or } x = \frac{a}{b}X,$$

which shows that the minor auxiliary circle will become the ellipse by increasing all its abscissas in the ratio $a:b$.

III. The angle $A'Cp$, measured on the major auxiliary circle, is called the *eccentric angle* of the corresponding point P on the ellipse.

If this angle is denoted by ϕ, then the coordinates of p are $(a\cos\phi, a\sin\phi)$; and the coordinates of P are $(a\cos\phi, \frac{b}{a}\cdot a\sin\phi)$, or $(a\cos\phi, b\sin\phi)$.

If the eccentric angle is measured on the minor auxiliary circle, then the coordinates of p' and P' are $(b\cos\phi, b\sin\phi)$ and $(a\cos\phi, b\sin\phi)$.

IV. Draw PR' parallel to pC; then $PR' = pC = a$. Also, $A'RP = A'Cp = \phi$; therefore $RP\sin\phi = PN = b\sin\phi$; therefore $RP = b$, $R'R = a - b$. If the points R' and R move on the axes, then the point P will describe the ellipse; and any other point on $R'P$ will also describe an ellipse, the difference of whose semi-axes equals $a - b$, except a point midway between R' and R, which will describe a circle. This is the principle of the elliptic compasses.

V. *To construct a point P on the ellipse for a given eccentric angle $A'Cp$.*

First draw pN perpendicular to AA', and through the point in which pC cuts the minor auxiliary circle draw a parallel to AA'; this parallel will meet pN in P, a point on the ellipse.

VI. *The tangents at the corresponding points $p(x'y'')$ and $P(x'y')$ meet on the axis of X; while tangents at the corresponding points p' and P' meet on the axis of Y.*

The tangents of the ellipse and major auxiliary circle are

$$\frac{yy'}{b^2} + \frac{xx'}{a^2} = 1, \text{ and } \frac{yy''}{a^2} + \frac{xx'}{a^2} = 1,$$

both of which, for $y = 0$, give $x = \dfrac{a^2}{x'} = CT$; while the tangents for the ellipse and minor auxiliary circle both give $y = \dfrac{b^2}{y'}$ for $x = 0$.

VII. *To express the equations of the tangent and normal at a given point $P(x'y')$ in terms of the eccentric angle of this point.*

For the point P, $x' = a\cos\phi$, $y' = b\sin\phi$, and the tangent at $P(x'y')$,

$$\frac{xx'}{a^2} + \frac{yy'}{b^2} = 1, \text{ becomes } \frac{x}{a}\cos\phi + \frac{y}{b}\sin\phi = 1.$$

The normal (b) (Art. 137), which may be written

$$\frac{a^2}{x'}x - \frac{b^2}{y'}y = a^2 - b^2, \text{ becomes } \frac{ax}{\cos\phi} - \frac{by}{\sin\phi} = a^2 e^2.$$

VIII. *If $P(x'y')$ and $D(x''y'')$ are the ends of a pair of conjugate diameters whose eccentric angles are ϕ and ϕ', then $\phi' - \phi = \dfrac{\pi}{2}$.* *(Fig. 49.)*

The equations of PP' and DD' are

$$y = \frac{y'}{x'}x, \text{ and } y = \frac{y''}{x''}x. \quad \therefore m' = \frac{y'}{x'}, \ m = \frac{y''}{x''}.$$

AUXILIARY CIRCLES AND ECCENTRIC ANGLE.

Substituting these values of m' and m in $m'm = -\dfrac{b^2}{a^2}$ (Art. 151, VI.), it becomes

$$\frac{y'y''}{x'x''} = -\frac{b^2}{a^2}, \text{ or } \frac{y'y''}{b^2} + \frac{x'x''}{a^2} = 0. \tag{a}$$

But $x' = a \cos \phi$, $y' = b \sin \phi$, $x'' = a \cos \phi'$, $y'' = b \sin \phi'$, and (a) becomes

$$\cos \phi' \cos \phi + \sin \phi' \sin \phi = 0. \quad \therefore \phi' - \phi = \frac{\pi}{2}.$$

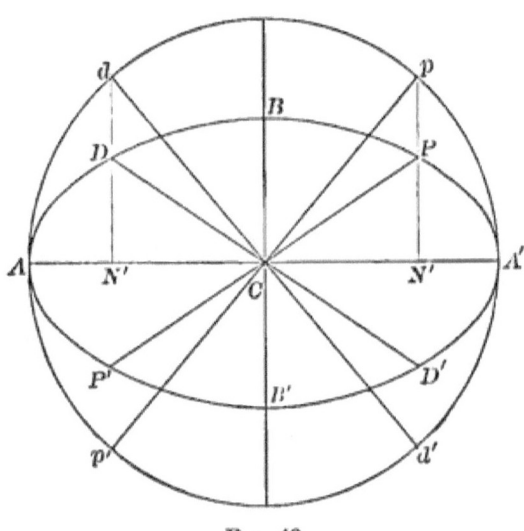

Fig. 49.

IX. *The coordinates of D and D' can be expressed in terms of the coordinates of P, their conjugate point.*

The coordinates of D and D' are

$$x'' = a \cos \phi' = a \cos\left(\phi \pm \frac{\pi}{2}\right) = \mp a \sin \phi = \mp \frac{ay'}{b},$$

$$y'' = b \sin \phi' = b \sin\left(\phi \pm \frac{\pi}{2}\right) = \pm b \cos \phi = \pm \frac{bx'}{a}.$$

The equations of the conjugate diameters PP' and DD' are, therefore,
$$y = \frac{b}{a} x \tan \phi, \quad y = -\frac{b}{a} x \cot \phi.$$

X. *The squares of the semi-conjugate diameters CP and CD, usually denoted by a' and b', are*
$$a'^2 = a^2 \cos^2 \phi + b^2 \sin^2 \phi = b^2 + (a^2 - b^2) \cos^2 \phi = b^2 + e^2 x'^2,$$
$$b'^2 = a^2 \sin^2 \phi + b^2 \cos^2 \phi = a^2 - (a^2 - b^2) \cos^2 \phi = a^2 - e^2 x'^2 = r \cdot r'.$$

XI. *The sum of the squares of any pair of semi-conjugate diameters is equal to the sum of the squares of the semi-axes.*

For $a'^2 + b'^2 = a^2 + b^2$, by VIII.

XII. *The area of any parallelogram formed by the tangents drawn at the ends of any pair of conjugate diameters equals the area of the rectangle described on the axes, and is therefore constant. (Fig. 47.)*

For $CD \cdot CK = ab$, since $\overline{CK}^2 = \dfrac{a^2 b^2}{rr'}$ (Art. 148), and $\overline{CD}^2 = rr'$, by VIII. $\therefore 4 CD \cdot CK = 4 ab$.

It also follows that the area of the triangle PCD, formed by joining the ends of any pair of conjugate diameters, is constant.

XIII. *To find the angle $PCD = \theta' - \theta$ between any pair of conjugate diameters.*

From X.,
$$CP \cdot CD \sin PCD = a'b' \sin(\theta' - \theta) = ab.$$
$$\therefore \sin(\theta' - \theta) = \frac{ab}{a'b'}.$$

XIV. *To find the position of a pair of equi-conjugate diameters.*

AUXILIARY CIRCLES AND ECCENTRIC ANGLE. 241

If $a' = b'$, then, from VIII.,

$$a'^2 - b'^2 = (a^2 - b^2)(\cos^2 \phi - \sin^2 \phi) = (a^2 - b^2)\cos 2\phi = 0.$$

$$\therefore \cos 2\phi = 0, \text{ or } 2\phi = \frac{\pi}{2}, \text{ or } \frac{3\pi}{2}. \quad \therefore \phi = \frac{\pi}{4}, \text{ or } \frac{3}{4}\pi.$$

The equations of the equi-conjugate diameters are therefore

$$y = \frac{b}{a}x, \text{ and } y = -\frac{b}{a}x \text{ (IX), since } \phi = \frac{\pi}{4};$$

and they coincide with the diagonals of the rectangle described on the axes.

Therefore, for a pair of equi-conjugate diameters,

$$\tan PCD' = \frac{2ab}{a^2 - b^2}, \text{ or } \sin PCD' = \frac{2ab}{a^2 + b^2}.$$

XV. *The acute angle PCD' made by two conjugate diameters is least when these diameters are equal.*

From XIII., $\sin PCD' = \frac{ab}{a'b'}$. Since ab is constant, PCD' is least when $a'b'$ is greatest; but $a'b'$ is greatest when $a' = b'$, since $a'^2 + b'^2 = a^2 + b^2$ is constant; that is, PCD' is least for equi-conjugate diameters, for which $\sin PCD' = \frac{2ab}{a^2 + b^2}$, by XIV.

XVI. *To construct a pair of conjugate diameters which make a given angle with each other.*

By XIII., this angle must be greater than PCD', or $\sin^{-1}\frac{2ab}{a^2 + b^2}$. On any diameter PCP' describe the segment of a circle which shall contain the given angle. The circle will cut the ellipse in some point D; DP and DP' are supplemental chords making the given angle with each other, and the diameters parallel to these chords are the required conjugates, by Art. 151, XI.

EXERCISES AND PROBLEMS ON THE ECCENTRIC ANGLE OF THE ELLIPSE.

Given the ellipse $3x^2 + 4y^2 = 12$ and the eccentric angle $\phi = 30°$, to find (see Fig. 47)

1. The equations of the conjugate diameters PP' and DD'.
   Ans. $y = \frac{1}{2}x, \ y = -\frac{3}{2}x$.

2. The lengths of the semi-conjugates CP and CD.
   Ans. $\frac{1}{2}\sqrt{15}, \ \frac{1}{2}\sqrt{13}$.

3. The equations of the tangents at P and D.
   Ans. $2y + 3x = 4\sqrt{3}, \ 2y - x = 4$.

4. The equations of the normals at P and D.
   Ans. $2x - 3y = \frac{1}{2}\sqrt{3}, \ 4x + 2y + 1 = 0$.

5. The perpendiculars from C on the tangent and normal at P.
   Ans. $\frac{4}{13}\sqrt{39}, \ \frac{1}{26}\sqrt{39}$.

6. The perpendiculars from the focus $S'(ae, 0)$ on the tangent and normal at P.
   Ans. $\dfrac{3 - 4\sqrt{3}}{\sqrt{13}}, \ \dfrac{4 - \sqrt{3}}{2\sqrt{13}}$.

Given the ellipse $x^2 + 5y^2 = 5$ and the abscissa $x = 1$, to find

7. The eccentric angle and the equations of the conjugate diameters.
   Ans. $\phi = \tan^{-1}2; \ y = \frac{2}{5}\sqrt{5}x; \ y = -\frac{1}{10}\sqrt{5}x$.

8. The lengths of the semi-conjugate diameters and the angle they make with each other.
   Ans. $a' = \frac{3}{5}\sqrt{5}; \ b' = \frac{1}{5}\sqrt{105}; \ \tan^{-1}(-\frac{5}{8}\sqrt{5})$

9. The length of the chord joining any two conjugate points on the ellipse is $\sqrt{a^2 + b^2 + a^2 e^2 \sin 2\phi}$.

10. The equation of the secant through any two points
$$(a \cos \phi, \ b \sin \phi), \ (a \cos \phi', \ b \sin \phi')$$
on the ellipse is
$$\frac{x}{a} \cos \tfrac{1}{2}(\phi' + \phi) + \frac{y}{b} \sin \tfrac{1}{2}(\phi' + \phi) = \cos \tfrac{1}{2}(\phi' - \phi).$$

EXERCISES ON THE ECCENTRIC ANGLE.

11. The equation of the secant through any two conjugate points on the ellipse is
$$\frac{x}{a}(\cos\phi - \sin\phi) + \frac{y}{b}(\cos\phi + \sin\phi) = 1.$$

12. The length of the perpendicular from the centre C to a chord joining any two conjugate points is
$$\frac{ab}{\sqrt{a^2 + b^2 + a^2 e^2 \sin 2\phi}}.$$

13. The angle θ which the chord joining any two conjugate points makes with the major axis of the ellipse is
$$\tan\theta = \frac{b}{a}\tan\left(\phi - \frac{\pi}{4}\right).$$

14. If a, b are the semi-axes, and a', b' the semi-conjugate diameters of the ellipse, then
$$\sin\phi = \frac{\sqrt{a^2 - a'^2}}{ae} = \frac{\sqrt{b'^2 - b^2}}{ae}.$$

15. The area of the parallelogram formed by tangents to the ellipse at points whose eccentric angles are ϕ, ϕ', $\phi + \pi$, $\phi' + \pi$, is
$$\frac{4ab}{\sin(\phi' - \phi)}.$$

16. If tangents at any two points P and D (Fig. 47) whose eccentric angles are ϕ and ϕ' meet in L, the area of the quadrilateral $LPCD$ is $ab \tan \frac{1}{2}(\phi' - \phi)$.

17. If tangents are drawn to the ellipse at the points P, P' (Fig. 49), and at the ends p and p' of the corresponding diameter of the auxiliary circle, show that the area of the parallelogram formed by these tangents is $\dfrac{8a^2b}{(a-b)\sin 2\phi}$, ϕ being the eccentric angle of P.

18. The area of a triangle inscribed in an ellipse is
$$\tfrac{1}{2}ab\left[\sin(\beta - \gamma) + \sin(\gamma - \alpha) + \sin(\alpha - \beta)\right]$$
if α, β, γ are the eccentric angles of its vertices.

19. Show that the locus of the point of intersection of tangents to an ellipse at two points whose eccentric angles differ by a constant angle a is the ellipse

$$\frac{x^2}{a^2}\cos^2 a + \frac{y^2}{b^2}\cos^2 a = 1.$$

The Ellipse referred to Conjugate Diameters as Axes.

153. *To find the equation of the ellipse when referred to any pair of conjugate diameters as coordinate axes.*

Let the diameters PP' and DD' make the angles $A'CP = \theta$, and $A'CD = \theta'$ with the axis of X.

The equations for transforming coordinates from a rectangular to an oblique system of axes are, (g) (p. 49),

$$y = x' \sin \theta + y' \sin \theta',$$
$$x = x' \cos \theta + y' \cos \theta'.$$

Substituting these values of x and y in the equation of the ellipse $\frac{x^2}{a^2} + \frac{y^2}{b^2} = 1$, it becomes, omitting the primes on the new coordinates,

$$\frac{(x \cos \theta + y \cos \theta')^2}{a^2} + \frac{(x \sin \theta + y \sin \theta')^2}{b^2} = 1;$$

or

$$x^2\left[\frac{\sin^2 \theta}{b^2} + \frac{\cos^2 \theta}{a^2}\right] + 2\left[\frac{\sin \theta \sin \theta'}{b^2} + \frac{\cos \theta \cos \theta'}{a^2}\right]xy$$
$$+ y^2\left[\frac{\sin^2 \theta'}{b^2} + \frac{\cos^2 \theta'}{a^2}\right] = 1;$$

or

$$x^2\frac{(a^2 \sin^2 \theta + b^2 \cos^2 \theta)}{a^2b^2} + y^2\frac{(a^2 \sin^2 \theta' + b^2 \cos^2 \theta')}{a^2b^2} = 1, \quad (a)$$

since the coefficient of xy equals zero, by Art. 151, VI.

By making $x = 0$ and $y = 0$ in succession, we get for the squares of the intercepts on the new axes,

ELLIPSE REFERRED TO CONJUGATE DIAMETERS.

$$y^2 = \frac{a^2b^2}{a^2 \sin^2 \theta' + b^2 \cos^2 \theta'} = \overline{CD}^2 = b'^2,$$

$$x^2 = \frac{a^2b^2}{a^2 \sin^2 \theta + b^2 \cos^2 \theta} = \overline{CP}^2 = a'^2.$$

Therefore (a) becomes

$$\frac{x^2}{a'^2} + \frac{y^2}{b'^2} = 1, \tag{a'}$$

the required equation of the ellipse referred to a pair of conjugate diameters as coordinate axes.

I. For $b' = a'$, or when the ellipse is referred to a pair of equi-conjugate diameters as axes, its equation is

$$x^2 + y^2 = a'^2 = \frac{a^2 + b^2}{2}. \tag{XI.} \text{ (Art. 152)}$$

This equation is now of the form of $x^2 + y^2 = r^2$, the equation of the circle, the ellipse being referred to oblique, and the circle to rectangular, axes.

II. It is obvious that the equation $\frac{x^2}{a^2} + \frac{y^2}{b^2} = 0$ of a pair of conjugate diameters, will, by the same transformation, become $\frac{x^2}{a'^2} + \frac{y^2}{b'^2} = 0$.

III. Any proposition relating to the ellipse which does not presuppose rectangular axes will hold when the curve is referred to a pair of conjugate diameters as axes. Thus the equation of a chord is

$$\frac{x(x' + x'')}{a'^2} + \frac{y(y' + y'')}{b'^2} = 1 + \frac{x'x''}{a'^2} + \frac{y'y''}{b'^2}.$$

A tangent at $(x'y')$ is $\frac{xx'}{a'^2} + \frac{yy'}{b'^2} = 1$.

The polar of (h, k) is $\frac{hx}{a'^2} + \frac{ky}{b'^2} = 1$.

EXERCISES AND PROBLEMS ON THE ELLIPSE.

Given the ellipse $9x^2 + 25y^2 = 225$, and $\phi = 60°$:

1. The equation referred to the corresponding conjugate diameters is $21x^2 + 13y^2 = 273$.

2. The length of the chord joining P and D, the ends of the congate diameters, is $\sqrt{34 + 8\sqrt{3}}$.

3. The perpendicular from C on this chord is $\dfrac{15}{\sqrt{34 + 8\sqrt{3}}}$.

4. The equation of the given ellipse referred to its equi-conjugate diameters is $x^2 + y^2 = 17$.

5. Tangents to the ellipse and auxiliary circle at any corresponding points P and p (Fig. 49) always meet in the fixed point $\left(\dfrac{a^2}{x''}, 0\right)$ or in $\left(-\dfrac{a^2}{x''}, 0\right)$.

6. Normals at any two corresponding points P and p on the ellipse and auxiliary circle always meet on the circle $x^2 + y^2 = (a+b)^2$.

7. The lines SR and CP'' (Fig. 46) meet on the left-hand directrix, and $S'R'$ and CP' meet on the right-hand directrix.

8. The ends of a straight line AB of given length move on a pair of rectangular axes; show that the locus of any point P on this line which divides it in the ratio $m:n$ is an ellipse. The locus of the middle point of this line is a circle.

9. An ellipse slides along a pair of rectangular axes; show that the locus of its centre is a circle.

10. The equations of the ellipse referred to the conjugate diameters PP' and DD' (Fig. 47) is $\dfrac{x^2}{a'^2} + \dfrac{y^2}{b'^2} = 1$; show that the tangents at the ends R and R' of any chord meet on the diameter which bisects this chord. Note that the coordinates of R' and R are $(x'y')$ and $(x', -y')$.

EXERCISES AND PROBLEMS ON THE ELLIPSE.

11. If p and P are any two points on the ellipse and auxiliary circle (Fig. 49), the perpendiculars from the foci S and S' on the tangent at p are equal to the focal distances SP and $S'P$ respectively.

12. If points S and S' be taken on the minor axis of an ellipse, at the same distance from the centre C as the foci are, then the sum of the squares of the perpendiculars from these points on any tangent to the ellipse is constant.

13. The tangent at P' (Fig. 46) meets the tangent at A in a point Y; show that CY is parallel to $A'P'$.

14. The equation of the ellipse when the right-hand focus is the origin, and the latus-rectum is the axis of Y, is

$$\frac{x^2}{a^2}+\frac{y^2}{b^2}+\frac{2ex}{a}=\frac{b^2}{a^2}.$$

15. If the chords of an ellipse pass through the point $Q(x'y')$, and tangents are drawn through the ends of these chords, the locus of these pairs of tangents is the line

$$\frac{xx'}{a^2}+\frac{yy'}{b^2}=1.$$

16. If the pole $P(h, k)$ is on the line $Ax + By + C = 0$, then the polar will pass through the point $\left(-\dfrac{Aa^2}{C}, -\dfrac{Bb^2}{C}\right)$.

17. The perpendicular SR (Fig. 46) produced will meet the line $S'P'$ produced in a point which is equal to $2SR$ from the focus S.

18. Find the locus of this point of intersection for a moving tangent.

19. The ratio of the subnormals for any two corresponding points p and P on the auxiliary circle and the ellipse is $\dfrac{a^2}{b^2}$.

20. The length of any focal chord which makes an angle θ with the major axis is $\dfrac{2b^2}{a(1-e^2\cos^2\theta)}$. The length of the diameter parallel to this chord is $\dfrac{2b}{\sqrt{1-e^2\cos^2\theta}}$.

21. If PNP' is a double ordinate of a point on an ellipse, and Q any point on the curve, then if QP and QP' meet the major axis in the points M and M', $CM \cdot CM' = \overline{CA}^2$.

22. The tangents at any point P on the ellipse cut the equi-conjugate diameters in T and T'; show that the triangles TCP, $T'CP$ are in the ratio of $\overline{CT}^2 : \overline{CT'}^2$.

23. If CQ be conjugate to the normal at P, then will CP be conjugate to the normal at Q.

24. If P, D (Fig. 47) are the ends of conjugate diameters, and the tangent at P cuts the major axis in T, and the tangent at D cuts the minor axis in T'; show that TT' will be parallel to one of the equi-conjugates.

25. Connect any point P on the ellipse with the vertices A and A'; at P draw perpendiculars to AP and $A'P$, meeting the major axis in M and N; show that MN equals the latus-rectum.

26. If P and D are the ends of conjugate diameters of an ellipse, show that the tangents at P and D meet on the ellipse $\frac{x^2}{a^2} + \frac{y^2}{b^2} = 2$, and that the locus of the middle point of PD is the ellipse $\frac{x^2}{a^2} + \frac{y^2}{b^2} = \frac{1}{2}$.

27. A line is drawn parallel to the minor axis at a point midway between the focus and corresponding directrix; show that the product of the perpendiculars on it from the ends of any chord passing through that focus is constant.

28. If P, D be the points of contact of perpendicular tangents to an ellipse, and p, d are the corresponding points on the auxiliary circle, then Cp and Cd are conjugate diameters.

29. Prove that the sum of the products of the perpendiculars from the ends P, P' and D, D' of a pair of conjugate diameters on any tangent to an ellipse is equal to the square of the perpendicular from the centre on that tangent.

EXERCISES AND PROBLEMS ON THE ELLIPSE.

30. If P, p are corresponding points on the ellipse and major auxiliary circle (Fig. 49), and CP be produced to meet the auxiliary circle in q, show that the tangent at P on the ellipse corresponding to q, is perpendicular to Cp, and cuts off from Cp a length equal to CP.

31. If a pair of tangents to an ellipse are perpendicular to each other, the product of the perpendiculars from the centre and from the intersection of the tangents on the chord of contact is constant.

32. Tangents to an ellipse are perpendicular to each other; find the locus of the middle point of the chord of contact.

33. Normals at the ends P and D of any pair of conjugate diameters intersect on the curve
$$2(a^2x^2 + b^2y^2)^3 = a^4e^4(a^2x^2 - b^2y^2)^2.$$

34. If tangents at any two points P and Q on the ellipse $\frac{x^2}{a^2} + \frac{y^2}{b^2} = 1$ meet in $T(h, k)$, then, if β is the angle between them,
$$\tan \beta = \frac{2\sqrt{a^2k^2 + b^2h^2 - a^2b^2}}{k^2 + h^2 - (a^2 + b^2)}.$$

If $\beta = 90$, then the locus of the intersection is the director circle
$$k^2 + h^2 = a^2 + b^2.$$

35. If the points T, P, Q in the last example are joined to the focus S, then
$$\frac{\overline{ST}^2}{SP \cdot SQ} = \frac{k^2}{b^2} + \frac{h^2}{a^2}.$$

36. If the normals at the ends of the focal chord $P'p$ (Fig. 46) meet in H, then the line HQ parallel to the major axis will bisect this chord.

CHAPTER VIII.

THE HYPERBOLA $\frac{x^2}{a^2} - \frac{y^2}{b^2} = 1$.

Secants, Tangents, Normals, and Polars.

154. As the equations of the hyperbola and ellipse only differ in the sign of the square of the semi-conjugate axis b^2, it follows that the proof of many of the propositions relating to the hyperbola is the same as for the ellipse, and in the results it is only necessary to change b^2 to $-b^2$.

155. The following results may be used for review:

I. The secant through $P'(x'y')$ and $P''(x''y'')$, and the tangent of the angle it makes with the axis of X.

(a) $\quad \dfrac{x(x'+x'')}{a^2} - \dfrac{y(y'+y'')}{b^2} = 1 + \dfrac{x'x''}{a^2} - \dfrac{y'y''}{b^2}.$ (Art. 136)

(b) $\quad m = \dfrac{b^2(x'+x'')}{a^2(y'+y'')}.$

II. The equations of the tangent at $(x'y')$, in terms of m, and in the normal form (Arts. 137, 139).

(a) $\quad \dfrac{xx'}{a^2} - \dfrac{yy'}{b^2} = 1.$

(b) $\quad y = mx \pm \sqrt{a^2 m^2 - b^2}.$

(c) $\quad x \cos a + y \sin a = a\sqrt{1 - e^2 \sin^2 a}.$

III. The values of m for tangents to the hyperbola which pass through the point (h, k) (Art. 140).

(a) $$m = \frac{kh \pm \sqrt{a^2k^2 - b^2h^2 + a^2b^2}}{h^2 - a^2}.$$

IV. The equations of the normal at $(x'y')$, in terms of m, and in the normal form (Arts. 137, 139).

(a) $$\frac{a^2(x-x')}{x'} = -\frac{b^2(y-y')}{y'}, \text{ or } \frac{a^2x}{x'} + \frac{b^2y}{y'} = a^2 + b^2.$$

(b) $$y = mx \pm \frac{(a^2+b^2)m}{\sqrt{a^2 - b^2m^2}}.$$

(c) $$x \cos a + y \sin a = \frac{ae^2 \sin a \cos a}{\sqrt{1 - e^2 \cos^2 a}}.$$

V. The polar of the point (h, k) (Art. 142).

(a) $$\frac{hx}{a^2} - \frac{ky}{b^2} = 1.$$

VI. The equation of the director circle (Art. 144).

(a) $$x^2 + y^2 = a^2 - b^2.$$

This circle is real for $a > b$, a point circle for $a = b$, and an imaginary circle for $b > a$.

VII. The equation of the major auxiliary circle (Art. 145).

(a) $$x^2 + y^2 = a^2.$$

VIII. The tangent bisects the angle $SP'S'$ between the focal distances of any point on the curve, and the normal bisects the supplementary angle (Fig. 50).

This proposition is proved in the same way as the corresponding one for the ellipse (Art. 146). A simple proof is as follows:

In the right triangles $P'SR$ and $P'S'R'$, the sines of the angles $SP'R$ and $S'P'R'$ are $\dfrac{SR}{P'S}$ and $\dfrac{S'R'}{P'S'}$. But by Art. 148,

$$\frac{\overline{SR}^2}{\overline{P'S}^2} = \frac{\overline{S'R'}^2}{\overline{P'S'}^2} = \frac{b^2}{rr'};$$

therefore $SP'T = S'P'T$, and the normal bisects the supplement of $SP'S'$.

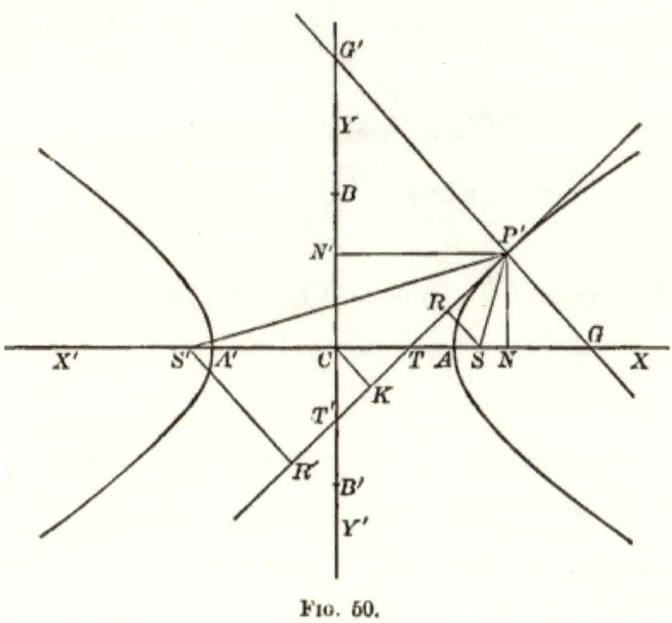

Fig. 60.

IX. All the results in Arts. 147, 148, 149, 150, are true for the hyperbola.

X. The locus of the middle points of a system of parallel chords, having a direction parameter m (Art. 151).

(a) $\qquad m = \dfrac{b^2 x}{a^2 y}$, or $y = \dfrac{b^2}{a^2 m} x = m'x.$

EXERCISES ON TANGENTS AND NORMALS.

XI. For a pair of conjugate diameters $mm' = \frac{b^2}{a^2}$ (Art. 151, VI.). Since the product mm' of the tangents of the angles which the conjugate diameters make with the axis of X is positive, it follows that both angles are acute or both obtuse, and that both diameters lie on the same side of the conjugate axis BB'.

XII. Since the product $mm' = \frac{b^2}{a^2}$ is constant, it follows that as one angle increases, the other diminishes, and the conjugate diameters will finally coincide with the diagonals of the rectangle described on the axes AA' and BB' of the curve. The equations of the diagonals are

$$y = \frac{b}{a}x \quad \text{and} \quad y = -\frac{b}{a}x.$$

EXERCISES ON TANGENTS AND NORMALS.

Given the equations

$$\frac{xx'}{a^2} - \frac{yy'}{b^2} = 1 \quad \text{and} \quad \frac{a^2x}{x'} + \frac{b^2y}{y'} = a^2e^2$$

of the tangent and normal at the point $P'(x'y')$ (Fig. 50), to find the following lines:

1. $CT = \frac{a^2}{x'}$.
2. $CT' = -\frac{b^2}{y'}$.
3. $CG = e^2x'$.
4. $CG' = \frac{a^2e^2}{b^2}y'$.
5. $TN = \frac{x'^2 - a^2}{x'}$.
6. $NG = \frac{b^2}{a^2}x'$.
7. $T'N' = \frac{y'^2 + b^2}{y'}$.
8. $N'G' = \frac{a^2}{b^2}y'$.
9. $TG = \frac{(ex' + a)(ex' - a)}{x'}$.
10. $T'G' = \frac{a^2e^2y'^2 + b^4}{b^2y'}$.

11. Given $P'(x'y')$, $S(ae, 0)$, $S(-ae, 0)$; show by (a) (Art. 4) that $P'S' = ex' + a = r'$, $P'S = ex' - a = r$.

254   PLANE ANALYTIC GEOMETRY.

12. The equations of $P'S'$ and $P'S$ are
$$y(x' + ae) - y'(x + ae) = 0, \quad y(x' - ae) - y'(x - ae) = 0.$$

13. The equations of the perpendiculars $S'R'$, SR, CK on the tangent are
$$b^2 x'y + a^2 y'x + a^3 ey' = 0, \quad b^2 x'y + a^2 y'x - a^3 ey' = 0, \quad b^2 x'y + a^2 y'x = 0.$$

14. Show by Art. 44 that
$$\tan SP''T = \tan S'P''T' = \frac{b^2}{aey'}.$$

15. $\overline{SR}^2 = \dfrac{b^2 r}{r'}$.   16. $\overline{S'R'}^2 = \dfrac{b^2 r'}{r}$.   17. $\overline{CK}^2 = \dfrac{a^2 b^2}{rr'}$.

18. $\overline{P''G}^2 = \dfrac{b^2}{a^2} rr'$.   19. $\overline{P''G'}^2 = \dfrac{a^2}{b^2} rr'$.

20. $SR \cdot S'R' = CK \cdot P'G = -b^2$.

21. $CK \cdot P''G' = a^2$.   22. $P''G \cdot P''G' = rr'$.

23. The equation of the hyperbola whose vertex bisects the distance between the centre and focus is $3x^2 - y^2 = 3a^2$.

24. If the distance between the foci is $2c$, and e is the eccentricity, the equation of the hyperbola is $\dfrac{e^2 x^2}{c^2} - \dfrac{e^2 y^2}{c^2(e^2 - 1)} = 1$.

25. The sum of the focal distances of any point $P'(x'y')$ on the hyperbola is $2ex'$.

26. If an ellipse and an hyperbola have the same foci, their tangents at the points of intersection are perpendicular to each other.

27. For the points $\left(\pm \dfrac{a^2}{\sqrt{a^2 - b^2}},\ \pm \dfrac{b^2}{\sqrt{a^2 - b^2}}\right)$ on the hyperbola, the subtangent is equal to the subnormal.

Diameters of the Hyperbola.

156. *If a diameter PP', $y = m'x$, meets the hyperbola $\dfrac{x^2}{a^2} - \dfrac{y^2}{b^2} = 1$ in real points, its conjugate DD', $y = mx$, will meet the curve in imaginary points. (Fig. 51.)*

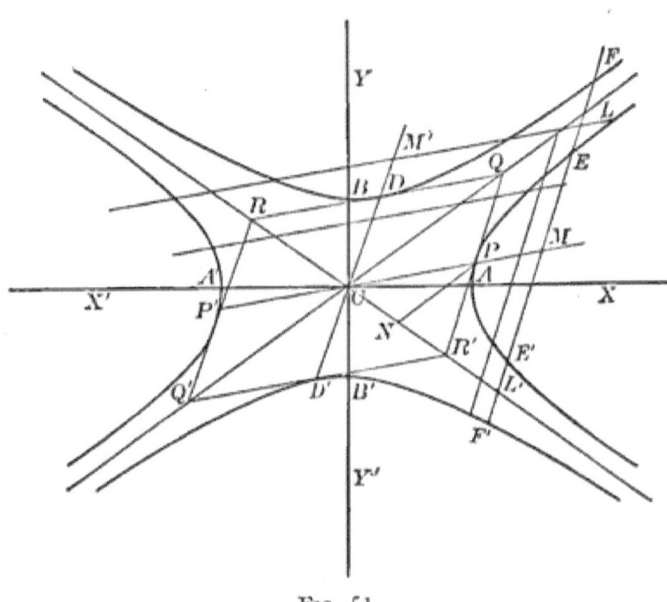

Fig. 51.

The abscissas of the points of intersection of the diameters $y = m'x$ and $y = mx$ with the hyperbola are

$$\frac{x^2}{a^2} - \frac{m'^2 x^2}{b^2} = 1, \text{ or } x = \pm \frac{ab}{\sqrt{b^2 - a^2 m'^2}}. \quad (a)$$

$$\frac{x^2}{a^2} - \frac{m^2 x^2}{b^2} = 1, \text{ or } x = \pm \frac{ab}{\sqrt{b^2 - a^2 m^2}}. \quad (b)$$

Since $m'm = \dfrac{b^2}{a^2}$, $m' < \dfrac{b}{a}$ when $m > \dfrac{b}{a}$. The roots of (a) are real for $m'^2 < \dfrac{b^2}{a^2}$, while the roots of (b) are imaginary for

$m^2 < \dfrac{b^2}{a^2}$, and conversely. Therefore one conjugate diameter meets the hyperbola in real points, and the other in imaginary points.

I. If $m' = \dfrac{b}{a}$, then $m = \dfrac{b}{a}$, and the roots of both (a) and (b) are infinite; that is, the hyperbola approaches the diagonals of the rectangle described on the axes as the values of the coordinates of the points of intersection approach infinity.

II. *Asymptotes.* — An asymptote of a conic is a straight line which the curve constantly approaches, to within any assignable distance however small, but never meets.

The diagonals of the rectangle described on the axes of the hyperbola are its asymptotes, since the curve meets these lines at infinity.

The equations of the asymptotes $y = \dfrac{b}{a}x$, $y = -\dfrac{b}{a}x$ may be found as follows:

The abscissas of the points in which the line $y = mx + c$ intersects the hyperbola $\dfrac{x^2}{a^2} - \dfrac{y^2}{b^2} = 1$ are given by the quadratic equation

$$\dfrac{x^2}{a^2} - \dfrac{(mx+c)^2}{b^2} = 1,$$

or
$$(b^2 - a^2 m^2)x^2 - 2a^2 mcx - a^2 c^2 - a^2 b^2 = 0. \qquad (a)$$

If both the intersections of the line $y = mx + c$ and the hyperbola are at infinity, then both roots of (a) must be infinite, and by Art. 20,

$$b^2 - a^2 m^2 = 0 \text{ and } 2a^2 mc = 0.$$

Therefore $m = \pm \dfrac{b}{a}$ and $c = 0$, and $y = mx + c$ becomes

$$y = \pm \frac{b}{a}x,$$

the equations of the asymptotes.

III. When the transverse axis AA' is *real*, and the conjugate axis BB' is *imaginary*, $\frac{x^2}{a^2} - \frac{y^2}{b^2} = 1$ is called the *primary* hyperbola; when BB' is the *real* or the transverse axis, and AA' is the *imaginary* or the conjugate axis, $\frac{y^2}{b^2} - \frac{x^2}{a^2} = 1$ is called the *conjugate* hyperbola; that is, these two hyperbolas are *conjugate to each other*.

IV. *If the diameter PP', $y = m'x$, meets the primary hyperbola in real points, its conjugate diameter DD', $y = mx$, will also meet the conjugate hyperbola in real points.* (*Fig. 51.*)

For the intersections of $y = m'x$ and the primary hyperbola, and for the intersections of $y = mx$ and the conjugate hyperbola, the respective roots are

$$x = \pm \frac{ab}{\sqrt{b^2 - a^2 m'^2}}, \qquad (a)$$

$$x = \pm \frac{ab}{\sqrt{a^2 m^2 - b^2}}. \qquad (b)$$

The roots of (*a*) are real for $m'^2 < \frac{b^2}{a^2}$, and the roots of (*b*) are also real for $m^2 > \frac{b^2}{a^2}$. Therefore a pair of conjugate diameters will meet both the primary and conjugate hyperbolas in real points.

If $m' = \frac{b}{a}$, then $m = \frac{b}{a}$, and the roots of both (*a*) and (*b*) are infinite, and $y = \frac{b}{a}x$, $y = -\frac{b}{a}x$ are the equations of the asymptotes of both the primary and its conjugate hyperbola.

V. *To find the coordinates of $D(x''y'')$ and $D'(-x'', -y'')$, points on the conjugate hyperbola, in terms of the coordinates $(x'y')$ of the conjugate point P on the primary hyperbola. (Fig. 51.)*

The equations of CP and CD are $y = \dfrac{y'}{x'}x$ and $y = \dfrac{y''}{x''}x$;

$\therefore m' = \dfrac{y'}{x'}$, $m = \dfrac{y''}{x''}$, and $m'm = \dfrac{b^2}{a^2}$ becomes

$$\dfrac{x'x''}{a^2} = \dfrac{y'y''}{b^2}. \tag{a}$$

Since $P(x'y')$ is on the primary hyperbola, and $D(x''y'')$ is on the conjugate hyperbola, we have the equations of condition

$$\dfrac{x'^2}{a^2} - \dfrac{y'^2}{b^2} = 1, \quad (b) \qquad \dfrac{y''^2}{b^2} - \dfrac{x''^2}{a^2} = 1. \tag{c}$$

But (a), combined with (b) and (c), may be written

$$\dfrac{y'^2}{b^2} \cdot \dfrac{y''^2}{b^2} = \dfrac{x'^2}{a^2} \cdot \dfrac{x''^2}{a^2}, \text{ or } \dfrac{y'^2}{b^2}\left(1 + \dfrac{x''^2}{a^2}\right) = \dfrac{x''^2}{a^2}\left(1 + \dfrac{y'^2}{b^2}\right). \tag{d}$$

Therefore from (d) and (a),

$$x'' = \pm \dfrac{a}{b}y', \quad y'' = \pm \dfrac{b}{a}x', \tag{e}$$

which are the coordinates of D and D' in terms of $P(x'y')$.

If the points D and D' refer to the primary hyperbola, then

$$x'' = \pm \dfrac{a}{b}y'\sqrt{-1}, \quad y'' = \pm \dfrac{b}{a}x'\sqrt{-1}.$$

VI. *If CP and CD are semi-conjugate diameters of the two conjugate hyperbolas, then $\overline{CP}^2 - \overline{CD}^2 = a^2 - b^2$.*

For $\overline{CP}^2 - \overline{CD}^2 = x'^2 + y'^2 - \dfrac{a^2}{b^2}y'^2 - \dfrac{b^2}{a^2}x'^2$ \hfill (e) (V.)

$$= a^2\left(\dfrac{x'^2}{a^2} - \dfrac{y'^2}{b^2}\right) - b^2\left(\dfrac{x'^2}{a^2} - \dfrac{y'^2}{b^2}\right) = a^2 - b^2.$$

DIAMETERS OF THE HYPERBOLA.

If CP and CD are semi-conjugate diameters of the primary hyperbola, then CD is imaginary; but if they are semi-conjugates of the conjugate hyperbola, then CP is imaginary.

If CD is imaginary, then $\overline{CP}^2 + \overline{CD}^2 = a^2 - b^2$, and the sum of the squares of two conjugate diameters is constant, as in the ellipse.

VII. *Tangents at the ends P, P' and D, D' of a pair of diameters form a parallelogram.*

The equations of the tangents at $P(x'y')$ and $P'(-x', -y')$ are
$$\frac{xx'}{a^2} - \frac{yy'}{b^2} = 1, \quad \frac{x(-x')}{a^2} - \frac{y(-y')}{b^2} = 1.$$

But for both of these tangents, $m = \dfrac{b^2 x'}{a^2 y'}$; therefore they are parallel. In the same way show that the tangents at D and D' are parallel.

VIII. *The area of the parallelogram formed by the tangents at P, P' and D, D', the ends of a pair of conjugate diameters, is constant and equal to $4ab$, the rectangle described on the axes.*

The tangent at $P(x'y')$ is $\dfrac{xx'}{a^2} - \dfrac{yy'}{b^2} = 1$, and the perpendicular CK on this tangent is (Art. 42)

$$\overline{CK}^2 = \frac{1}{\dfrac{x'^2}{a^4} + \dfrac{y'^2}{b^4}}.$$

Also, $\quad \overline{CD}^2 = \dfrac{a^2}{b^2} y'^2 + \dfrac{b^2}{a^2} x'^2 = a^2 b^2 \left(\dfrac{x'^2}{a^4} + \dfrac{y'^2}{b^4} \right).$ \hfill (e) (V.)

$\therefore CK \cdot CD = ab,$ or $4 CK \cdot CD = 4ab.$

IX. *The asymptotes bisect the lines PD and PD'. (Fig. 51.)*
For (xy), the middle point of PD or PD' (e) (Art. 5),

$$2x = x' \pm \frac{a}{b} y', \quad 2y = y' \pm \frac{b}{a} x'. \hfill (e)\ (V.)$$

$$\therefore \frac{y}{x} = \frac{y' \pm \frac{b}{a}x'}{x' \pm \frac{a}{b}y'} = \pm \frac{b}{a},$$

which are the equations of the asymptotes.

X. *The diagonals of the parallelogram formed by the tangents at the ends P, P'' and D, D' of a pair of conjugate diameters are the asymptotes of the hyperbola.*

For in the parallelograms $PQDC$ and $PR'D'C$, CQ and CR' bisect PD and PD'; and therefore, by IX., QQ' and RR' are asymptotes.

XI. *The lines PD and PD' are parallel to the asymptotes.*

The coordinates of P, D, D' are

$$(x'y'), \quad \left(\frac{a}{b}y', \frac{b}{a}x'\right), \quad \left(-\frac{a}{b}y', -\frac{b}{a}x'\right); \qquad \text{(V.)}$$

and therefore the tangents of the angles which PD and PD' make with the axis of X are

$$\frac{y' \mp \frac{b}{a}x'}{x' \mp \frac{a}{b}y'} = \mp \frac{b}{a}. \qquad (b) \text{ (Art. 4)}$$

Therefore PD and PD' are parallel to RR' and QQ', respectively, since $\mp \frac{b}{a}$ are the tangents of the angles which the asymptotes make with the axis of X (II.).

XII. *Tangents at the ends $P(x'y')$ and $D\left(\frac{a}{b}y', \frac{b}{a}x'\right)$ of a pair of conjugate diameters meet on the asymptotes.*

Tangents at P and D, on the conjugate hyperbolas

$$\frac{x^2}{a^2} - \frac{y^2}{b^2} = 1, \quad \frac{x^2}{a^2} - \frac{y^2}{b^2} = -1,$$

are
$$\frac{xx'}{a^2} - \frac{yy'}{b^2} = 1, \quad (a) \qquad \frac{xy'}{ab} - \frac{yx'}{ab} = -1. \qquad (b)$$

By combining (a) and (b) we get for $Q(xy)$, the point of intersection of these tangents,

$$y = y' + \frac{b}{a}x', \quad x = x' + \frac{a}{b}y'.$$

But the equation of the line through C and Q is

$$\frac{y}{x} = \frac{y' + \frac{b}{a}x'}{x' + \frac{a}{b}y'} = \frac{b}{a},$$

which is the equation of the asymptote; therefore Q is on the asymptote.

XIII. *If $y = m'x$, $y = mx$ are conjugate diameters of the primary hyperbola, they are also conjugate diameters of the conjugate hyperbola*, since the relation $m'm = \frac{b^2}{a^2}$ is not changed when the axis AA' becomes imaginary and the axis BB' real.

XIV. *The portion QR' of any tangent intercepted between the asymptotes is bisected at the point of tangency, and is equal to the parallel diameter DD'.*

For in the equal parallelograms $PQDC$ and $PR'D'C$,

$$CD = CD'. \quad \therefore PQ = PR', \text{ and } QR' = DD'.$$

XV. *The portions of a chord parallel to any tangent QR', intercepted between the conjugate hyperbolas, are equal; that is, $EF = E'F'$.*

Since M is the middle point of the common chord, we have $MF = MF'$ and $ME = ME'$.

$$\therefore MF - ME = EF = MF' - ME' = E'F'.$$

XVI. *The portions EL and E'L' of the chord intercepted between the primary hyperbola and its asymptotes are equal.*

Since QR' is bisected at P, its parallel LL', intercepted between the asymptotes, is bisected at M. But $ME = ME'$;

$$\therefore ML - ME = EL = ML' - ME' = E'L'.$$

Also $\quad MF - ML = FL = MF' - ML' = FL'.$

XVII. *The polars of any point (h, k) with respect to the conjugate hyperbolas $\dfrac{x^2}{a^2} - \dfrac{y^2}{b^2} = 1$, $\dfrac{y^2}{b^2} - \dfrac{x^2}{a^2} = 1$ are parallel.*

The polars are

$$\frac{hx}{a^2} - \frac{ky}{b^2} = 1, \quad \frac{ky}{b^2} - \frac{hx}{a^2} = 1,$$

and have the same direction parameter $m = \dfrac{b^2 h}{a^2 k}$.

XVIII. *The polar of a point $P(x'y')$ on the primary hyperbola with respect to the conjugate hyperbola is a tangent to the primary hyperbola at P', the other end of the diameter PP'.*

The polar of $P(x'y')$ with respect to the conjugate hyperbola is

$$\frac{yy'}{b^2} - \frac{xx'}{a^2} = 1, \text{ or } \frac{x(-x')}{a^2} - \frac{y(-y')}{b^2} = 1,$$

which is the tangent at $P'(-x', -y')$.

From this it follows that if from P, any point on the primary hyperbola, tangents PQ and PQ' are drawn to the conjugate hyperbola, the line QQ' is a tangent to the primary hyperbola at P', the other end of the diameter PP'.

The Hyperbola and the Eccentric Angle.

157. I. The coordinates of any point $P(xy)$ on the hyperbola $\dfrac{x^2}{a^2} - \dfrac{y^2}{b^2} = 1$ can be expressed in terms of the eccentric angle ϕ, as in the ellipse. In the ellipse (Art. 152, III.) the coordinates of any point $P(xy)$ were found to be

$$x = a \cos \phi, \quad y = b \sin \phi,$$

since $\quad \dfrac{x^2}{a^2} + \dfrac{y^2}{b^2} = \cos^2 \phi + \sin^2 \phi = 1.$

In the same way any point $P(xy)$ on the hyperbola may be expressed by $x = a \sec \phi$, $y = b \tan \phi$, since

$$\dfrac{x^2}{a^2} - \dfrac{y^2}{b^2} = \sec^2 \phi - \tan^2 \phi = 1.$$

Also any point $P(xy)$ on the conjugate hyperbola may be represented by $x = a \cot \phi'$, $y = b \operatorname{cosec} \phi'$, since

$$\dfrac{y^2}{b^2} - \dfrac{x^2}{a^2} = \operatorname{cosec}^2 \phi' - \cot^2 \phi' = 1.$$

II. *To construct the point $P(a \sec \phi, b \tan \phi)$ on the hyperbola for a given value of ϕ. (Fig. 52.)*

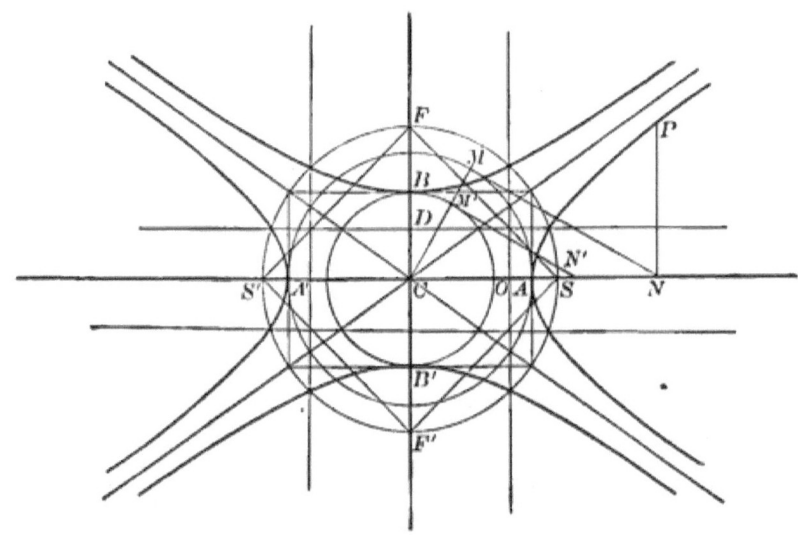

Fig. 52.

On the axes describe the auxiliary circles as in Fig. 52. Let ACM be the given eccentric angle, and at the points M and M' draw the tangents MN and $M'N'$.

Let $CN = x$; then in the right triangle NCM, we have $x \cos \phi = a$, or $x = a \sec \phi$. Again, in the right triangle $N'CM'$, $N'M' = b \tan \phi = y$. On the perpendicular to CN at N lay off $PN = N'M'$, and $P(a \sec \phi, b \tan \phi)$ is the required point.

III. *If $P(x'y')$ and $D(x''y'')$ are the ends of a pair of conjugate diameters whose eccentric angles are ϕ and ϕ', then $\phi' + \phi = \dfrac{\pi}{2}$, or ϕ' and ϕ are complementary angles. (Fig. 51.)*

For the conjugate diameters PP' and DD',

$$m' = \frac{y'}{x'} = \frac{b \tan \phi}{a \sec \phi} = \frac{b}{a} \sin \phi,$$

$$m = \frac{y''}{x''} = \frac{b \csc \phi'}{a \cot \phi'} = \frac{b}{a} \sec \phi'.$$

But for conjugate diameters,

$$m'm = \frac{b^2}{a^2} \sin \phi \sec \phi' = \frac{b^2}{a^2}. \qquad \text{(Art. 155, XI.)}$$

$\therefore \sin \phi \sec \phi' = 1$, or $\sin \phi = \cos \phi'$. $\therefore \phi' + \phi = \dfrac{\pi}{2}$.

Therefore the coordinates $(a \cot \phi', b \csc \phi')$ of any point D on the conjugate hyperbola may also be written $(a \tan \phi, b \sec \phi)$ in terms of ϕ, the angle of the conjugate point P on the primary hyperbola; and these coordinates also satisfy the equation of the conjugate hyperbola, since

$$\frac{y^2}{b^2} - \frac{x^2}{a^2} = \sec^2 \phi - \tan^2 \phi = 1.$$

The equations of the conjugate diameters PP' and DD' are

$$y = \frac{b}{a} x \sin \phi \quad \text{and} \quad y = \frac{b}{a} x \csc \phi.$$

IV. The lengths of the semi-conjugate diameters CP and CD are

$$\overline{CP}^2 = a^2 \sec^2 \phi + b^2 \tan^2 \phi,$$

$$\overline{CD}^2 = a^2 \tan^2 \phi + b^2 \sec^2 \phi.$$

$$\therefore \overline{CP}^2 - \overline{CD}^2 = (a^2 - b^2)(\sec^2 - \tan^2 \phi) = a^2 - b^2,$$

since $\sec^2 \phi - \tan^2 \phi = 1$;

that is, the difference of the squares of a pair of semi-conjugate diameters is equal to the difference of the squares of the semi-axes.

V. The equations of the tangents at the conjugate points P and D, in terms of the eccentric angle ϕ, are

$$\frac{x}{a} \sec \phi - \frac{y}{b} \tan \phi = 1, \quad \text{since} \quad \frac{xx'}{a^2} - \frac{yy'}{b^2} = 1; \quad (a)$$

$$\frac{y}{b} \sec \phi - \frac{x}{a} \tan \phi = 1, \quad \text{since} \quad \frac{yy''}{b^2} - \frac{xx''}{a^2} = 1. \quad (b)$$

VI. The equations of the normals at the conjugate points P and D, in terms of ϕ, are

$$\frac{ax}{\sec \phi} + \frac{by}{\tan \phi} = a^2 + b^2, \quad \text{since} \quad \frac{a^2 x}{x'} + \frac{b^2 y}{y'} = a^2 + b^2; \quad (a)$$

$$\frac{ax}{\tan \phi} + \frac{by}{\sec \phi} = a^2 + b^2, \quad \text{since} \quad \frac{a^2 x}{x''} + \frac{b^2 y}{y''} = a^2 + b^2. \quad (b)$$

VII. *The locus of the intersections of tangents at conjugate points on conjugate hyperbolas is an asymptote.*

The difference of the squares of (a) and (b) in V. gives $y^2 = \dfrac{b^2}{a^2} x^2$, the equations of the asymptotes, since

$$\frac{y^2}{b^2} - \frac{x^2}{a^2} = 0, \quad \text{or} \quad \left(\frac{y}{b} + \frac{x}{a}\right)\left(\frac{y}{b} - \frac{x}{a}\right) = 0,$$

or
$$y = \frac{b}{a}x, \quad y = -\frac{b}{a}x.$$

Therefore the asymptotes are the diagonals of the parallelogram formed by tangents at the vertices of conjugate diameters.

VIII. *The loci of the intersections of the normals at the ends P, D, and P', D' of conjugate diameters are the lines*

$$y = \frac{a}{b}x, \quad y = -\frac{a}{b}x.$$

The difference of the squares of (*a*) and (*b*) (VI.) is

$$(a^2x^2 - b^2y^2)(\sec^2\phi - \tan^2\phi) = 0.$$

$$\therefore a^2x^2 - b^2y^2 = 0, \text{ or } y = \frac{a}{b}x, \quad y = -\frac{a}{b}x,$$

since $\sec^2\phi - \tan^2\phi = 1$.

IX. *The tangent at P, Fig. 51, intersects the asymptotes QQ' and RR' in the points*

$$Q\,[a(\sec\phi + \tan\phi), \quad b(\sec\phi + \tan\phi)]$$

and
$$R'\,[a(\sec\phi - \tan\phi), \quad -b(\sec\phi - \tan\phi)],$$

as may readily be found by combining $y = \frac{b}{a}x$ and $y = -\frac{b}{a}x$, the equations of the asymptotes, with the tangent at P (*a*) (V.).

In the same way find the coordinates of

$$Q'\,[-a(\sec\phi + \tan\phi), \quad -b(\sec\phi + \tan\phi)],$$
$$R\,[-a(\sec\phi - \tan\phi), \quad b(\sec\phi - \tan\phi)],$$

by combining $y = \frac{b}{a}x$ and $y = -\frac{b}{a}x$ with $\frac{x}{a}\sec\phi - \frac{y}{b}\tan\phi = -1$, the tangent at P'.

Combining the equations of the asymptotes with

$$\frac{y}{b}\sec\phi - \frac{x}{a}\tan\phi = -1,$$

the tangent at D', we shall find the same coordinates of Q' and R'. It follows that the products of the abscissas of Q and R' $= a^2$, and the product of their ordinates $= -b^2$, since

$$\sec^2 \phi - \tan^2 \phi = 1.$$

The same is true of the coordinates of Q' and R.

X. *The lengths of CQ and CR' are*

$$CQ = \frac{\sqrt{a^2 + b^2}}{\sec \phi - \tan \phi}, \quad CR' = \frac{\sqrt{a^2 + b^2}}{\sec \phi + \tan \phi},$$

since the distances of the points Q and R' from the centre C are the square roots of the sums of the squares of their coordinates.

XI. *The product of the intercepts made by any tangent upon the asymptotes is constant.*

For by X., $CQ \cdot CR' = a^2 + b^2$.

XII. *The area of the triangle made by any tangent and its intercepts on the asymptotes is constant.*

The area of the triangle QCR' is $\tfrac{1}{2} CQ \cdot CR \sin QCR'$. But the product $CQ \cdot CR'$ is constant by XI., as well as $\sin QCR'$, since the angle between the asymptotes is constant.

XIII. *Prove that $\overline{CQ}^2 + \overline{CR'}^2 = 2(\overline{CP}^2 + \overline{CD}^2).$*

This follows from X. and IV.

XIV. *The foci S, S', F, F' of the primary and conjugate hyperbolas, and the vertices of the rectangle on the axes, are all equidistant from the centre C.*

For by Art. 101, XII.,

$$CS = CS' = ae = \sqrt{a^2 + b^2}, \quad CF = CF' = be' = \sqrt{a^2 + b^2},$$

which are also the lengths of the semi-diagonals of the rectangle described on the axes.

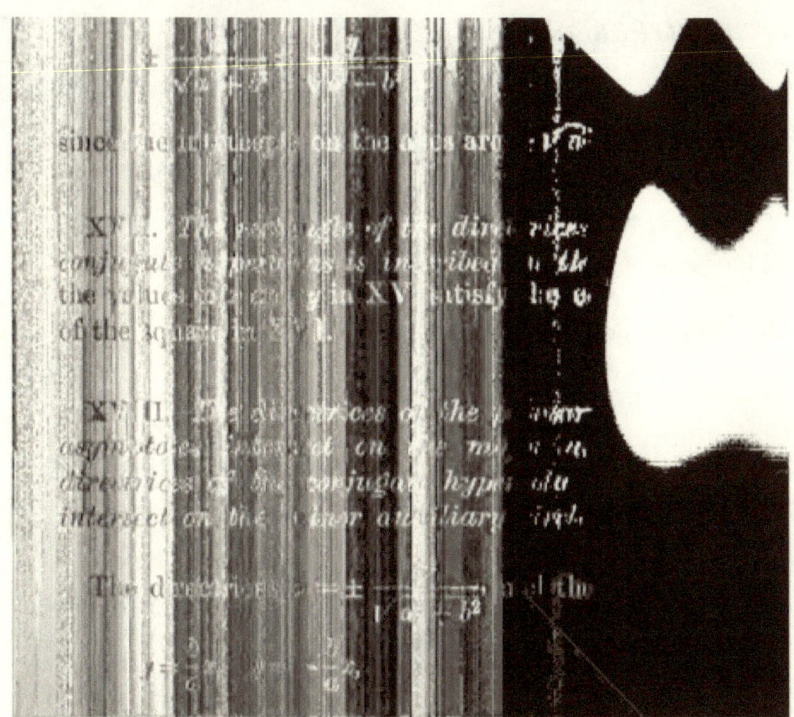

intersect in the four points

$$\left(\pm \frac{a^2}{\sqrt{a^2+b^2}},\ \pm \frac{ab}{\sqrt{a^2+b^2}} \right).$$

But these points are on the major auxiliary circle, since

$$x^2 + y^2 = \frac{a^4}{a^2+b^2} + \frac{a^2 b^2}{a^2+b^2} = a^2.$$

The directrices $y = \pm \dfrac{b^2}{\sqrt{a^2+b^2}}$ of the conjugate hyperbola and the asymptotes intersect in the four points

$$\left(\pm \dfrac{ab}{\sqrt{a^2+b^2}},\ \pm \dfrac{b^2}{\sqrt{a^2+b^2}}\right),$$

which are on

$$x^2 + y^2 = \dfrac{a^2 b^2}{a^2+b^2} + \dfrac{b^4}{a^2+b^2} = b^2,$$

the minor auxiliary circle.

EXERCISES ON THE ECCENTRIC ANGLE.

1. The equations of the conjugate diameters PP' and DD', Fig. 51, are

$$y = \dfrac{b}{a}x \sin \phi, \quad y = \dfrac{b}{a}x \csc \phi.$$

2. Find the equations and lengths of the conjugate diameters for $\phi = 0,\ 30°,\ 45°$.

3. The equations of the lines PD and PD' are

$$ay + bx = ab(\tan \phi + \sec \phi), \quad ay - bx = ab(\tan \phi - \sec \phi).$$

4. The lines PD and $P'D'$ are parallel to the asymptote RR'; the lines PD' and $D'P''$ are parallel to the asymptote QQ'.

5. The lengths of PD and PD' are

$$ae(\sec \phi - \tan \phi) \quad \text{and} \quad ae(\sec \phi + \tan \phi).$$

6. Show by Art. 4 that the lengths of the focal distances PS' and PS are

$$a(e \sec \phi + 1), \quad a(e \sec \phi - 1).$$

7. The equations of PS' and PS are

$$ya(\sec \phi + e) - xb \tan \phi - abe \tan \phi = 0,$$
$$ya(\sec \phi - e) - xb \tan \phi + abe \tan \phi = 0.$$

8. The equations of the internal and external bisectors of the angle between the focal distances PS' and PS are, by Art. 52,

$$\frac{x}{a}\sec\phi - \frac{y}{b}\tan\phi = 1, \quad \frac{ax}{\sec\phi} + \frac{by}{\tan\phi} = a^2e^2,$$

the equations of the tangent and normal at P.

9. Find the equations of the conjugate diameters PP' and DD', and show that they are parallel respectively to the tangents at D and P.

10. Show that the value of the eccentric angle of the point P is $\phi = \tan^{-1}\dfrac{\sqrt{e^2-1}}{e}$ when $CP = ae$; also, for the point D, $\phi' = \cot^{-1}\dfrac{\sqrt{e'^2-1}}{e'}$ when $CD = be'$.

11. Find the value of ϕ for the upper end of the latus-rectum of the right-hand branch of the primary hyperbola; also, the value of ϕ for the right-hand end of the latus-rectum of the upper branch of the conjugate hyperbola.

12. Prove that the tangents to the conjugate hyperbola at the points where it is cut by the tangent at the vertex A of the primary hyperbola, pass through the other vertex A'.

13. Find the tangents to the primary and conjugate hyperbolas for $\phi = 90°$ and $270°$.

14. If the normal at P meets the coordinate axes in G and G', show that $GG' = \dfrac{a}{b}e^2 CD$.

15. If the tangent at P meets the coordinate axes in T and T', show that $TT' = CD \cos\phi \cot\phi$.

16. Show that $\overline{PG}^2 = b^2(e^2\sec^2\phi - 1) = \dfrac{b^2}{a^2}\overline{CD}^2$.

17. Prove that a tangent at P, Fig. 52, will pass through the foot of the ordinate of M.

The Hyperbola referred to Conjugate Diameters.

158. *To find the equation of the hyperbola referred to a pair of conjugate diameters as coordinate axes. (Fig. 51.)*

The solution is the same as for the ellipse (Art. 153). We shall find for the squares of the new semi-axes,

$$\frac{a^2 b^2}{b^2 \cos^2 \theta - a^2 \sin^2 \theta} = \overline{CP}^2 = a'^2, \text{ real for } \tan \theta < \frac{b}{a};$$

$$\frac{a^2 b^2}{b^2 \cos^2 \theta' - a^2 \sin^2 \theta'} = -\overline{CD}^2 = -b'^2, \text{ imaginary for } \tan \theta' > \frac{b}{a}.$$

Therefore $\frac{x^2}{a'^2} - \frac{y^2}{b'^2} = 1$ is the required equation of the hyperbola referred to a pair of conjugate diameters as axes.

I. The same transformation will give the equation $\frac{y^2}{b'^2} - \frac{x^2}{a'^2} = 1$ of the conjugate hyperbola, and $\frac{x^2}{a'^2} - \frac{y^2}{b'^2} = 0$, the equations of the asymptotes.

II. Since these equations of the conjugate hyperbolas and the asymptotes are of the same form as when referred to the axes, it follows that all propositions upon the hyperbola which have not assumed rectangular axes will still hold when a pair of conjugate diameters are the coordinate axes.

The Hyperbola referred to its Asymptotes.

159. *To find the equation of the hyperbola referred to the asymptotes as coordinate axes. (Fig. 51.)*

For the asymptotes $m = \frac{b}{a}$ and $m' = -\frac{b}{a}$; and therefore, if $ACQ = \theta$ and $ACR' = \theta' = -\theta$,

$$\sin \theta = \frac{b}{\sqrt{a^2 + b^2}}, \quad \cos \theta = \frac{a}{\sqrt{a^2 + b^2}};$$

$$\sin \theta' = -\frac{b}{\sqrt{a^2+b^2}}, \quad \cos \theta' = \frac{a}{\sqrt{a^2+b^2}}.$$

The equations of transformation (g) (p. 49) become

$$y = \frac{bx'}{\sqrt{a^2+b^2}} - \frac{by'}{\sqrt{a^2+b^2}} = \frac{b}{\sqrt{a^2+b^2}}(x'-y'),$$

$$x = \frac{ax'}{\sqrt{a^2+b^2}} + \frac{ay'}{\sqrt{a^2+b^2}} = \frac{a}{\sqrt{a^2+b^2}}(x'+y').$$

Substituting these values of x and y in $\frac{x^2}{a^2} - \frac{y^2}{b^2} = 1$, and omitting the primes, we get

$$\frac{(x+y)^2}{a^2+b^2} - \frac{(x-y)^2}{a^2+b^2} = 1, \text{ or } xy = \frac{a^2+b^2}{4},$$

the required equation.

The equation of the conjugate hyperbola referred to the asymptotes as coordinate axes is $xy = -\frac{a^2+b^2}{4}$.

160. *To find the equations of the secant and tangent for the hyperbola* $xy = \frac{a^2+b^2}{4}$.

I. *The Secant.* — For two points $(x'y')$ and $(x''y'')$ on the hyperbola,

$$x'y' = \frac{a^2+b^2}{4} = x''y''.$$

$$\therefore y' - y'' = \frac{x''y''}{x'} - y'' = -\frac{y''}{x'}(x'-x'').$$

$$\therefore \frac{y'-y''}{x'-x''} = -\frac{y''}{x'} = m,$$

and $\qquad y - y' = -\frac{y''}{x'}(x-x'), \text{ or } \frac{y-y'}{y''} + \frac{x-x'}{x'} = 0,$ \qquad (a)

is the equation of the secant.

II. *The Tangent.* — For tangency $x'' = x'$, $y'' = y'$, and (a) reduces to

$$\frac{x}{x'} + \frac{y}{y'} = 2,$$

the required equation of the tangent.

III. *The intercepts of the tangent on the asymptotes are* $x = 2x'$, $y = 2y'$; therefore the part of the tangent intercepted between the asymptotes is bisected at the point of tangency.

Equilateral or Rectangular Hyperbolas.

161. An hyperbola is *equilateral* when its axes are equal; that is, for $b = a$. The equilateral conjugate hyperbolas are

$$x^2 - y^2 = a^2, \quad y^2 - x^2 = a^2.$$

These hyperbolas referred to the asymptotes as coordinate axes are

$$xy = \frac{a^2}{2} \text{ and } xy = -\frac{a^2}{2}.$$

When $b = a$, the angle between the asymptotes is

$$\theta = 2 \tan^{-1} 1 = \frac{\pi}{2};$$

that is, the asymptotes are at right angles to each other. For this reason these hyperbolas are also called rectangular hyperbolas.

I. Any two conjugate diameters of conjugate equilateral hyperbolas are equal; for

$$\overline{CP}^2 - \overline{CD}^2 = a^2 - b^2 = 0. \qquad \text{(Art. 156, VI.)}$$

II. These equi-conjugate diameters CP, CD make equal angles with the asymptotes; for CPD is an isosceles triangle, and the asymptote bisects PD (Art. 156, IX.).

III. The eccentricity of an equilateral hyperbola is $\sqrt{2}$; for $e^2 = \dfrac{a^2 + b^2}{a^2} = 2$ when $b = a$.

IV. The latus-rectum $\dfrac{2b^2}{a} = 2a$ when $b = a$.

Problems on the Hyperbola.

1. If the straight line $y = mx + c$ meets the hyperbola $\dfrac{x^2}{a^2} - \dfrac{y^2}{b^2} = 1$ in two points, one of which is at a finite and the other at an infinite distance from the origin, show that this straight line is parallel to an asymptote.

2. If a straight line cuts an hyperbola in the points P and P', and its asymptotes is Q and Q', show that the middle of PP' is also the middle of QQ'.

3. If PN and PM are the perpendiculars from P on any coordinate axes OA and OB, find the locus of P when the area of $OMPN$ is constant.

4. The ordinate PN of an hyperbola produced meets the asymptote in Q; the normal at P meets the axis of X in G; show that QG is perpendicular to the asymptote.

5. If e and e' are the eccentricities of the primary and conjugate hyperbolas, then $\dfrac{1}{e^2} + \dfrac{1}{e'^2} = 1$.

6. If any number of hyperbolas have the same transverse axis, show that the tangents to these hyperbolas having the same abscissa all pass through the same point on the transverse axis.

7. If the tangent at any point P on the hyperbola intersects the axis in T, and CP meets the tangent at the vertex A in E, then ET is parallel to AP.

8. A and A' are the vertices of a rectangular hyperbola, P is any point on the curve; show that the internal and external bisectors of the angle APA' are parallel to the asymptotes.

9. Show that the coordinates of the points of intersection of two tangents to an hyperbola referred to its asymptotes as axes are harmonic means between the coordinates of the points of contact.

10. The straight lines drawn from any point of an equilateral hyperbola to the extremities of any diameter are equally inclined to the asymptotes.

11. Find the condition that the line $lx + my = 1$ should touch the hyperbola $\dfrac{x^2}{a^2} - \dfrac{y^2}{b^2} = 1$.

12. Find the points in which the tangents from O, the foot of the directrix, touch the hyperbola, and the angle they make with the transverse axis.

13. A tangent at the end of the latus-rectum through the focus S meets any ordinate PN produced in R; show that $PS = NR$.

14. If a tangent at any point P on the hyperbola cuts the tangents at the vertices A and A' in the points T and T', then $AT \cdot A'T' = b^2$.

15. If in an hyperbola $3\,CA = 2\,CS$, find the angles of the asymptotes with the transverse axis.

16. If from a point P on an hyperbola PK is drawn parallel to the transverse axis, cutting the asymptotes in Q and R, then $PQ \cdot PR = a^2$; if PK is drawn parallel to the conjugate axis, then $PQ \cdot PR = b^2$.

17. If from a point P on an hyperbola PN be drawn parallel to the asymptote meeting the directrix in N, then $PS = PN$.

18. The tangent at P meets the transverse axis of the hyperbola in T; connect P with the vertices A, A'; the perpendicular to AA' at T meets PA and PA' in Q and R; show that QR is bisected at T.

19. Find the eccentricity and latus-rectum of the hyperbola
$$y^2 = 4(x^2 + a^2).$$

20. In an equilateral hyperbola the eccentricity is the ratio of the diagonal of a square to its side.

21. If a tangent at P on an hyperbola is intersected by the tangents at the vertices in the points Q and R, the circle described on QR as a diameter will pass through the foci.

22. At what point on an hyperbola are the subtangent and subnormal equal to each other?

23. The line $x = 3y$ is a diameter of the hyperbola $\dfrac{x^2}{16} - \dfrac{y^2}{25} = 1$; find the equation of the conjugate diameter.

24. Find the condition that the line $\dfrac{x}{b} + \dfrac{y}{a} = 1$ shall touch the hyperbola $\dfrac{x^2}{a^2} - \dfrac{y^2}{b^2} = 1$. *Ans.* $e^4 - e^2 - 1 = 0$.

25. Show that the linear equation of the right-hand branch of the hyperbola when a focus is the origin is $r = ex \pm a(1 - e^2)$.

26. A perpendicular is drawn from the focus of an hyperbola to an asymptote; show that its foot is at the distances a and b from the centre and focus respectively.

27. Show that the distances cut off from the normal to an hyperbola by the axes are in the ratio of $a^2 : b^2$.

28. A, A' are the ends of a fixed diameter of a circle, and P, P' are the ends of any chord perpendicular to this diameter; show that the locus of the point of intersection of AO, AP, and $A'P'$ is a rectangular hyperbola.

29. A series of chords of the hyperbola $\dfrac{x^2}{a^2} - \dfrac{y^2}{b^2} = 1$ are tangents to the circle described on the distance between the foci of the hyperbola as a diameter; show that the locus of their poles with reference to the hyperbola is $\dfrac{x^2}{a^4} + \dfrac{y^2}{b^4} = \dfrac{1}{a^2 + b^2}$.

CHAPTER IX.

THE GENERAL EQUATION OF THE SECOND DEGREE.

$$Ax^2 + 2Hxy + By^2 + 2Gx + 2Fy + C = 0.$$

162. In the preceding chapters we have found that the equations of the conic sections are always of the second degree. In this chapter it is proposed to show that the general equation of the second degree, which for brevity in writing we shall usually denote by $S = 0$ and designate by the conic $S = 0$, always represents a conic section for all real values of the arbitrary constants A, H, B, G, F, C.

We may assume that the locus represented by $S = 0$ is referred to a pair of rectangular axes; for if not, we may transform to a pair of rectangular axes without changing the degree of the equation (Art. 22).

163. *To show that $S = 0$ always represents a conic section; or that the conic is the only curve of the second order.*

I. If $H = 0$, then $S = 0$ becomes

$$Ax^2 + By^2 + 2Gx + 2Fy + C = 0, \tag{a}$$

or
$$A\left(x + \frac{G}{A}\right)^2 + B\left(y + \frac{F}{B}\right)^2 = \frac{AF^2 + BG^2 - ABC}{ABC} = K$$

by completing the squares (Art. 112).

If now we take the point $\left(-\dfrac{G}{A}, -\dfrac{F}{B}\right)$ as the origin, then (a) reduces to

$$Ax^2 + By^2 = K,$$

or
$$\frac{x^2}{\frac{K}{A}} + \frac{y^2}{\frac{K}{B}} = 1. \tag{a'}$$

If A, B, K, have like signs, then (a') denotes an ellipse; but if A or B has a sign different from K, then (a') denotes an hyperbola.

It follows, then, that *the axes of the conic (a) are respectively parallel to the coordinate axes.*

II. If $K = 0$, then $Ax^2 + By^2 = 0$ denotes two intersecting straight lines passing through the point $\left(-\dfrac{G}{A}, -\dfrac{F}{B}\right)$, *real lines* when A and B have unlike signs, and *imaginary lines* when A and B have like signs.

III. In addition to $H = 0$, A and B cannot both be zero; for this would reduce $S = 0$ to an equation of the first degree.

But suppose that $H = 0$ and $A = 0$, then (a) becomes

$$By^2 + 2Fy + 2Gx + C = 0, \qquad (a'')$$

or
$$B\left(y + \frac{F}{B}\right)^2 = -2Gx + \frac{F^2}{B} - C,$$

or
$$\left(y + \frac{F}{B}\right)^2 = -\frac{2G}{B}\left(x - \frac{F^2 - BC}{2BG}\right),$$

and denotes a parabola whose axis is parallel to the axis of X, and whose vertex is $\left(\dfrac{F^2 - BC}{2BG}, -\dfrac{F}{B}\right)$ (Art. 112).

If G also equals zero, then (a'') becomes

$$\left(y + \frac{F}{B}\right)^2 = \frac{F^2 - BC}{B^2},$$

and represents two parallel straight lines.

If now we can show that the term $2Hxy$ can be made to disappear by referring the locus to a new pair of rectangular axes, the origin remaining the same, then it will follow that $S = 0$ always represents a conic section.

164. *To transform $S = 0$ to a new pair of rectangular axes, the origin remaining the same, such that the term containing the product xy of the coordinates shall disappear.*

The formulas for transforming to a new pair of rectangular axes, in which θ denotes the angle which the new axis of X makes with the old one, are (Art. 21, c)

$$x = x' \cos\theta - y' \sin\theta, \quad y = x' \sin\theta + y' \cos\theta.$$

Substitute these values of x and y in $S = 0$, and it becomes, omitting primes, and reducing,

$$(A \cos^2\theta + 2H \sin\theta \cos\theta + B \sin^2\theta) x^2$$
$$+ 2[H(\cos^2\theta - \sin^2\theta) - (A - B) \sin\theta \cos\theta] xy$$
$$+ (A \sin^2\theta - 2H \sin\theta \cos\theta + B \cos^2\theta) y^2$$
$$+ 2(G \cos\theta + F \sin\theta) x + 2(F \cos\theta - G \sin\theta) y + C = 0. \quad (a')$$

But the term containing xy will disappear if

$$H(\cos^2\theta - \sin^2\theta) - (A - B) \sin\theta \cos\theta = 0,$$

or $\quad \tan 2\theta = \dfrac{2H}{A - B}$, or $\theta = \tfrac{1}{2} \tan^{-1} \dfrac{2H}{A - B}$.

This value of θ is always real, and therefore *it is always possible to remove the term from $S = 0$ which contains the product of the coordinates.*

If now we substitute this value of θ in the coefficients of x^2, y^2, x, y, and denote the results by A', B', $2G'$, $2F'$, then equation (a') becomes

$$A'x^2 + B'y^2 + 2G'x + 2F'y + C = 0.$$

But this equation is now of the same form as (a) in Art. 163, and *therefore $S = 0$ always represents a conic.*

We can most readily find the functions of the coefficients which *characterize* the different conics represented by $S = 0$ by solving the following problem.

165. *To find the points in which the conic $S=0$ is cut by any straight line drawn through the origin.*

I. It will be simplest to use polar coordinates. Place the pole at the origin, and let the axis of X be the polar axis; then, Art. 12, $x = r\cos\theta$, $y = r\sin\theta$, and $S = 0$ becomes the quadratic

$$(A\cos^2\theta + 2H\sin\theta\cos\theta + B\sin^2\theta)r^2$$
$$+ 2(G\cos\theta + F\sin\theta)r + C = 0. \qquad (a')$$

Since θ can have all values from 0 to 2π, the radius vector, r, must cut the conic $S = 0$; but for no value of θ can it ever cut $S = 0$ in more than two points. But these points may be *real* or *imaginary*, and at *finite* or *infinite* distances from the origin.

II. *To find for what values of θ the radius vector, r, will cut $S = 0$ in one finite point, and in one point at infinity.*

If (a') has *one* finite root and *one* infinite root, then, Art. 20, the finite root is given by the equation

$$2(G\cos\theta + F\sin\theta)r + C = 0, \qquad (b)$$

and the directions, or values of θ for which r is infinite, are given by the equation

$$A\cos^2\theta + 2H\sin\theta\cos\theta + B\sin^2\theta = 0, \qquad (c)$$

or $\qquad B\tan^2\theta + 2H\tan\theta + A = 0.$

Solving, we get

$$\tan\theta = \frac{-H \pm \sqrt{H^2 - AB}}{B} = \frac{y}{x},$$

which shows that there can be drawn through the origin *two real different straight lines*, or *two coincident ones*, or *two imaginary ones*, which will cut the conic $S = 0$ at *infinity*, according as $H^2 - AB > 0$, $= 0$, or < 0; and that each of these lines will also meet the conic $S = 0$ in the finite point given by (b).

It appears, then, that the conic $S = 0$ has *two* infinite branches when $H^2 - AB > 0$, *one* infinite branch when $H^2 - AB = 0$, and *no* infinite branch when $H^2 - AB < 0$. Therefore

$H^2 - AB > 0$ *is the characteristic of the hyperbola;*
$H^2 - AB = 0$ *is the characteristic of the parabola;*
$H^2 - AB < 0$ *is the characteristic of the ellipse.*

It follows that A and B must have like signs for the ellipse and parabola, and unlike signs for the hyperbola.

If we multiply (c) by r^2, and introduce x and y, we get

$$Ax^2 + 2Hxy + By^2 = 0, \qquad (d)$$

the equation of the two straight lines, drawn through the origin, which meet the conic $S = 0$ in points at infinity.

III. *To find the position of the origin which will make the terms in $S = 0$ of the first degree in x and y disappear.*

If the roots of (a') are equal and with opposite signs for all values of θ, then by the theory of equations the coefficient of the first power of r must disappear; that is, $G\cos\theta + F\sin\theta = 0$ for all values of θ. Therefore $G = 0$, $F = 0$, are the required conditions. It follows that the origin now bisects all chords drawn through it, and is therefore the centre of the conic.

But these conditions also make the root r, of (a'), which is determined by equation (b), *infinite;* hence two lines respectively parallel to the lines (d) can be drawn through the centre which will meet the conic in *two coincident points* at infinity; that is, two tangents whose points of contact are at infinity. These lines, called the *asymptotes* of the conic (Art. 156, II.), are real for the hyperbola, imaginary for the ellipse, and lie altogether at infinity for the parabola, since its centre is at infinity.

166. *To find by construction (Art. 18) the shapes and positions of the curves represented by $S = 0$.*

I. *The solution of $S = 0$.* Suppose that B is positive and we solve for y; then, by Art. 64,

$$By + Hx + F = \pm \sqrt{[(H^2 - AB)x^2 - 2(BG - HF)x + F^2 - BC]} \quad (b)$$

$$= \pm \sqrt{\left[(H^2 - AB)\left\{x^2 - \frac{2(BG - HF)}{H^2 - AB}x + \frac{F^2 - BC}{H^2 - AB}\right\}\right]}$$

$$= \pm \sqrt{[(H^2 - AB)(x - x')(x - x'')]},$$

x' and x'' denoting the roots of the quadratic

$$x^2 - \frac{2(BG - HF)}{H^2 - AB}x + \frac{F^2 - BC}{H^2 - AB} = 0.$$

Solving this quadratic, we get

$$x' = \frac{BG - HF}{H^2 - AB} + \frac{\sqrt{B\Delta}}{H^2 - AB},$$

$$x'' = \frac{BG - HF}{H^2 - AB} - \frac{\sqrt{B\Delta}}{H^2 - AB},$$

in which

$$B\Delta = (BG - HF)^2 - (H^2 - AB)(F^2 - BC)$$

$$= B(AF^2 + BG^2 + CH^2 - ABC - 2FGH),$$

or

$$\Delta = AF^2 + BG^2 + CH^2 - ABC - 2FGH.$$

The roots x' and x'' are *real and unequal* for $\Delta > 0$, *real and equal* for $\Delta = 0$, and *imaginary* for $\Delta < 0$. This function of the coefficients of $S = 0$, which is denoted by Δ, and which determines the character of the roots, is called the *Discriminant* of the equation $S = 0$.

II. *The Construction.* Equation (b) may be separated into the three parts

$$y_1 = -\frac{Hx + F}{B} - \frac{1}{B}\sqrt{(H^2 - AB)(x - x')(x - x'')}, \quad (d)$$

$$y_2 = -\frac{Hx + F}{B}, \quad (e)$$

$$y_3 = -\frac{Hx + F}{B} + \frac{1}{B}\sqrt{(H^2 - AB)(x - x')(x - x'')}. \quad (f)$$

GENERAL EQUATION OF SECOND DEGREE. 283

On any rectangular axes plot the straight line (e) AB (Fig. 53), and also lay off $ON' = x'$, $ON'' = x''$, and draw $N'P$, $N''P'$, parallel to the axis of y. Any point (xy_2) which lies on AB, and the corresponding points (xy_1), (xy_3), on the conic,

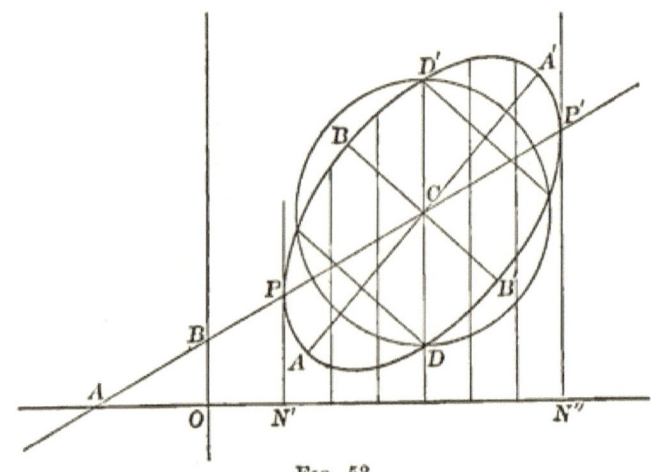

FIG. 53.

all lie on a parallel to the axis of Y, since all three have the same abscissa X. The distance between any two corresponding points (xy_1), (xy_3), on the curve, that is, the *length* of any chord which is parallel to the axis of Y, is

$$y_3 - y_1 = \frac{2}{B}\sqrt{(H^2 - AB)(x - x')(x - x'')}.$$

Also, since $y_2 = \frac{y_1 + y_3}{2}$ for all values of x, the straight line (e) AB bisects this system of parallel chords, and is therefore *a diameter of the conic*.

For $H^2 - AB < 0$ and $\Delta > 0$, all chords which lie between the tangents $N'P$ and $N''P'$ are *real*, since the product $(x - x')(x - x'')$ is *minus* for all values of x between these limits; therefore the curve is an *ellipse*. For all values of x without these limits the curve is an *imaginary* ellipse.

For $H^2 - AB > 0$ and $\Delta > 0$, all chords between $N'P$ and $N''P'$ are *imaginary;* but without these limits they are all *real to infinity*, and the curve is the hyperbola whose vertices on the diameter AB are at P, P' (Fig. 54).

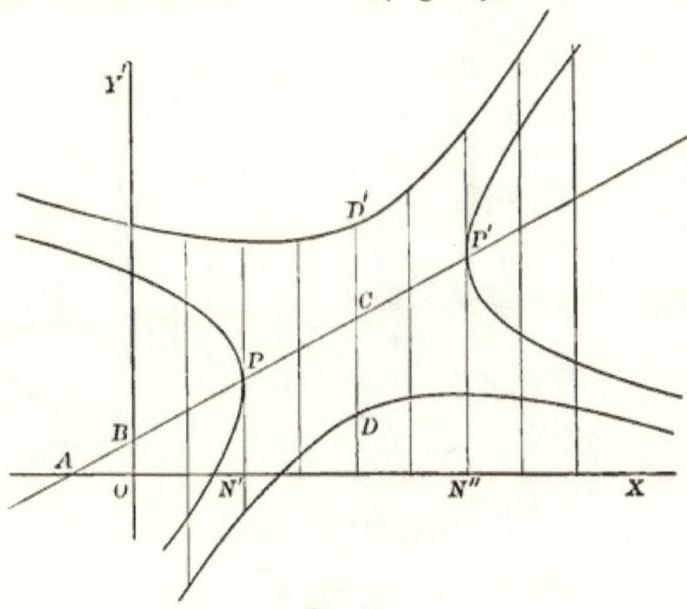

Fig. 54.

If $\Delta < 0$, that is, if the roots x' and x'' are imaginary, they have the form $x' = a + b\sqrt{-1}$, $x'' = a - b\sqrt{-1}$, and the product $(x - x')(x - x'') = (x - a)^2 + b^2$ is positive for all values of x. Therefore the chords

$$\frac{2}{B}\sqrt{(H^2 - AB)(x - x')(x - x'')}$$

are all imaginary for $H^2 - AB < 0$, and the ellipse is *imaginary;* but for $H^2 - AB > 0$ these chords are *real* for all values of x, and we get the conjugate hyperbola whose vertices on the diameter parallel to the axis Y are at DD'.

To construct the axes of the ellipse or hyperbola, draw a concentric circle (Fig. 53), and join the points of intersection by

GENERAL EQUATION OF SECOND DEGREE.

parallel chords; lines drawn through the middle points of these chords are the axes.

For $H^2 - AB = 0$, one root is infinite and one finite (Art. 20), and equation (b) becomes

$$y = -\frac{Hx + F}{B} \pm \frac{1}{B}\sqrt{2(HF - BG)(x - x')},$$

if the finite root $x' = \dfrac{BC - F^2}{2(HF - BG)}.$

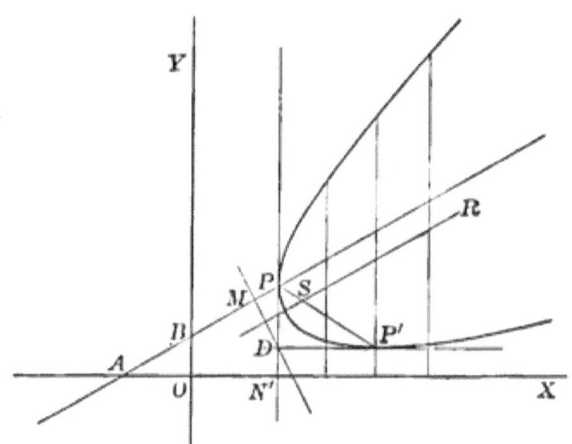

Fig. 55.

The straight line (e) AB (Fig. 55) still bisects all the chords

$$y_3 - y_1 = \frac{2}{B}\sqrt{2(HF - BG)(x - x')}$$

parallel to the axis of Y, and is a diameter.

Lay off $ON' = x'$, and draw $N'P$ parallel to the axis of Y. For all values of x greater than x' the chords are *real*, and the parabola extends to infinity on the positive side of $N'P$. If the chords are imaginary on the positive side of the tangent $N'P$, they will be real on the negative side, and the parabola will extend to infinity in this direction.

To construct the axis of the parabola, solve $S = 0$ for x, make $H^2 - AB = 0$, and find the root $y = y' = N'D$, the equation of the tangent parallel to the axis of X. The directrix passes through D, the intersection of the perpendicular tangents $N'P$ and DP' (Art. 120), and is perpendicular to the diameter AB. The focus S is on the chord PP' (Art. 130), and $PS = PM$ (Art. 99). The axis SR is drawn parallel to AB.

For $\Delta = 0$, or $x' = x''$, equation (b) becomes

$$By + Hx + F = \pm (x - x') \sqrt{H^2 - AB},$$

two *real* or two *imaginary* straight lines, according as $H^2 - AB > 0$, or < 0.

When $\Delta = 0$, and $H^2 - AB = 0$, then $BG - HF = 0$, and (b) reduces to the two parallel straight lines

$$By + Hx + F = \pm \sqrt{F^2 - BC}.$$

EXERCISES.

1. Show that $Ax + Hy + G = 0$ is a diameter of $S = 0$.

2. Find the expression for the length of any chord of the system parallel to the axis of X, which the diameter $Ax + Hy + G = 0$ bisects.

3. Find the tangents to $S = 0$ which are parallel to the axis of X.

4. Find the equation of the diameter DD'.

5. The length of $DD' = 2\sqrt{\dfrac{B\Delta}{AB - H^2}}$.

6. The length of $PP' = \dfrac{2\sqrt{B(H^2 + B^2)\Delta}}{B(AB - H^2)}$.

7. The coordinates of the centre are $\left(\dfrac{BG - HF}{H^2 - AB}, \dfrac{AF - HG}{H^2 - AB}\right)$.

8. Show that for the parabola the three second-degree terms form a perfect square.

9. Show that if $S = 0$ is a tangent to both coordinate axes, then $G^2 = AC$, $F^2 = BC$.

GENERAL EQUATION OF SECOND DEGREE.

10. What curve is represented by

$$\frac{x^2}{a^2} - \frac{2xy}{ab} + \frac{y^2}{b^2} - \frac{2x}{a} - \frac{2y}{b} + 1 = 0?$$

11. What conics are represented by the following equations? First determine the kind by computing the values of $H^2 - AB$ and Δ; then compute a diameter, and the tangents at the vertices on this diameter, and find the limits within which the curve must lie.

1. $3x^2 + 4xy + y^2 - 3x - 2y + 21 = 0$.   An hyperbola.
2. $5x^2 + 4xy + 2y^2 - 5x - 2y - 19 = 0$.   An ellipse.
3. $4x^2 + 4xy + y^2 - 5x - 2y - 10 = 0$.   A parabola.
4. $3x^2 + 4xy - 4y^2 - 7x + 2y + 2 = 0$.   Two straight lines.
5. $x^2 + 4xy + 4y^2 + 6x + 12y + 9 = 0$.   Two coincident straight lines.

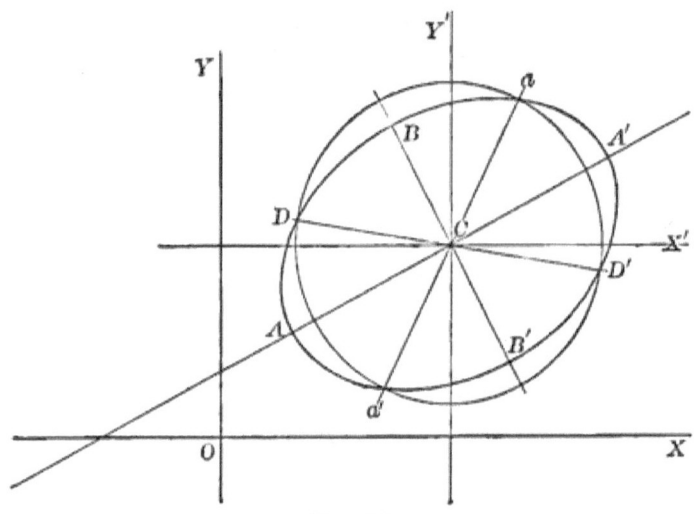

Fig. 56.

167. *To find the form of equation $S = 0$ when the centre $C(h, k)$ of the conic is taken as the origin, and the new coordinate axes CX', CY', are respectively parallel to the old ones. (Fig. 56.)*

For this purpose change x and y into $x+h$ and $x+k$ respectively (Art. 21, a); then $S=0$ becomes

$$A(x+h)^2 + 2H(x+h)(y+k) + B(y+k)^2 \\ + 2G(x+h) + 2F(y+k) + C = 0,$$

or $\qquad Ax^2 + 2Hxy + By^2 + 2G'x + 2F'y + C' = 0, \qquad (a')$

in which

$$G' = Ah + Hk + G,$$
$$F' = Bk + Hh + F,$$
$$C' = Ah^2 + 2Hhk + Bk^2 + 2Gh + 2Fk + C.$$

Since (h, k) is the centre of the conic, we have (Art. 165, III.)

$$G' = Ah + Hk + G = 0, \quad F' = Bk + Hh + F = 0. \qquad (b)$$

Solving these equations for (h, k), we get the single values

$$h = \frac{BG - HF}{H^2 - AB}, \quad k = \frac{AF - HG}{H^2 - AB},$$

and

$$\begin{aligned}
C' &= (Ah + Hk + G)h + (Bk + Hh + F)k \\
&\qquad + Gh + Fk + C \qquad (b') \\
&= Gh + Fk + C \qquad \text{by } (b) \\
&= G \cdot \frac{BG - HF}{H^2 - AB} + F \cdot \frac{AF - HG}{H^2 - AB} + C \\
&= \frac{AF^2 + BG^2 + CH^2 - ABC - 2FGH}{H^2 - AB} = \frac{\Delta}{H^2 - AB},
\end{aligned}$$

and (a') becomes

$$Ax^2 + 2Hxy + By^2 + C' = 0, \qquad (a'')$$

or $\qquad Ax^2 + 2Hxy + By^2 + \dfrac{\Delta}{H^2 - AB} = 0.$

It will be noticed that by this transformation the coefficients A, $2H$, B, of the second-degree terms, as well as the discriminant Δ, and the characteristic $H^2 - AB$, remain unchanged; that the new independent term C' is the result of the substitution of the coordinates (h, k) of the centre for x and y, respectively, in $S = 0$.

When $H^2 - AB$ is not zero, (h, k) denotes a single finite point, and the ellipse and hyperbola are therefore called *central curves*; but for $H^2 - AB = 0$, h and k are both infinite, and the parabola is called a *non-central* curve.

If $\Delta = 0$, then $S = 0$ becomes

$$Ax^2 + 2Hxy + By^2 = 0, \qquad (a''')$$

and represents two straight lines drawn through $C(h, k)$, *real* for the hyperbola, and *imaginary* for the ellipse, which meet the conic (a'') in two points at infinity (Art. 164, III.), and are therefore its asymptotes.

But the axes of the conic bisect the angles between the asymptotes (Art. 156, I.), and therefore (Art. 67),

$$Hy^2 + (A - B)xy - Hx^2 = 0,$$

or

$$\frac{y}{x} = \frac{-(A-B) \pm \sqrt{(A-B)^2 + (2H)^2}}{2H}, \qquad (b)$$

are the equations of the axes AA', BB', which can readily be drawn. Combining the equation of each axis with that of the curve (a''), we get the coordinates of the points, A, A', B, B', of intersection, and can then compute the lengths AA', BB', of the axes.

168. We can still further simplify the central equation of the conic $Ax^2 + 2Hxy + By^2 + C'' = 0$, (a''), (Art. 167), by adopting the axes AA', BB', as coordinate axes, when it reduces to $A'x^2 + B'y^2 + C'' = 0$ (Fig. 56).

I. *Transform equation (a'') to a new pair of rectangular axes, the new axis of x making an angle θ with the old one.*

Since equation (a'') does not contain the first powers of x and y, it becomes at once, by the same transformation as in Art. 164,

290   PLANE ANALYTIC GEOMETRY.

$$\left.\begin{array}{l}(A\cos^2\theta + 2H\sin\theta\cos\theta + B\sin^2\theta)x^2 \\ + 2[H(\cos^2\theta - \sin^2\theta) - (A-B)\sin\theta\cos\theta]xy \\ + (A\sin^2\theta - 2H\sin\theta\cos\theta + B\cos^2\theta)y^2\end{array}\right\} + C' = 0. \quad (a''')$$

If we denote the coefficients of x^2, xy, y^2, by A', $2H'$, B', and remember, from Trigonometry, that

$$2\sin^2\theta = 1 - \cos 2\theta, \qquad 2\cos^2\theta = 1 + \cos 2\theta,$$
$$\cos 2\theta = \cos^2\theta - \sin^2\theta, \quad \sin 2\theta = 2\sin\theta\cos\theta,$$

we have

$$2A' = A + B + 2H\sin 2\theta + (A-B)\cos 2\theta, \qquad (b)$$
$$2B' = A + B - 2H\sin 2\theta - (A-B)\cos 2\theta, \qquad (c)$$
$$2H' = 2H\cos 2\theta - (A-B)\sin 2\theta. \qquad (d)$$

II. *Show that the values of the two functions $A + B$ and $H^2 - AB$ of the coefficients of the second-degree terms of $S = 0$ do not change by transforming the equation from one pair of rectangular axes to another.*

By adding (b) and (c), we get at once

$$A' + B' = A + B. \qquad (e)$$

The product of (b) and (c) and the square of (d) are

$$4A'B' = (A+B)^2 - [2H\sin 2\theta + (A-B)\cos 2\theta]^2,$$
$$4H'^2 = [2H\cos 2\theta - (A-B)\sin 2\theta]^2.$$
$$\therefore 4(H'^2 - A'B') = 4H^2 + (A-B)^2 - (A+B)^2$$
$$= 4(H^2 - AB). \qquad (e)$$

On account of the invariability of these functions they are called *Invariants* in the Modern Geometry.

III. If we make $\theta = X'CA'$ (Fig. 56), then the new pair of rectangular axes will coincide with the axes of the conic, and the term in (a''') containing xy must disappear.

GENERAL EQUATION OF SECOND DEGREE. 291

$$\therefore 2H' = 2H\cos 2\theta - (A-B)\sin 2\theta = 0,$$

or
$$\tan 2\theta = \frac{2H}{A-B} = \frac{2\tan\theta}{1-\tan^2\theta}. \tag{f}$$

$$\therefore \tan\theta = \frac{-(A-B) \pm \sqrt{(A-B)^2 + (2H)^2}}{2H},$$

as found in Art. 167 (b), for determining the angles which the axes of the conic make with CX', the axis of X.

IV. *To find the values of A', B', when the coordinate axes coincide with the axes of the conic.*

In this case $H' = 0$, and equations (e) II. become

$$A' + B' = A + B, \qquad -A'B' = H^2 - AB,$$

for which we readily get

$$A' = \tfrac{1}{2}(A+B \pm Q), \quad B' = \tfrac{1}{2}(A+B \mp Q),$$

if
$$Q = \pm \sqrt{(A-B)^2 + (2H)^2} = \pm\sqrt{(A+B)^2 + 4(H^2-AB)}.$$

We see that Q is less than $A+B$ for the ellipse, greater for the hyperbola, and equal to $A+B$ for the parabola; and therefore it follows that A', B', have like signs for the ellipse, unlike signs for the hyperbola, and that either A' or B' must equal zero for the parabola. Therefore

$$A'x^2 + B'y^2 + C'' = 0 \tag{a^{iv}}$$

is the equation of the central conic referred to its centre as origin, and to its axes as coordinate axes.

The signs of the radical in these values of A' and B' must be so taken that the sign of $A' - B' = Q$ shall be the same as that of $2H$ in the general equation $S = 0$. The difference of A' and B' from (b), (c), I. is

$$A' - B' = 2H\sin 2\theta + (A-B)\cos 2\theta,$$

and from (f), III.

$$2H = (A-B)\tan 2\theta.$$

$$\therefore 2H(A'-B') = [2H\sin 2\theta + (A-B)\cos 2\theta](A-B)\tan 2\theta$$
$$= [(2H)^2 + (A-B)^2]\sin 2\theta.$$

Therefore $2H$ and $A'-B'$, or Q, always have the same sign, since $\sin 2\theta$ is always positive, 2θ being less than π.

When $A' + B' = A + B = 0$, then $S = 0$,

or
$$A'x^2 - A'y^2 + C'' = 0,$$

represents an *equilateral or rectangular* hyperbola (Art. 161).

V. *To express the equation of the conic a^{iv} in terms of the semi-axes CA, CB. (Fig. 56.)*

For $x = 0$, $y^2 = -\dfrac{C''}{B'} = \pm \overline{CB}^2 = \pm b^2$.

For $y = 0$, $x^2 = -\dfrac{C''}{A'} = \pm \overline{CA}^2 = \pm a^2$.

$$\therefore \frac{x^2}{-\dfrac{C''}{A'}} + \frac{y^2}{-\dfrac{C''}{B'}} = 1, \text{ or } \frac{x^2}{\pm a^2} + \frac{y^2}{\pm b^2} = 1. \qquad (a^v)$$

For $+a^2, +b^2$, the conic is a *real* ellipse; for $-a^2, -b^2$, it is an *imaginary* ellipse; for $+a^2, -b^2$, it is the *primary* hyperbola; for $-a^2, b^2$, it is the *conjugate* hyperbola; and for $C'' = 0$, (a^v) represents *two real* or *two imaginary* straight lines, according as the signs of A', B', are unlike or like.

VI. *To express a^2, b^2, e^2, the squares of the semi-axes and eccentricity in terms of the coefficients of $S = 0$.*

From the values of A', B', IV., and $C'' = \dfrac{\Delta}{H^2 - AB}$ (Art. 167, II.), we readily get

$$a^2 = -\frac{2C''}{2A'} = -\frac{2C''}{A+B\pm Q} = \frac{2\Delta}{(AB-H^2)(A+B\pm Q)},$$

$$b^2 = -\frac{2C''}{2B'} = -\frac{2C''}{A+B\mp Q} = \frac{2\Delta}{(AB-H^2)(A+B\mp Q)}.$$

If $a > b$, then for the ellipse

$$e^2 = 1 - \frac{b^2}{a^2} = 1 - \frac{A+B-Q}{A+B+Q} = \frac{2Q}{A+B+Q};$$

and for the hyperbola, if $2a$ is the real axis,

$$e^2 = 1 + \frac{b^2}{a^2} = 1 + \frac{A+B-Q}{A+B+Q} = \frac{2(A+B)}{A+B+Q}.$$

169. A direct and simple method for determining the equations and lengths of the axes of the central conic

$$Ax^2 + 2Hxy + By^2 + C'' = 0 \qquad (a^{iv}), \text{ Art. 168}$$

is to cut it by the concentric circle

$$x^2 + y^2 = r^2$$

in the points D, D', E, E' (Fig. 56).

By combining these two equations we get

$$Ax^2 + 2Hxy + By^2 + C''\frac{(x^2+y^2)}{r^2} = 0,$$

or
$$\left(A + \frac{C''}{r^2}\right)x^2 + 2Hxy + \left(B + \frac{C''}{r^2}\right)y^2 = 0, \qquad (a)$$

which represents the two straight lines DD', EE', passing through the centre and the points of intersection of the conic and circle.

If the points D and E become coincident, the radius r will then equal the semi-conjugate axis CB of the conic; and if the points E and D' become coincident, r will equal the semi-transverse axis CA of the conic.

But for equal roots in (a), or for coincidence of the lines DD' and EE', we must have (Art. 20),

$$\left(A + \frac{C''}{r^2}\right)\left(B + \frac{C''}{r^2}\right) = H^2, \qquad (b)$$

in which the roots, or values of r^2, are the squares of the semi-axes of the conic. Solving, we get

$$\frac{1}{r^2} = \frac{-(A+B) \pm Q}{2C''}, \tag{b'}$$

or
$$= \frac{(AB - H^2)(A + B \mp Q)}{2\Delta},$$

in which both values of r^2 are positive for the ellipse, but have opposite signs for the hyperbola.

The equations of the axes are now readily found. Multiply (a) by $A + \dfrac{C''}{r^2}$; then by (b).

$$\left(A + \frac{C''}{r^2}\right)^2 x^2 + 2H\left(A + \frac{C''}{r^2}\right)xy + H^2 y^2 = 0;$$

or, extracting the square root,

$$\left(A + \frac{C''}{r^2}\right)x + Hy = 0,$$

or
$$\frac{y}{x} = -\frac{(A-B) \pm Q}{2H}, \tag{c}$$

are the required equations of the axes of the conic as already found in (b), Art. 167.

170. *To reduce $S = 0$ to its simplest form when the locus is a parabola; that is, when $H^2 - AB = 0$.*

When $H^2 = AB$, A and B must have the same sign, and both can be taken as positive. We cannot now transform $S = 0$ to the centre of the conic as an origin, as in Art. 166, since the centre of the parabola is at infinity. But without changing the origin we may readily transform $S = 0$ to a new pair of rectangular axes, the new axis of x being parallel to the axis of the parabola.

GENERAL EQUATION OF SECOND DEGREE.   295

If we multiply $S = 0$ by B, and put $AB = H^2$, it becomes

$$(By + Hx)^2 + 2BGx + 2BFy + BC = 0; \qquad (a)$$

which shows that when $S = 0$ represents a parabola, *its second-degree terms form a perfect square.*

The equation of the diameter through the origin, and parallel to diameter (e), Art. 165, is

$$By + Hx = 0. \qquad (b)$$

If we take this diameter as the new axis of x, then

$$\tan\theta = -\frac{H}{B}, \quad \therefore \sin\theta = -\frac{H}{\sqrt{H^2+B^2}}, \quad \cos\theta = \frac{B}{\sqrt{H^2+B^2}}.$$

But for transformation to the new rectangular axes (Art. 21 (c)),

$$x = x'\cos\theta - y'\sin\theta = \frac{Bx' + Hy'}{\sqrt{H^2+B^2}},$$

$$y = x'\sin\theta + y'\cos\theta = \frac{By' - Hx'}{\sqrt{H^2+B^2}},$$

$$(By + Hx)^2 = (H^2 + B^2)y'^2 = B(A + B)y'^2,$$

and (a) reduces to

$$y'^2 + 2F'y' + 2G'x' + C' = 0, \qquad (a')$$

in which

$$F' = \frac{BF + GH}{\sqrt{B(A+B)^3}}, \quad G' = \frac{BG - FH}{\sqrt{B(A+B)^3}}, \quad C' = \frac{C}{A+B}.$$

Now it is easy to reduce (a') to the form $y^2 = 4px$ (Art. 102). Completing the square, (a') becomes

$$(y' + F')^2 = -2G'\left(x' - \frac{F'^2 - C'}{2G'}\right),$$

or $\quad y'^2 = -2G'x'$,

when the vertex V (Fig. 57) is $\left(\dfrac{F'^2-C'}{2G'},\ -F'\right)$ and

$$-2G = \dfrac{2(FH-BG)}{\sqrt{B(A+B)^3}} = 4p = LL',\ \text{the latus-rectum.}$$

If we had taken diameter (b) as the axis of Y, then $\tan\theta = \dfrac{B}{H}$, and we should have found that (a) reduces to

$$x'^2 + 2F'x' + 2G'y' + C' = 0,$$

the axis of the parabola being now parallel to the axis of Y.

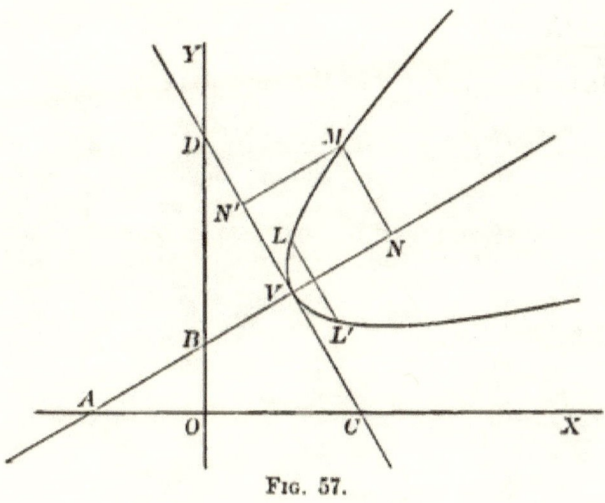

Fig. 57.

171. The equation (a), Art. 170, of the parabola can be reduced to the form $y^2 = 4px$, as follows:

If λ is an arbitrary constant, then

$$By + Hx + \lambda = 0$$

represents a system of straight lines (Art. 30) parallel to the diameter $By + Hx = 0$, one of which is the axis of the curve.

By adding $2(By + Hx)\lambda + \lambda^2$ to both sides of

$$(By + Hx)^2 = -2BGx - 2BFy - BC \qquad (a),\ \text{Art. 170}$$

it becomes
$$(By+Hx+\lambda)^2 = 2(H\lambda-BG)x+2(B\lambda-BF)y+\lambda^2-BC, \quad (a')$$
and the lines (Fig. 57)

$AB, \quad By+Hx+\lambda = 0,$ \hfill (b)

$CD, \quad 2(H\lambda-BG)x+2(B\lambda-BF)y+\lambda^2-BC = 0,$ \hfill (b')

will be perpendicular to each other if
$$\frac{H}{B} = -\frac{B\lambda-BF}{H\lambda-BG}, \text{ or } \lambda = \frac{BF+GH}{A+B}.$$

If we take AB and CD as new rectangular coordinate axes of X and Y respectively, then for any point $M(xy)$ on the curve, we get (Art. 42)
$$\frac{(By+Hx+\lambda)^2}{H^2+B^2} = MN^2 = y^2,$$
$$\frac{2(H\lambda-BG)x+2(B\lambda-BF)y+\lambda^2-BC}{2\sqrt{(H\lambda-BG)^2+(B\lambda-BF)^2}} = MN' = x,$$

and (a') becomes
$$(H^2+B^2)y^2 = 2[(H\lambda-BG)^2+(B\lambda-BF)^2]^{\frac{1}{2}}x,$$
or $\quad y^2 = 4px,$

if $\quad 4p = \dfrac{2[(H\lambda-BG)^2+(B\lambda-BF)^2]^{\frac{1}{2}}}{H^2+B^2} = \dfrac{2(FH-BG)}{\sqrt{B(A+B)^3}}.$

Therefore (b), (b') and $4p$ are the axis, the tangent at the vertex V, and the latus-rectum of the parabola after substituting the value of λ: and the coordinates of V are found by solving equations (b) and (b').

If $4p = 0$, or $BG = FH$, then equation (a), Art. 170, represents the two parallel straight lines
$$By+Hx+F = \pm \sqrt{F^2-BC},$$
as in Art. 166.

EXERCISES ON THE CONIC $S = 0$.

I. What conics do the following equations represent? Compute the values of the characteristic $H^2 - AB$, and of the discriminant Δ of each.

1. $3x^2 - 5xy + 6y^2 + 11x - 17y + 13 = 0.$ An ellipse.
2. $3x^2 + 4xy + y^2 - 3x - 2y = 0.$ An hyperbola.
3. $5x^2 + 4xy + 2y^2 - 5x - 2y - 19 = 0.$ An ellipse.
4. $36x^2 + 24xy + 29y^2 - 72x + 126y + 81 = 0.$ An ellipse.
5. $x^2 - 5xy + y^2 + 8x - 20y + 15 = 0.$ An hyperbola.
6. $7x^2 - 17xy + 6y^2 + 23x - 2y - 20 = 0.$ Two straight lines.
7. $3x^2 - 7xy - 6y^2 + 3x - 9y + 5 = 0.$ An hyperbola.
8. $(2x - 3y)^2 - 2x + 4y - 2 = 0.$ A parabola.
9. $x^2 - 4xy - 2y^2 + 10x + 4y = 0.$ An hyperbola.
10. $41x^2 + 24xy + 9y^2 - 130ax - 60ay + 116a^2 = 0.$ An ellipse.
11. $x^2 - 2xy + y^2 - 6x - 6y + 9 = 0.$ A parabola.
12. $6x^2 - 5xy + y^2 - 14x + 5y + 4 = 0.$ Two intersecting lines.
13. $x^2 - xy + y^2 - 3(x + y) = 0.$ An ellipse.
14. $25x^2 - 8xy + y^2 - 42cx + 6cy + 9c^2 = 0.$ An ellipse.
15. $4x^2 + 4xy + x^2 - 5x - 24 - 10 = 0.$ A parabola.
16. $25x^2 - 120xy + 144y^2 - 2x - 29y - 1 = 0.$ A parabola.
17. $4(x + 2y)^2 + (y - 2x)^2 - 5a^2 = 0.$ An ellipse.
18. $9x^2 + 24xy + 16y^2 + 22x + 46y + 9 = 0.$ A parabola.
19. $16x^2 - 24xy + 9y^2 - 75x - 100y = 0.$ A parabola.
20. $14x^2 - 4xy + 11y^2 - 44x - 58y + 71 = 0.$ An ellipse.
21. $11x^2 + 84xy - 24y^2 - 146x + 180y + 409 = 0.$ An hyperbola.
22. $2x^2 + xy - 6y^2 - 7y - 2 = 0.$ Two straight lines.
23. $9x^2 + 12xy + 4y^2 - 3x - 2y - 2 = 0.$ Two parallel lines.
24. $4x^2 - 24xy + 9y^2 + 4x - 6y + 1 = 0.$ An hyperbola.
25. $(x + y)^2 - a(x - y) = 0.$ A parabola.

EXERCISES ON THE CONIC $S=0$.

26. $xy - 2x + y = 0$.   Two straight lines.
27. $xy + 3ax - 3ay = 0$.   An hyperbola.
28. $x^2 + 4xy + 4y^2 + 4x - 6y + 2 = 0$.   A parabola.
29. $x^2 + 4xy + 4y^2 - 6x + 8y - 1 = 0$.   A parabola.
30. $x^2 - xy + 2y^2 - 2x - 6y + 7 = 0$.   An ellipse.

II. *Central Conics.* (*a*) Find the coordinates (h, k) of the centres of the above central conics; (*b*) find their equations referred to a new pair of coordinate axes CX', CY' (Fig. 56), drawn through the centre, respectively parallel to the old ones; (*c*) find the squares of their semi-axes, $\pm a^2$, $\pm b^2$; (*d*) find the equations of their axes.

Ex. 1. $3x^2 - 5xy + 6y^2 + 11x - 17y + 13 = 0$.   An ellipse.

Solution. (*a*) The equations for determining the centre are (Art. 167),
$$6k - \tfrac{5}{2}h - \tfrac{17}{2} = 0, \quad 3h - \tfrac{5}{2}k + \tfrac{11}{2} = 0. \quad \therefore (h, k) = (-1, 1).$$

(*b*) $C'' = Gh + Fk + C = -\tfrac{11}{2} - \tfrac{17}{2} + 13 = -1$.   (Art. 167)
$$\therefore 3x^2 - 5xy + 6y^2 - 1 = 0$$
is the equation of the ellipse referred to its centre as an origin.

(*c*) From Art. 167, IV., $Q = \pm \sqrt{(A-B)^2 + (2H)^2}$,
and $A' = \tfrac{1}{2}[(A+B) \pm Q]$, $B' = \tfrac{1}{2}[(A+B) \mp Q]$.
$$\therefore Q = \pm\sqrt{34}, \quad A' = \tfrac{1}{2}(9 - \sqrt{34}), \quad B' = \tfrac{1}{2}(9 + \sqrt{34});$$
and the squares of the semi-axes are
$$a^2 = \frac{2}{9 - \sqrt{34}}, \quad b^2 = \frac{2}{9 + \sqrt{34}} \cdot \therefore \frac{9 - \sqrt{34}}{2} x^2 + \frac{9 + \sqrt{34}}{2} y^2 = 1.$$

In the values of A' and B', the sine of Q must be so taken that the sine of $A' - B' = Q$ shall be the same as $2H$.

The squares of the semi-axes are readily computed from (*b'*), Art. 169.
$$\frac{1}{r^2} = \frac{-(A+B) \pm Q}{2C''} = \frac{-9 \pm \sqrt{34}}{-2} = \frac{9 \mp \sqrt{34}}{2}.$$

300   PLANE ANALYTIC GEOMETRY.

(d) $\dfrac{y}{x} = -\dfrac{(A-B) \pm Q}{2H}$

$= \dfrac{-3 \pm \sqrt{34}}{5}$.   (b), Art. 167, or (c), Art. 169.

Ex. 5. $x^2 - 5xy + y^2 + 8x - 20y + 15 = 0$.   An hyperbola.

(a) $2k - 5h - 20 = 0$, $2h - 5k + 8 = 0$. $\therefore (h, k) = (-4, 0)$.

(b) $C' = -16 + 15 = -1$. $\therefore x^2 - 5xy + y^2 - 1 = 0$.

(c) $Q = \pm 5$, $\dfrac{1}{r^2} = \dfrac{-2 \pm 5}{-2} = -\dfrac{3}{2}$ and $\dfrac{7}{2}$. $\therefore \dfrac{2y^2}{7} - \dfrac{2x^2}{3} = 1$.

(d) $\dfrac{y}{x} = -\dfrac{\pm 5}{-5} = \pm 1$. $\therefore y - x = 0$ is the equation of the imaginary axis, and $y + x = 0$ of the real axis.

The following are the results for the central conics in I.:

Ex.	(h, k)	Reduced Equations.	$\pm a^2, \pm b^2$.	Equations of Axes.
1.	$(-1, 1)$,	$3x^2 - 5xy + 6y^2 - 1 = 0$,	$\dfrac{2}{9 \mp \sqrt{34}}$,	$y = \tfrac{1}{5}(-3 \pm \sqrt{34})x$.
2.	$(\tfrac{1}{2}, 0)$,	$12x^2 + 16xy + 4y^2 - 3 = 0$,	$\dfrac{3}{8 \pm 4\sqrt{5}}$,	$y = \tfrac{1}{2}(-1 \pm \sqrt{5})x$.
3.	$(2, -3)$,	$20x^2 + 16xy + 8y^2 - 81 = 0$,	$\tfrac{81}{24}, \tfrac{81}{4}$,	$2y - x = 0,\ y + 2x = 0$.
4.	$(2, -3)$,	$36x^2 + 24xy + 29y^2 - 180 = 0$,	$4, 9$,	$4y - 3x = 0,\ 3y + 4x = 0$.
5.	$(-4, 0)$,	$x^2 - 5xy + y^2 - 1 = 0$,	$-\tfrac{2}{3}, \tfrac{2}{7}$,	$y - x = 0,\ y + x = 0$.
6.	$(2, 3)$,	$7x^2 - 17xy + 6y^2 = 0$,	Two straight lines through $(2, 3)$.	
7.	$(-\tfrac{9}{11}, -\tfrac{3}{11})$,	$3x^2 - 7xy - 6y^2 + 5 = 0$,	$\dfrac{10}{3 \pm \sqrt{130}}$,	$y = \tfrac{1}{6}(9 \pm \sqrt{130})x$.
9.	$(-1, 2)$,	$x^2 - 4xy - 2y^2 - 1 = 0$,	$-\tfrac{1}{3}, \tfrac{1}{2}$,	$y - 2x = 0,\ 2y + x = 0$.
10.	$(a, 2a)$,	$41x^2 + 24xy + 9y^2 - 9a^2 = 0$,	$\dfrac{a^2}{5}, \dfrac{9a^2}{5}$,	$3y - x = 0,\ y + 3x = 0$.
12.	$(-3, -10)$,	$6x^2 - 5xy + y^2 = 0$,	Two straight lines through $(-3, -10)$.	
13.	$(3, 3)$,	$x^2 - xy + y^2 - 9 = 0$,	$18, 6$,	$y - x = 0,\ y + x = 0$.
14.	(c, c),	$25x^2 - 8xy + y^2 - 9c^2 = 0$,	$\dfrac{9c^2}{13 \mp 4\sqrt{10}}$,	$y = (3 \pm \sqrt{10})x$.
17.	$(0, 0)$,	$8x^2 + 12xy + 17y^2 - 5a^2 = 0$,	$\tfrac{a^2}{4}, a^2$,	$y - 2x = 0,\ 2y + x = 0$.
20.	$(2, 3)$,	$14x^2 - 4xy + 11y^2 - 60 = 0$,	$6, 4$,	$y - 2x = 0,\ 2y + x = 0$.

EXERCISES ON THE CONIC $S=0$.

Ex.	(h, k)	Reduced Equations.	$\pm a^2, \pm b^2$.	Equations of Axes.
21.	$(-1, 2),$	$11x^2+84xy-24y^2+662=0,$	$-\frac{662}{39}, \frac{331}{26},$	$3y-2x=0, 2y+3x=0.$
22.	$(\frac{1}{2}, -\frac{4}{3}),$	$2x^2+xy-6y^2=0,$	Two straight lines through $(\frac{1}{2}, -\frac{4}{3}).$	
24.	$(-\frac{1}{6}, -\frac{1}{3}),$	$12x^2-72xy+27y^2+1=0,$	$\dfrac{2}{-39 \pm 3\sqrt{601}},$	$y = \tfrac{1}{24}(-5 \pm \sqrt{601})x.$
26.	$(-1, 2),$	$xy=0,$	The coordinate axes.	
27.	$(-3a, 3a),$	$xy-9a^2=0,$	$18a^2, -18a^2,$	$y-x=0, y+x=0.$
30.	$(2, 2),$	$x^2-xy+2y^2-1=0,$	$\dfrac{2}{3 \mp \sqrt{2}},$	$y = (-1 \pm \sqrt{2})x.$

III. *The Parabola.* (a) Find the equations of the axes of the parabolas in I.; (b) find the equations of their tangents at the vertex; (c) reduce their equations to the form $y^2 = 4px$.

Ex. 8. $4x^2 - 12xy + 9y^2 - 2x + 4y - 2 = 0.$ A parabola.

Solution. From Art. 171, $\lambda = \dfrac{BF+GH}{A+B} = \dfrac{24}{13}$, and (b), (b′), $4p$, become

$$39y - 26x + 8 = 0, \quad 26y - 39x - 201 = 0, \quad 4p = -\dfrac{2\sqrt{13}}{169};$$

therefore $169 y^2 = -2\sqrt{13}\, x$ is the reduced equation of the parabola, and shows that the curve lies on the negative side of the tangent at V.

Ex.	Equation of Axis.	Equation of Tan at V.	$y^2 = 4px.$
8.	$39y - 26x + 8 = 0,$	$26y + 39x - 201 = 0,$	$169 y^2 = -2\sqrt{13}\, x.$
11.	$y - x = 0,$	$y + x = 0,$	$y^2 = 3\sqrt{2}\, x.$
15.	$5y + 10x - 6 = 0,$	$5x - 10y + 286 = 0,$	$25 y^2 = \sqrt{5}\, x.$
16.	$12y - 5x - 1 = 0,$	$12x + 5y + 2 = 0,$	$y^2 = \tfrac{1}{13} x.$
18.	$4y + 3x + 5 = 0,$	$4x - 3y + 8 = 0,$	$y^2 = \tfrac{2}{5} x.$
19.	$3y - 4x = 0,$	$3x + 4y = 0,$	$y^2 = 5x.$
23.	The parallel lines	$3x + 2y - 2 = 0,$	$3x + 2y + 1 = 0.$
25.	$ay + x + \tfrac{a}{2} = 0.$	$y - x = 0,$	$y^2 = \tfrac{1}{2} a\sqrt{2}\, x.$
28.	$10y + 5x - 4 = 9,$	$35y - 70x - 17 = 0,$	$25 y^2 = 2\sqrt{5}\, x.$
29.	$2y + x + 1 = 0,$	$4x - 2y + 1 = 0,$	$5 y^2 = 4\sqrt{5}\, x.$

31. If a and b are the semi-axes of a central conic, show that
$$a^2 b^2 = \frac{C'^2}{AB - H^2}, \quad a^2 + b^2 = \frac{(A+B)C'}{AB - H^2}, \quad a^2 - b^2 = \frac{QC'}{AB - H^2}.$$

32. The equation of the conic $Ax^2 + 2Hxy + By^2 = C'$ transformed to its axes is
$$(A + B + Q)x^2 + (A + B - Q)y^2 = 2C'.$$

33. Show that the product of the semi-axes of the conic
$(x - 2y + 1)^2 + (4x + 2y - 3)^2 - 10 = 0$ is 1.

34. Show that the product of the semi-axes of the ellipse
$x^2 - xy + 2y^2 - 2x - 6y + 7 = 0$ is $\dfrac{2}{\sqrt{7}}$,
and that the equation of its axes is $x^2 - 2xy - y^2 + 8y - 8 = 0$.

35. Show that, if
$$Ax^2 + 2Hxy + By^2 = 1, \text{ and } A'x^2 + 2H'xy + B'y^2 = 1$$
represent the same conic, and the axes are rectangular, then
$$(A - B)^2 + (2H)^2 = (A' - B')^2 + (2H')^2.$$

36. Show that for $\lambda = 5$, or $= -3$, the equation
$$2x^2 + \lambda xy - y^2 - 3x + 6y - 9 = 0$$
will represent pairs of straight lines.

37. Show that for all positions of the axes, so long as they remain rectangular, and the origin is unchanged, the value of $G^2 + F^2$, in the equation $S = 0$, is constant.

38. Show that $AF - HG = 0$, when $H^2 - AB = 0$ and $BG - HF = 0$.

39. When $\Delta = 0$, and the origin is at the centre of the conic, show that $S = 0$ becomes the product of the two straight lines
$$By + (H + \sqrt{H^2 - AB})x = 0, \quad By + (H - \sqrt{H^2 - AB})x = 0.$$

40. Find the conditions which will make
$$(A + B + Q)x^2 + (A + B - Q)y^2 = 2C'$$
represent a circle.

CHAPTER X.

GENERAL PROPERTIES OF THE CONIC $S=0$.

172. *To find the equations of the asymptotes of $S=0$.*

The equations (a'') and (a''') (Art. 167) of the conic and its asymptotes only differ by the constant term $\dfrac{\Delta}{H^2-AB}$, and therefore the equations of the asymptotes of $S=0$ are

$$S-\dfrac{\Delta}{H^2-AB}=0.$$

Or determine λ by making the discriminant of the conic $S+\lambda=0$ equal to zero; that is,

$$AF^2+BG^2+(C+\lambda)H^2-AB(C+\lambda)-2FGH=0,$$

or $\quad AF^2+BG^2+CH^2-ABC-2FGH+\lambda(H^2-AB)=0.$

Therefore $\lambda=-\dfrac{\Delta}{H^2-AB}=0$, and $S-\dfrac{\Delta}{H^2-AB}=0$ are the equations of the asymptotes of $S=0$.

These equations are real for the hyperbola and imaginary for the ellipse.

173. *To find the equation of the hyperbola conjugate to $S=0$.*

The equations of the primary hyperbola, of the asymptotes, and of the conjugate hyperbola are (Art. 156, II., III.),

$$\dfrac{x^2}{a^2}-\dfrac{y^2}{b^2}=1,\ \dfrac{x^2}{a^2}-\dfrac{y^2}{b^2}=0,\ \dfrac{x^2}{a^2}-\dfrac{y^2}{b^2}=-1,$$

in which the independent terms of the equations of the hyperbolas are obtained from the independent term of the equations of the asymptotes by adding and subtracting the same constant quantity; therefore

$$S = 0, \quad S - \frac{\Delta}{H^2 - AB} = 0, \quad S - \frac{2\Delta}{H^2 - AB} = 0,$$

are the equations of the primary hyperbola, of the asymptotes, and of the conjugate hyperbola.

174. *To find the condition that $S = 0$ shall be a rectangular hyperbola.*

The two lines

$$Ax^2 + 2Hxy + By^2 = 0 \qquad (d), \text{ Art. 165, II.}$$

are parallel to the asymptotes of $S = 0$ (Art. 166 (a''')). But for the rectangular hyperbola the asymptotes are perpendicular to each other; and the required condition is $A + B = 0$ for rectangular axes, and $A + B - 2H\cos\theta = 0$ for oblique axes (Art. 66).

175. *To find the locus of the middle points of a system of parallel chords of the conic $S = 0$.*

If the origin is at the middle point of any chord which makes an angle θ with the axis of x, this chord will meet the conic $S = 0$ in the points (r, θ) and $(-r, \theta)$, and therefore the values of r in equation (a') (Art. 164) will be equal, with opposite signs. But for this $(G\cos\theta + F\sin\theta)r = 0$; and since this equation is satisfied by the coordinates (r, θ) and $(-r, \theta)$, it is the equation of the bisected chord.

Therefore *any given point will in general bisect one chord of the conic $S = 0$.*

If now we transform $S = 0$ to a new origin $(x'y')$ and to a new pair of parallel axes through this point, we shall get (Art. 167) for the new coefficients of x and y,

$$G' = Ax' + By' + G, \quad F' = By' + Hx' + F,$$

GENERAL PROPERTIES OF THE CONIC $S=0$.   305

and the equation of the parallel chord through $(x'y')$ is

$$(Ax' + Hy' + G)\cos\theta + (By' + Hx' + F)\sin\theta = 0,$$

after dividing by r. Since, then, this equation is satisfied by the coordinates $(0, 0)$ of the origin and any other point $(x'y')$, the middle point of any parallel chord must lie on the straight line

$$(Ax + Hy + G)\cos\theta + (By + Hx + F)\sin\theta = 0, \qquad (b)$$

which is the required locus.

Hence (b) is the equation of *a diameter*, and the chords which it bisects are called its *ordinates*. But this equation is of the form $lL + mM = 0$ (Art. 53), and therefore (b) always passes through the point determined by the equations

$$Ax + Hy + G = 0, \quad By + Hx + F = 0;$$

that is, the centre of $S = 0$.

If $\theta = \frac{\pi}{2}$, (b) becomes $By + Hx + F = 0$, a diameter which bisects the system of chords parallel to the axis of Y; if $\theta = 0$, (b) becomes $Ax + Hy + G = 0$, a diameter which bisects the system of chords parallel to the axis of X.

176. *To find the condition that any two diameters of $S = 0$ shall be conjugates.*

The diameter (b) (Art. 175) can be written

$$(H + B\tan\theta)y + (A + H\tan\theta)x + G + F\tan\theta = 0,$$

or

$$y = -\frac{A + H\tan\theta}{H + B\tan\theta}x - \frac{G + F\tan\theta}{H + B\tan\theta}.$$

If this diameter makes an angle θ' with the axis of X, then

$$\tan\theta' = -\frac{A + H\tan\theta}{H + B\tan\theta},$$

or

$$B\tan\theta\tan\theta' + H(\tan\theta + \tan\theta') + A = 0, \qquad (b)$$

is the required condition. Since θ and θ' can be interchanged, it follows that each diameter bisects the system of chords which is parallel to the other. The two diameters are therefore conjugates (Art. 151, VII.).

177. *The axes of $S=0$ are the only pair of conjugate diameters which are perpendicular to each other.*

For perpendicular diameters $\tan\theta \tan\theta' + 1 = 0$ (Art. 46), which reduces condition (b) (Art. 176) to

$$H \tan^2\theta + (A-B)\tan\theta - H = 0,$$

or $\qquad Hy^2 + (A-B)xy - Hx^2 = 0,$

the equations of the axes (Art. 167 (b)).

178. *To find the condition that the two lines*
$$A'x^2 + 2H'xy + B'y^2 = 0$$
shall be parallel to a pair of conjugate diameters of $S=0$.

Write this equation in the form

$$\frac{y^2}{x^2} + \frac{2H'}{B'}\cdot\frac{y}{x} + \frac{A'}{B'} = 0;$$

the sum and product of its roots (Art. 20, IV., V.) are

$$-\frac{2H'}{B'} \text{ and } \frac{A'}{B'},$$

and these must satisfy (b) (Art. 176). Therefore

$$AB' + A'B = 2HH'$$

is the required condition.

179. *If any two chords be drawn in given directions through the origin, the ratio of the products of the segments into which they are cut by the conic $S=0$ is constant.*

For brevity, put $A_\theta = A\cos^2\theta + 2H\sin\theta\cos\theta + B\sin^2\theta$; then equation (a') (Art. 165) becomes

$$A_\theta r^2 + 2(G\cos\theta + F\sin\theta)r + C = 0. \qquad (a')$$

GENERAL PROPERTIES OF THE CONIC $S=0$.

Suppose that the line drawn through the origin in the direction θ cuts $S=0$ in the points P and P', and that the line drawn in the direction θ' cuts $S=0$ in the points Q and Q'; then OP, OP', and OQ, OQ', are the corresponding roots of (a').

The products of these roots are (Art. 20)

$$OP \cdot OP' = \frac{C}{A_\theta}, \quad OQ \cdot OQ' = \frac{C}{A_{\theta'}}.$$

$$\therefore \frac{OP \cdot OP'}{OQ \cdot OQ'} = \frac{A_{\theta'}}{A_\theta} = \text{a constant, for given directions.}$$

180. *The ratio of the products of the segments of any two chords of $S=0$ is equal to the ratio of the squares of the diameters respectively parallel to these chords.*

If we take the centre of the conic as the origin, and a new pair of parallel axes, its equation (a'') (Art. 167), with the notation of Art. 179, becomes

$$A_\theta r^2 + C' = 0. \tag{b}$$

Now draw two diameters through the centre C, in the directions θ and θ', the first cutting the conic in the points p, p', and the second in the points q, q'; then from (b)

$$Cp \cdot Cp' = -\frac{C'}{A_\theta}, \quad Cq \cdot Cq' = -\frac{C'}{A_{\theta'}}.$$

$$\therefore \frac{Cp \cdot Cp'}{Cq \cdot Cq'} = \frac{\overline{pp'}^2}{\overline{qq'}^2} = \frac{A_{\theta'}}{A_\theta} = \frac{OP \cdot OP'}{OQ \cdot OQ'}. \tag{Art. 179}$$

181. *Two tangents drawn through any point to $S=0$ have the same ratio as the diameters to which they are parallel.*

If T and T' are the points of tangency, then $OP = OP' = OT$ and $OQ = OQ' = OT'$, and (Art. 180)

$$\frac{OP \cdot OP'}{OQ \cdot OQ'} = \frac{\overline{OT}^2}{\overline{OT'}^2} = \frac{\overline{pp'}^2}{\overline{qq'}^2}. \quad \therefore \frac{OT}{OT'} = \frac{pp'}{qq'}.$$

If a concentric circle cuts the conic $S=0$ in the points P, P', and Q, Q', then, by Geometry, $OP \cdot OP' = OQ \cdot OQ'$; therefore the respectively parallel diameters pp' and qq' are equal, and make equal angles with either axis of the conic.

182. *To find the equation of the tangent to $S=0$ at the point $(x'y')$.*

The equation is (Art. 80)

$$Axx' + H(x'y + xy') + Byy' + G(x+x') + F(y+y') + C = T' = 0.$$

Any point (xy) on the tangent $T'=0$ lies without the conic $S=0$.

From the sum $S + S' = 0$ subtract $2T' = 0$; then

$$A(x-x')^2 + 2H(x-x')(y-y') + B(y-y')^2 = S - 2T' + S',$$

and $S - 2T' + S'$ is never zero, except for $(x'y')$, the point of contact.

183. *To find the equations of the two tangents drawn through any point to the conic $S=0$.*

In this case the roots of equation (a') (Art. 165) are equal; therefore (Art. 20) the required equations are

$$(A\cos^2\theta + 2H\sin\theta\cos\theta + B\sin^2\theta)C = (G\cos\theta + F\sin\theta)^2,$$

or

$$(AC - G^2)x^2 + 2(CH - GF)xy + (BC - F^2)y^2 = 0,$$

by introducing x and y.

184. *To find the equation of the chord of contact of the two tangents to the conic $S=0$, which meet in the point (h, k).*

Let $T'=0$ and $T''=0$ denote the equations of the two tangents to $S=0$ at the points $(x'y')$ and $(x''y'')$. If these tangents meet in the point (h, k), then $T'(h, k) = 0$ and $T''(h, k) = 0$ are the two equations of condition which show that the points of contact $(x'y')$ and $(x''y'')$ are on the line whose equation is $T(h, k) = 0$. Therefore $T(h, k) = 0$ is the required equation of the chord of contact. There is, therefore, a fixed relation between the point (h, k) and the line $T(h, k) = 0$ with respect to the conic $S = 0$, which is called the *polar relation* of the point and line with respect to the conic. The point is called the *pole* of the line, and the line is called the *polar* of the point, with respect to the conic (Art. 86).

185. *To find the equation of the polar of any point $(x'y')$ with respect to the conic $S = 0$.*

As we have just seen, the equations of the tangent and polar are of the same form; that is, $T(x'y') = 0$ is the equation of the tangent to $S = 0$ at any point $(x'y')$ on the conic, and it is also the equation of the polar of any point $(x'y')$ in the plane of the conic. It follows that the point of tangency and the tangent at this point are respectively the pole and polar of each other with respect to the conic $S = 0$.

The equation of the polar of the origin with respect to $S = 0$ is found to be $Gx + Fy + C = 0$ by making $x' = 0$, $y' = 0$ in the equation of the polar $T(x'y') = 0$. Propositions VIII., IX., X. (Art. 86), on poles and polars with respect to the circle, apply equally to the conic $S = 0$.

186. *To find what the equation $S = 0$ becomes when the coordinate axes of Y and X are a tangent and a normal at any given point on the conic.*

Since the origin is on the curve, its coordinates $(0, 0)$ reduce $S = 0$ to $C = 0$. The axis of Y meets the conic in the point for which $x = 0$, or $By^2 + 2Fy = 0$; but since the axis of Y is a tangent, the roots of this equation must be equal; therefore

$F = 0$. We find, then, that $C = 0$ and $F = 0$ are the necessary and sufficient conditions, and $S = 0$ reduces to

$$Ax^2 + 2Hxy + By^2 + 2Gx = 0.$$

187. *To find the equation $S = 0$ of the conic which passes through any five given points* $(x'y')$, $(x''y'')$, $(x'''y''')$, (x^{iv}, y^{iv}), (x^v, y^v).

If the values of A, H, B, G, F, C, derived from the five equations of condition

$$Ax'^2 + 2Hx'y' + By'^2 + 2Gx' + 2Fy' + C = 0,$$
$$Ax''^2 + 2Hx''y'' + By''^2 + 2Gx'' + 2Fy'' + C = 0,$$
$$Ax'''^2 + 2Hx'''y''' + By'''^2 + 2Gx''' + 2Fy''' + C = 0,$$
$$Ax^{iv\,2} + 2Hx^{iv}y^{iv} + By^{iv\,2} + 2Gx^{iv} + 2Fy^{iv} + C = 0,$$
$$Ax^{v\,2} + 2Hx^v y^v + By^{v\,2} + 2Gx^v + 2Fy^v + C = 0,$$

are substituted in $S = 0$, it will be the required equation.

These five equations of condition contain six arbitrary constants; but they are not independent, for by dividing each equation by the same coefficient, as A, we get the five independent ratios which are determined by the five equations. It will be noticed that if the origin $(0, 0)$ is one of the five given points, then $C = 0$, and cannot be used as a divisor.

These equations are of the first degree, or linear, with reference to these ratios, and give but one solution, which can be readily effected for special numerical points.

I. If all five points are on the same straight line, then the two coincident straight lines through these points is a solution.

II. If four of these points are on the same straight line, then this line and any second line through the fifth point is a solution.

III. If three of the points are on the same straight line, then this line and the straight line passing through the remaining two points is a solution.

IV. Since a straight line cannot cut a curve in more than two points, it follows that no three of the given points can lie on the same straight line if the required conic is a curve.

188. *To find the foci, the equations of the directrices, and the eccentricity of the conic $S = 0$.*

It will be best to take the equation of the conic
$$A'x^2 + B'y^2 + C' = 0 \qquad (a^{iv})$$
referred to its centre and axes. If $(x'y')$ are the coordinates of a focus, e the eccentricity, and $x \cos a + y \sin a - p = 0$ the equation of a directrix, then the squares of the distances of any point (xy) on the conic, from the focus and directrix, are
$$(x - x')^2 + (y - y')^2 \quad \text{and} \quad (x \cos a + y \sin a - p)^2,$$
and by Art. 99 the equation of the conic is
$$(x - x')^2 + (y - y')^2 - e^2(x \cos a + y \sin a - p)^2 = 0,$$
or $(1 - e^2 \cos^2 a)x^2 - 2e^2 xy \sin a \cos a + (1 - e^2 \sin^2 a)y^2$
$$- 2(x' - e^2 p \cos a)x - 2(y' - e^2 p \sin a)y + x'^2 + y'^2 - e^2 p^2 = 0. \quad (a^v)$$

If (a^{iv}) and (a^v) represent the same conic, we must have, since the coefficients of xy, x, y, are zero in (a^{iv}),
$$e^2 \sin a \cos a = 0, \quad x' - e^2 p \cos a = 0, \quad y' - e^2 p \sin a = 0, \qquad (b)$$
and also the equal ratios
$$\frac{A'}{1 - e^2 \cos^2 a} = \frac{B'}{1 - e^2 \sin^2 a} = \frac{C'}{x'^2 + y'^2 - e^2 p^2}. \qquad (c)$$

I. *The foci and the directrices.* — From $e^2 \sin a \cos a = 0$ we get $\sin a = 0$ or $\cos a = 0$. For $\sin a = 0$, $a = 0$ or $= \pi$, and from

(b) the coordinates $(x'y')$ of the foci are $(\pm e^2p, 0)$, two points on the axis of X equidistant from the origin; and the equations of the corresponding directrices are $x = \pm p$, two lines equidistant from the origin and parallel to the axis of Y.

For $\cos a = 0$, $a = \dfrac{\pi}{2}$ or $= \dfrac{3\pi}{2}$, and the coordinates of the foci $(x'y')$ are $(0, \pm e^2p)$, two points on the axis of Y equidistant from the origin; and the equations of the corresponding directrices are $y = \pm p$, parallel to the axis of X.

The polar of the focus $(e^2p, 0)$ with reference to the conic (a^{iv}) is $A'e^2px + C' = 0$, which is the equation of the directrix of the corresponding focus.

For $a = 0$ or $= \pi$, equations (c) become

$$\frac{A'}{1-e^2} = \frac{B'}{1} = \frac{C'}{x'^2 - e^2p^2}.$$

Since $x' = e^2p$, by solving these equations, we get

$$e^2 = 1 - \frac{A'}{B'}, \quad x'^2 = C'\left(\frac{1}{B'} - \frac{1}{A'}\right), \quad p^2 = -\frac{\dfrac{C'}{A'}}{1 - \dfrac{A'}{B'}}.$$

For C' negative, and A', B', both positive, (a^{iv}) represents an ellipse (Art. 168); if $A' < B'$, both values of e, x', p are *real*, but if $A' > B'$ they are *imaginary*.

For C' negative, and A', B', of different signs, (a^{iv}) represents an hyperbola; if A' is plus and B' minus, both values of e, x', p are *real*, but if A' is minus and B' plus, both values of e are still real, while both values of x' and p are imaginary.

For $a = \dfrac{\pi}{2}$ or $= \dfrac{3\pi}{2}$, A', B', and x', y', will interchange; the results just found for the axis of X will then be true of e, y', p with reference to the axis of Y.

The central conic has then four foci, two real and two imaginary, with the corresponding directrices real or imaginary.

GENERAL PROPERTIES OF THE CONIC $S = 0$. 313

II. *To express the values of the eccentricity e in terms of the coefficients of $S = 0$.*

To find e we have
$$A' = B'(1 - e^2),$$
and $\quad A' + B' = A + B, \quad -A'B' = H^2 - AB.$ (Art. 168, IV.)

Eliminating A', B' from these equations, we get
$$\frac{(2 - e^2)^2}{1 - e^2} = \frac{(A + B)^2}{AB - H^2},$$
or $\quad e^4 + \dfrac{(A - B)^2 + (2H)^2}{AB - H^2}(e^2 - 1) = 0.$ $\qquad(d)$

If we put $Q^2 = (A - B)^2 + (2H)^2$ for brevity, and solve, we get
$$e^2 = -\frac{Q^2}{2(AB - H^2)} \pm \frac{\sqrt{Q^4 + 4Q^2(AB - H^2)}}{2(AB - H^2)}.$$

For the ellipse, $AB - H^2$ is positive. In this case one value of e^2 is positive and the other negative; the two real values of e correspond to the real foci, and the two imaginary values to the imaginary foci.

For the hyperbola, $AB - H^2$ is negative. In this case both values of e^2 are positive, and all four values of e are real.

For the parabola, $AB - H^2 = 0$. Introducing this condition in (d), it becomes
$$(AB - H^2)e^4 + (A + B)^2(e^2 - 1) = 0,$$
or $\quad (A + B)^2(e^2 - 1) = 0.$

Therefore (Art. 20), $e^2 = \infty$, and $e^2 = 1$ are the four roots.

EXERCISES AND PROBLEMS.

1. Find the asymptotes of the hyperbola
$$3x^2 + 4xy + y^2 - 3x - 2y + 21 = 0.$$
$\qquad$ Ans. $12x^2 + 16xy + 4y^2 - 12x - 8y + 3 = 0.$

2. Show that the equation of the conic whose asymptotes are $2x+3y-5=0$ and $5x+3y-8=0$, and which passes through the point $(1, -1)$, is
$$10x^2 + 21xy + 9xy^2 - 41x - 39y + 4 = 0.$$

3. Find the equation of the asymptotes of the conic
$$3x^2 - 2xy - 5y^2 + 7x - 9y = 0;$$
and find the equation of the conic which has the same asymptotes and passes through the point $(2, 2)$.

Ans. $3x^2 - 2xy - 5y^2 + 7x - 9y + 2 = 0,$
$3x^2 - 2xy - 5y^2 + 7x - 9y + 20 = 0.$

4. Find the asymptotes of the hyperbola
$$6x^2 - 7xy - 3y^2 - 2x - 8y - 6 = 0;$$
and also the equation of the conjugate hyperbola.

Ans. $6x^2 - 7xy - 3y^2 - 2x - 8y - 4 = 0,$
$6x^2 - 7xy - 3y^2 - 2x - 8y - 2 = 0.$

5. Find the equation of the locus of the middle points of a system of parallel chords of the conic
$$41x^2 + 24xy + 9y^2 - 130ax - 60ay + 116a^2 = 0,$$
which make an angle of 45° with the axis of x.

Ans. $53x + 21y - 100a = 0.$

6. A diameter of the conic
$$36x^2 + 24xy + 29y^2 - 72x + 126y + 81 = 0$$
makes an angle of 45° with the axis of x; find the equation of its conjugate. *Ans.* $x - 4 - 5 = 0.$

7. Find the equation of the equi-conjugate diameters of the conic
$$Ax^2 + 2Hxy + By^2 = 1. \qquad (a)$$

Let $x^2 + y^2 = r^2$ be the equation of any concentric circle (Fig 56);
then $\left(A - \dfrac{1}{r^2}\right)x^2 + 2Hxy + \left(B - \dfrac{1}{r^2}\right)y^2 = 0 \qquad (b)$

is the equation of the two diameters which pass through the points of common intersection of the conic and circle. But these diameters are equal, and they will be conjugates if

$$B\left(A-\frac{1}{r^2}\right)+A\left(B-\frac{1}{r^2}\right)=2H^2. \qquad \text{(Art. 177)}$$

Therefore,

$$\frac{1}{r^2}=\frac{2(AB-H^2)}{A+B},$$

and $\quad Ax^2+2Hxy+By^2-\dfrac{2(AB-H^2)}{A+B}(x^2+y^2)=0$

is the equation of the required equi-conjugate diameters.

8. Show that any two concentric conics

$$Ax^2+2Hxy+By^2=1,\quad A'x^2+2H'xy+B'y^2=1$$

have one, and only one, pair of common conjugate diameters.

The diameters $ax^2+2hxy+by^2=0$ are conjugate to both conics if

$$aB+bA-2hH=0,$$

and $\quad aB'+bA'-2hH'=0. \qquad \text{(Art. 178)}$

Therefore,

$$\frac{a}{A'H-AH'}=\frac{-2h}{AB'-A'B}=\frac{b}{BH'-B'H}.$$

Therefore,

$$(A'H-AH')x^2-(AB'-A'B)xy+(BH'-B'H)y^2=0$$

is the equation of the common conjugate diameters.

9. The polar of the origin with respect to the conic

$$3x^2-5xy+6y^2+11x-17y+13=0$$

is $\quad 11x-17y+26=0.$

10. Find the equation of the conic which passes through the five points $(2, 1)$, $(0, 3)$, $(-1, -3)$, $(1, 0)$, $(3, -3)$.

Ans. $17y^2+11xy-48x^2-24y+129x-81=0.$

11. Find the equation of the conic which passes through the five points $(2, 1)$, $(1, 0)$, $(3, -1)$, $(-1, 0)$, $(3, -2)$.

Ans. $x^2+19xy+4y^2-45y-1=0.$

316   PLANE ANALYTIC GEOMETRY.

12. Find the equation of the axes of the conic $S = 0$.

If $P(x'y')$ is a point on the axis of a conic, then the polar of P is perpendicular to the line joining P and the centre. The polar of $P(x'y')$, with respect to $S = 0$, is

$$(Ax' + Hy' + G)x + (Hx' + By' + F)y + Gx + Fy' + C = 0, \quad (a)$$

and the equation of any diameter is

$$Ax + Hy + G + \lambda(Hx + By + F) = 0. \quad (b)$$

Since (a) and (b) are perpendicular, we have

$$(A + \lambda H)(Ax' + Hy' + G) + (H + \lambda B)(Hx' + By' + F) = 0; \quad (c)$$

and, since (b) passes through $(x'y')$, we have

$$Ax' + Hy' + G + \lambda(Hx' + By' + F) = 0. \quad (d)$$

By eliminating λ from (c) and (d), we see that $(x'y')$ must be on the conic

$$\frac{(Ax+Hy+G)^2 - (Hx+By+F)^2}{A-B} = \frac{(Ax+Hy+G)(Hx+By+F)}{H},$$

which is the required equation of the axes.

13. Show that the foci of the conic

$$Ax^2 + 2Hxy + By^2 + C' = 0$$

lie on the curves

$$\frac{x^2 - y^2}{A - B} = \frac{xy}{H} = \frac{-C'}{H^2 - AB}.$$

14. Find the equations of the asymptotes of the hyperbolas
(a) $x^2 - 2xy + 4x + 3 = 0$; (b) $y^2 - xy + ax = 0$.

 Ans. (a) $x = 0$, $x = 2y - 4$; (b) $y = a$, $y = x - a$.

15. Find the eccentricities and asymptotes of the parabolas
(a) $4xy - 3x^2 - 2ay = 0$; (b) $y^2 - 4x^2 + 2y - 4x - 9 = 0$.

 Ans. (a) $e^2 = \tfrac{5}{4}$ and 5, $2x = a$, $8y - 6x = 3a$;
 (b) $e^2 = \tfrac{5}{4}$ and 5, $y = 2x$, $y + 2x + 2 = 0$.

16. Show that the coordinates of the real foci of the conic

$$x^2 - 6xy + y^2 - 2x - 2y + 5 = 0$$

are $(1, 1)$ and $(-2, -2)$.

17. Show that the coordinates of the real foci of the conic
$$2x^2 - 8xy - 4y^2 - 4y + 1 = 0$$
are $(0, -\tfrac{1}{2})$ and $(-\tfrac{2}{3}, -\tfrac{5}{6})$.

18. The focus of the parabola
$$x^2 + 2xy + y^2 - 4x + 8y - 6 = 0$$
is the point $(-\tfrac{1}{4}, -\tfrac{2}{3})$.

19. All chords of a conic which subtend a right angle at a fixed point on a conic, cut the normal at O in a fixed point.

Solution. Take the tangent and normal at O for the coordinate axes; then (Art. 185),
$$Ax^2 + 2Hxy + By^2 + 2Gx = 0$$
is the equation of the conic.

Let the equation of any chord PQ be $\dfrac{x}{a} + \dfrac{y}{b} = 1$; then the equation
$$Ax^2 + 2Hxy + 2Gx\left(\dfrac{x}{a} + \dfrac{y}{b}\right) = 0,$$
or
$$\left(A + \dfrac{2G}{a}\right)x^2 + 2\left(H + \dfrac{Y}{b}\right)xy + By^2 = 0$$
is the equation of the lines OP and OQ, since it is satisfied for $(0, 0)$ and for the points (x, y) common to the chord and conic.

If OP and OQ are at right angles to each other, then (Art. 174),
$$A + B + \dfrac{2G}{a} = 0, \text{ or } a = -\dfrac{A+B}{2G};$$
that is, the intercept on the normal is constant.

20. If any two chords OP, OQ of a conic make equal angles with the tangent at O, the line PQ will cut this tangent in a fixed point.

21. Find the eccentricity of the conic whose equation is
$$x^2 - 4xy - 2y^2 + 10x + 4y = 0.$$
Ans. $e = \pm\sqrt{\tfrac{5}{3}}$ with reference to foci on the real axis, and $\pm\sqrt{\tfrac{5}{2}}$ with reference to foci on the imaginary axis.

22. Find the eccentricity of the conic
$$14x^2 - 4xy + 11y^2 - 44x - 58y + 71 = 0.$$
Ans. $e^2 = \frac{1}{3}$ and $-\frac{1}{12}$.

23. If POP', QOQ' are any two chords of a conic at right angles to one another through a fixed point O, then

$$\frac{1}{OP \cdot OP'} + \frac{1}{OQ \cdot OQ'} = \text{a constant.}$$

24. If a point be taken on the major-axis of an ellipse, whose abscissa is equal to $\left[\dfrac{a^4 - a^2 b^2}{a^2 + b^2}\right]^{\frac{1}{2}}$, prove that the sum of the squares of the reciprocals of the segments of any chord drawn through this point is constant.

25. From a fixed point on a conic chords are drawn making equal intercepts, measured from the centre, on a fixed diameter; find the locus of the point of intersection of the tangents at their other extremities.

26. If PSP' be any focal chord of a parabola, and PM, $P'M'$ be perpendiculars on a fixed line, then will

$$\frac{PM}{PS} + \frac{P'M'}{P'S} = \text{a constant.}$$

27. Two parallel chords of an ellipse drawn through the foci intersect the curve in the points P, P' on the same side of the major-axis; a line through P, P' intersects the semi-axes CA, CB, in M and N, respectively; prove that

$$\frac{\overline{CA}^4}{\overline{CM}^2} + \frac{\overline{CB}^4}{\overline{CN}^2} = \text{a constant.}$$

28. Through O, the middle point of a chord AB of an ellipse, any chord POP is drawn. The tangents at P and P' meet AB in S and T, respectively; prove that $AS = BT$.

29. Pairs of tangents are drawn to the conic $Ax^2 + By^2 = 1$, so as to be always parallel to conjugate diameters of the conic

$$ax^2 + 2hxy + by^2 = 1;$$

show that the locus of their intersection is

$$ax^2 + 2hxy + by^2 = \frac{a}{A} + \frac{h}{B}.$$

30. O is a fixed point on the tangent at the vertex A of a conic, and P, P' are points on this tangent equally distant from O; show that the locus of the point of intersection of the other tangents from P and P' is a straight line.

31. If OP and OQ are two tangents to an ellipse, and CP', CQ' are parallel semi-diameters, show that

$$OP \cdot OQ + CP' \cdot CQ' = OS \cdot OS',$$

S and S' being the foci.

32. In the conic $Ax^2 + 2Hxy + By^2 - 2y = 0$, the product of the focal distances of the origin is $\dfrac{1}{AB - H^2}$.

33. Find the value of c in order that the hyperbola $2xy - c = 0$ may touch the ellipse $\dfrac{x^2}{a^2} + \dfrac{y^2}{b^2} = 1$, and show that the point of contact will be at one extremity of one of the equi-conjugate diameters of the ellipse. Show also that the polars of any point with respect to the two curves will meet on that diameter.

34. A circle intersects the hyperbola in four points; prove that the product of the distances of the four points of intersection from one asymptote is equal to the product of their distances from the other.

35. Tangents to the primary hyperbola are drawn from a point P on its conjugate; show that the chord of contact will be a tangent to the other branch of the conjugate.

CHAPTER XI.

SYSTEMS OF CONICS.

189. Similar Conics. — *Two conics are similar when their corresponding axes are proportional.*

Two similar conics whose corresponding axes are parallel are similarly situated.

I. *Similar conics have the same eccentricity* e.

The eccentricity $e^2 = 1 - \dfrac{b^2}{a^2} = 1 - \dfrac{b'^2}{a'^2}$, since $\dfrac{b}{a} = \dfrac{b'}{a'}$ by definition.

II. *Two conics $S = 0$ and $S' = 0$ are similar and similarly situated when the coefficients of their second degree terms are proportional; that is, when* $\dfrac{A}{A'} = \dfrac{B}{B'} = \dfrac{H}{H'}$.

Since their eccentricities must be equal, we have, by Art. 188, III.,
$$\frac{(2-e^2)^2}{1-e^2} = \frac{(A+B)^2}{AB-H^2} = \frac{(A'+B')^2}{A'B'-H'^2},$$
from which we readily get
$$\frac{A}{A'} = \frac{B}{B'} = \frac{H}{H'}.$$

From these ratios we also get $\dfrac{2H}{A-B} = \dfrac{2H'}{A'-B'}$, and therefore the corresponding axes of the conics are parallel.

III. *The equation*
$$Ax^2 + 2Hxy + By^2 + 2Gx + 2FG + C = S,$$
where S is arbitrary, represents a system of similar, similarly situated, and concentric conics.

Since the coefficients of the second degree terms are the same for all the conics of the system, they are all similar and similarly situated, by II., and the equations $Ax + Hy + G = 0$, $By + Hx + F = 0$, which determine the centre, are also the same: therefore the conics are concentric.

190. Confocal Conics. — If $a > b$ are the semi-axes of the conic $S = 0$, the distance between the foci on the axis of x is $2ae = 2\sqrt{a^2 - b^2}$. If λ is any arbitrary constant, then $a^2 + \lambda$, $b^2 + \lambda$, are the squares of the semi-axes of a system of conics confocal with $S = 0$, since $2\sqrt{a^2 + \lambda - (b^2 + \lambda)} = 2\sqrt{a^2 - b^2}$. Therefore

$$\frac{x^2}{a^2 + \lambda} + \frac{y^2}{b^2 + \lambda} = 1 \qquad (a)$$

is the equation of the confocal system.

I. *For all positive values of λ, the confocals lie without $S = 0$; and as λ approaches the limit infinity, the confocal approaches a circle;* since its eccentricity $e^2 = 1 - \dfrac{b^2 + \lambda}{a^2 + \lambda}$ approaches zero.

II. For all negative values of λ numerically less than b^2, the confocals

$$\frac{x^2}{a^2 - \lambda} + \frac{y^2}{b^2 - \lambda} = 1 \qquad (b)$$

are ellipses, both axes diminish as λ increases, and when the conjugate axis $2\sqrt{b^2 - \lambda}$ reaches its limit zero, the transverse axis will reach its limit $2\sqrt{a^2 - b^2}$, the distance between the foci.

Therefore the limiting ellipse of the system is this straight line, and equation (b) becomes $y = 0$, the equation of this line.

III. When λ is numerically greater than b^2 and less than a^2, the system of confocals (b) become hyperbolas; as λ approaches a^2, the conjugate axis approaches $2\sqrt{b^2 - a^2}$, the distance between the foci on the axis of y, and the equation of the confocal becomes $x = 0$. When λ is greater than a^2, the confocals become imaginary.

191. *Two confocals of the system pass through any given point $(x'y')$.*

If any confocal passes through the point $(x'y')$, then

$$\frac{x'^2}{a^2+\lambda} + \frac{y'^2}{b^2+\lambda} = 1.$$

If we put $b^2 + \lambda = \lambda'$, then since $a^2 e^2 = a^2 - b^2$,

$$\frac{x'^2}{a^2 - b^2 + \lambda'} + \frac{y'^2}{\lambda'} = 1,$$

or
$$\lambda'^2 - (x'^2 + y'^2 - a^2 e^2)\lambda' - a^2 e^2 y'^2 = 0.$$

The roots of this quadratic in λ' are real and of opposite signs. Therefore $b^2 + \lambda$ is positive for one conic, the ellipse, and negative for the other, the hyperbola.

192. *Two confocal conics cut each other at right angles at all points of intersection.*

The tangents of the two confocals

$$\frac{x^2}{a^2} + \frac{y^2}{b^2} = 1, \quad \frac{x^2}{a^2+\lambda} + \frac{y^2}{b^2+\lambda} = 1, \qquad (a)$$

at the point $(x'y')$ of intersection are

$$\frac{xx'}{a^2} + \frac{yy'}{b^2} = 1, \quad \frac{xx'}{a^2+\lambda} + \frac{yy'}{b^2+\lambda} = 1,$$

and are at right angles to each other if

$$\frac{x'^2}{a^2(a^2+\lambda)} + \frac{y'^2}{b^2(b^2+\lambda)} = 0.$$

But this is the same relation which is obtained by taking the difference of the confocals (a) for the point $(x'y')$.

SYSTEMS OF CONICS.

193. *Only one of a confocal system of conics will touch a given straight line.*

The line $x \cos a + y \sin a - p = 0$ will touch the confocal

$$\frac{x^2}{a^2+\lambda} + \frac{y^2}{b^2+\lambda} = 1$$

if $\qquad p^2 = (a^2+\lambda)\cos^2 a + (b^2+\lambda)\sin^2 a.$ (Art. 139)

But this condition gives but one value of λ; hence only one confocal will touch the given straight line.

194. *The difference of the squares of the perpendiculars drawn from the centre to any two parallel tangents to two given confocal conics is constant.*

If equations (a), Art. 192, are the two given confocals, then

$$p^2 = a^2 \cos^2 a + b^2 \sin^2 a, \qquad \text{(Art. 139)}$$

and $\qquad p'^2 = (a^2+\lambda)\cos^2 a + (b^2+\lambda)\sin^2 a.$

$$\therefore p'^2 - p^2 = \lambda.$$

195. *If the tangents to any two confocal conics of a system are at right angles to each other, the locus of their intersection is a circle.*

The perpendicular tangents of the two confocals (a), Art. 192, are

$$x \cos a + y \sin a = \sqrt{a^2 \cos^2 a + b^2 \sin^2 a},$$
$$x \sin a - y \cos a = \sqrt{(a^2+\lambda)\sin^2 a + (b^2+\lambda)\cos^2 a}.$$

Adding the squares of these equations, we get

$$x^2 + y^2 = a^2 + b^2 + \lambda,$$

the equation of the required locus, the director circle of the system of confocals.

196. *The locus of the pole of a given straight line with respect to a system of confocal conics is a straight line.*

Let $lx + my - 1 = 0$ (*a*) be the given straight line. If this line is the same as

$$\frac{xx'}{a^2 + \lambda} + \frac{yy'}{b^2 + \lambda} = 1,$$

the polar of the point $(x'y')$ with respect to the confocals

$$\frac{x^2}{a^2 + \lambda} + \frac{y^2}{b^2 + \lambda} = 1,$$

then

$$\frac{x'}{a^2 + \lambda} = l, \text{ and } \frac{y'}{b^2 + \lambda} = m.$$

$$\therefore \frac{x'}{l} - a^2 = \frac{y'}{m} - b^2 = \lambda,$$

and, omitting primes,

$$\frac{x}{l} - \frac{y}{m} = a^2 - b^2$$

is the equation of the required locus.

This locus is a straight line perpendicular to the given line (*a*) at the point where it touches a confocal; and as the pole moves on a straight line, the corresponding polars are all parallel to each other.

197. *To find the coordinates of the points in which the two given conics $S = 0$ and $S' = 0$ intersect.*

The constants in $S' = 0$ are A', H', B', G', F', C'.

The values of x and y which make $S = 0$ and $S' = 0$ simultaneously are the required values, and are the only ones which belong to points of intersection.

If we eliminate, say y, from $S = 0$, $S' = 0$, we know from Algebra that the resulting equation will be of the fourth degree in x, and these four values of x are the abscissas of all the points common to the two given conics. The corresponding values of y are readily found.

These four points $P(x'y')$, $P'(x''y'')$, $Q(x'''y''')$, $Q'(x^{iv}y^{iv})$ may all be real; or all imaginary; or two real and two imaginary; or one pair coincident; or both pairs coincident.

It appears, then, that if the conic $S = 0$ and $S' = 0$ are tangents at any point P, they will meet again in two real or two imaginary points; but if they are tangents at both P and Q, they cannot meet again in any other point.

198. *To find the equation of the conic which passes through the four points of intersection of $S = 0$ and $S' = 0$.*

If k is an arbitrary quantity which can have any positive or negative value, then $S - kS' = 0$ is the required equation; for it is a general equation of the second degree and represents a conic, and is satisfied by the coordinates of the four points which make $S = 0$, $S' = 0$ simultaneously.

Also, since k is arbitrary, $S - kS' = 0$ represents a *system of conics* all passing through the four points of intersection of $S = 0$, $S' = 0$.

199. *One conic of the system $S - kS' = 0$, and only one, will pass through any fifth point $(x'y')$.*

If we substitute $(x'y')$ for (xy) in $S - kS' = 0$, we get $S(x'y') - kS'(x'y') = 0$, which determines a single value of k. This value of k substituted in $S - kS' = 0$ gives the equation of the only conic of the system which will pass through the fifth point $(x'y')$.

200. *Only two conics of the system $S - kS' = 0$ are parabolas.*

The characteristic of the parabola $H^2 - AB = 0$ becomes

$$(H - kH')^2 - (A - kA')(B - kB') = 0,$$

which is a quadratic in k; there are only two values of k which satisfy this condition, and therefore only two conics of the system are parabolas.

201. *Only three conics of the system $S - kS' = 0$ are pairs of straight lines.*

Since the discriminant Δ of any cubic is of the third degree in terms of its coefficients A, H, B, G, F, C, and since the coefficients in $S - kS' = 0$ are $A - kA'$, $H - kH'$, etc., it follows that $\Delta = 0$ is a cubic in k, giving three, and only three, values of k which will make $S - kS' = 0$ represent pairs of straight lines.

We know from the theory of equations that one value of k will always be real, and that the other two may be.

202. If the discriminant Δ' of $S' = 0$ is zero, then $S' = 0$ represents two straight lines, which we will denote by

$$\frac{x}{a} + \frac{y}{b} - 1 = 0, \qquad \frac{x}{a'} + \frac{y}{b'} - 1 = 0,$$

or $\qquad lx + my - 1 = 0, \quad (a) \qquad l'x + m'y - 1 = 0, \qquad (b)$

in which, as we know, l, m, l', m', are the reciprocals of the intercepts a, b, a', b', of the lines (a), (b) on the coordinate axes. The conic $S - kS' = 0$ now becomes

$$S - k(lx + my - 1)(l'x + m'y - 1) = 0. \qquad (c)$$

Suppose that the line (a) cuts $S = 0$ in the points P, P', and that the line (b) cuts $S = 0$ in the points Q, Q'; suppose, further, that the lines PP' and QQ' meet in O, and that we take the lines OPP', OQQ', as the coordinate axes of X and Y. Then $P(l, 0)$, $P'(l', 0)$ and $Q(0, m)$, $Q'(0, m')$ are the coordinates of the four points in which the straight lines (a) and (b) cut the conic $S = 0$, and (a) and (b) are the equations of PQ and $P'Q'$.

If the points P, P' become coincident, then $l' = l$, the axis of X, is a tangent to the system at the point $P(l, 0)$, and the equation (c) becomes

$$S - k(lx + my - 1)(lx + m'y - 1) = 0.$$

If Q, Q' also become coincident, then $m' = m$, the axis of Y, is a tangent to the system at the point $Q(0, m)$, and its equation is
$$S - k(lx + my - 1)^2 = 0.$$
It is obvious that $lx + my - 1 = 0$ is the equation of the common chord which cuts $S = 0$ in the points P and Q, at which all the conics of the system are tangents to $S = 0$ and to each other. The length of this chord can be easily computed.

203. Retaining the coordinate axes of Art. 202, the equations of the opposite sides and diagonals of the given quadrilateral are

for the sides PP', QQ',
$$y = 0, \qquad x = 0;$$
for the sides PQ, $P'Q'$,
$$lx + my - 1 = 0, \qquad l'x + m'y - 1 = 0;$$
for the diagonals PQ', $P'Q$,
$$lx + m'y - 1 = 0, \qquad l'x + my - 1 = 0.$$

Each pair of these straight lines
$$xy = 0,$$
$$(lx + my - 1)(l'x + m'y - 1) = 0,$$
$$(lx + m'y - 1)(l'x + my - 1) = 0,$$
is a conic which passes through the four given points. Hence
$$(lx + my - 1)(l'x + m'y - 1) - kxy = 0 \qquad (a)$$
is the equation of the system of conics which pass through the four given points; and combining any other two of these conics will give the equation of the same system.

If the points $P(x'y')$, $P'(x''y'')$, $Q(x'''y''')$, $Q'(x^{iv}y^{iv})$, are given either in rectangular or oblique coordinates, the equations of the sides and diagonals can be found, and also the equations of the required system of conics.

204. If the equation (*a*) of the system of conics (Art. 203)

$$(lx + my - 1)(l'x + m'y - 1) - kxy = 0,$$

or
$$ll'x^2 + (lm' + l'm - k)xy + mm'y^2$$
$$- (l + l')x - (m + m')y + 1 = 0, \quad (a)$$

represents a parabola, then the characteristic is

$$(lm' + l'm - k)^2 = 4\,ll'mm'. \quad (b)$$

As this condition gives only two values of k, only two conics of the system are parabolas. These values of k are real when all the intercepts have the same sign or when an even number of them are negative.

If only one intercept is negative, the quadrilateral $PP'Q'Q$ has a re-entrant angle, and cannot be inscribed in a real parabola.

The equations of these two parabolas are, by condition (*b*),

$$(x\sqrt{ll'} \pm y\sqrt{mm'})^2 - (l + l')x - (m + m')y + 1 = 0,$$

and their axes are parallel to the lines $ll'x^2 - mm'y^2 = 0$ (Art. 169).

But the coefficients of the second degree terms of

$$ll'x^2 - mm'y^2 = 0$$

and of equation (*a*) satisfy the condition of Art. 178 for all values of k.

Hence all the conics of a system passing through four given points have a pair of conjugate diameters respectively parallel to the axes of the two parabolas of the system.

205. *To find the locus of the centres of the system of conics which pass through four given points.*

The equation of the system is (Art. 204)

$$ll'x^2 + (lm' + l'm - k)xy + mm'y^2$$
$$- (l + l')x - (m + m')y + 1 = 0. \quad (a)$$

SYSTEMS OF CONICS.

If $(x'y')$ are the coordinates of the centre of any conic of the system, let us transform equation (a) to this centre as an origin, and to a new pair of parallel coordinate axes. For x and y substitute $x + x'$ and $y + y'$, respectively; then (Art. 167 (b)) the coefficients of x and y in the transformed equation, that is,

$$ky' - l(l'x' + m'y' - 1) - l'(lx' + my' - 1) = 0, \qquad (b)$$

$$kx' - m(l'x' + m'y' - 1) - m'(lx' + my' - 1) = 0, \qquad (c)$$

determine the relations of the coordinates $(x'y')$ of all the centres. Now eliminate the arbitrary quantity k by multiplying (b) and (c) by x' and y' respectively; then subtracting and omitting primes, and

$$(lx - my)(l'x + m'y - 1) + (l'x - m'y)(lx + my - 1) = 0,$$

or $\qquad 2ll'x^2 - 2mm'y^2 - (l + l')x + (m + m')y = 0, \qquad (d)$

is the equation of the required locus.

I. This centre-locus is a conic whose asymptotes are parallel to the lines given by the equations $ll'x^2 - mm'y^2 = 0$ (Art. 204), or to the axes of the two parabolas through the four given points. The centres of the two parabolas of the system (a) are points at infinity on the centre-locus, whose directions are determined by the equations $ll'x^2 - mm'y^2 = 0$.

II. *To find the points in which the centre-locus meets the sides of the quadrilateral $PP'Q'Q$.*

For $\quad y = 0, \qquad 2ll'x^2 - (l + l')x = 0;$

$\therefore\ x = 0, \qquad x = \dfrac{l + l'}{2ll'} = \dfrac{a + a'}{2}.$

For $\quad x = 0, \qquad 2mm'y^2 - (m + m')y = 0;$

$\therefore\ y = 0, \qquad y = \dfrac{m + m'}{2mm'} = \dfrac{b + b'}{2}.$

Therefore the centre-locus passes through the origin, that is, the intersection of the sides PP' and QQ' of the quadrilateral, and also through the middle points of these sides.

But we might have taken the sides PQ and $P'Q'$ as coordinate axes, and their point of intersection O' as an origin; or the diagonals PQ', $P'Q$ as coordinate axes, and their point of intersection O'' as an origin; and therefore the point O' and the middle points of the sides PQ, $P'Q'$, as well as O'' and the middle points of the diagonals PQ', $P'Q$, are on the centre-locus.

It follows, then, that the four given points through which a system of conics pass, and the nine points through which the centre-locus passes, all lie on the complete quadrilateral of $PP'Q'Q$ (see Fig. 30, p. 122).

III. If ll' and mm' have the same sign, the centre-locus (d) is an hyperbola, but an ellipse of ll' and mm' have opposite signs.

If $ll' = mm'$, the four given points P, P', Q, Q', are on a circle, and the centre-locus is an equilateral hyperbola.

If $ll' = -mm'$, and the coordinate axes are rectangular; that is, if the line joining any two of the four points is perpendicular to the line joining the other two, say the diagonals PQ', $P'Q$, then all the conics of system (a) are equilateral hyperbolas, and the centre-locus (d) is a circle passing through the nine points found in II., and is called the *nine-point circle*.

206. The asymptotes of the system of conics

$$(lx + my - 1)(l'x + m'y - 1) - kxy = 0 \qquad (a)$$

are parallel to the lines given by its second-degree terms (Art. 166); that is, by

$$(lx + my)(l'x + m'y) - kxy = 0,$$

or
$$ll'x^2 + (lm' + l'm - k)xy + mm'y^2 = 0; \qquad (b)$$

and these lines are parallel to the conjugate diameters of the centre-locus, since the coefficients of (b) and of the second-degree terms of the centre-locus (d) satisfy the condition

$$AB' + A'B = 2HH'. \qquad \text{(Art. 178)}$$

Hence *the asymptotes of the system of conics (a) are parallel to a pair of conjugate diameters of the centre-locus of this system.*

207. *To find the equation of the system of conics which touch the axes of coordinates.*

The conic

$$(lx + my - 1)(l'x + m'y - 1) - kxy = 0 \qquad (a)$$

will touch the coordinate axes if $l' = l$, $m' = m$, and

$$(lx + my - 1)^2 - kxy = 0 \qquad (b)$$

is the required equation, in which $lx + my - 1 = 0$ is the equation of the chord which passes through the points of tangency. The conics of this system are parabolas when the second-degree terms form a perfect square, that is, when

$$l^2 m^2 = (lm - \tfrac{1}{2}k)^2.$$

Therefore $k = 0$, or $4lm$. For $k = 0$ (b) becomes

$$(lx + my - 1)^2 = 0,$$

two coincident lines, the chord of contact of the parabola and the coordinate axes. For $k = 4lm$ (b) becomes

$$(lx + my - 1)^2 - 4lmxy = 0,$$

or $\qquad \sqrt{lx} + \sqrt{my} = 1.$

208. *To find the equation of the tangent to the parabola $\sqrt{lx} + \sqrt{my} = 1$ at the point $(x'y')$.*

The equation of the secant through the points $(x'y')$, $(x''y'')$,

is $\qquad \dfrac{x - x'}{x - x''} = \dfrac{y - y'}{y' - y''} \qquad (a)$

if
$$\sqrt{lx'} + \sqrt{my'} = 1 = \sqrt{lx''} + \sqrt{my''},$$

or
$$\sqrt{l}(\sqrt{x'} - \sqrt{x''}) = -\sqrt{m}(\sqrt{y'} - \sqrt{y''}). \qquad (b)$$

The products of the corresponding terms of (a) and (b) are

$$\frac{\sqrt{l}}{\sqrt{x'} + \sqrt{x''}}(x - x') = -\frac{\sqrt{m}}{\sqrt{y'} + \sqrt{y''}}(y - y').$$

For the tangent at $(x'y')$,

$$\sqrt{x''} = \sqrt{x'}, \quad \sqrt{y''} = \sqrt{y'},$$

and
$$\frac{\sqrt{l}}{\sqrt{x'}}(x - x') + \frac{\sqrt{m}}{\sqrt{y'}}(y - y') = 0,$$

or
$$x\sqrt{\frac{l}{x'}} + y\sqrt{\frac{m}{y'}} = \sqrt{lx'} + \sqrt{my'} = 1$$

is the required equation.

The polar of $(x'y')$ with respect to the conic (b) (Art. 207) is

$$(lx + my - 1)(lx' + my' - 1) - \tfrac{1}{2}k(x'y + xy') = 0.$$

The polar of the origin is $lx + my - 1 = 0$.

209. *To find the equation of the system of conics which touch four given straight lines.*

The conic (Art. 207, (b))

$$(lx + my - 1)^2 - kxy = 0 \qquad (b)$$

already touches two of the lines, which are taken for coordinate axes; and it only remains to find the values of the parameters l, m, k, when (b) also touches the two given straight lines

$$ax + by - 1 = 0, \qquad (c)$$
$$a'x + b'y - 1 = 0. \qquad (d)$$

Combine (b) and (c) by substituting $ax + by$ for the minus unity in (b); then

$$(lx + my - ax - by)^2 - kxy = 0$$

represents the straight lines passing through the origin $(0, 0)$ and the points in which the line (c) intersects the conic (b).

If the conic (b) touches the line (c), then the points in which they intersect become coincident, and the condition of equal roots is

$$(l-a)^2(m-b)^2 = [(l-a)(m-b) - \tfrac{1}{2}k]^2,$$

or $\quad k = 4(l-a)(m-b).$

In the same way find that

$$k = 4(l-a')(m-b')$$

is the condition that the conic shall touch the line (d).

Therefore the conic (b) will touch the four straight lines

$$x = 0, \quad y = 0, \quad ax + by - 1 = 0, \quad a'x + b'y - 1 = 0,$$

when the values of l, m, k, are determined from (b) and the two equations

$$k = 4(l-a)(m-b) = 4(l-a')(m-b').$$

These equations will give the values of l, m, and k, in terms of the four parameters of the lines (c) and (d), which will make the system of conics (b) touch the four given lines.

210. *To find the equation of the locus of the centres of the conics which touch four given straight lines.*

Transform the equation

$$(lx + my - 1)^2 - kxy = 0 \qquad (b)$$

to $(x'y')$, the centre of the system, and to a new pair of parallel axes; then the coefficients of x and y in the transformed equation, omitting primes, are

$$2l(lx + my - 1) - ky = 0, \quad 2m(lx + my - 1) - kx = 0,$$

or $\quad lx = my, \quad \text{and} \quad 2l(2lx - 1) = ky, \qquad (c)$

the equations which determine the relations between the coordinates of the centre.

But the equations are also subject to the conditions
$$k = 4(l-a)(m-b), \qquad (d)$$
$$k = 4(l-a')(m-b'). \qquad (e)$$

From (c), (d), (e), we can now eliminate l, m, k, and thus obtain an equation in (xy), the locus of the centres of the system of conics touching the four given straight lines.

But ky from (d) equals ky from (c); and since $lx = my$,

$$\therefore\ 2l(2lx - 1) = 4y(l-a)(m-b) = 4(l-a)(lx - by);$$
$$\therefore\ l(2ax + 2by - 1) = 2aby,$$

and $\qquad\therefore\ l(2a'x + 2b'y - 1) = 2a'b'y$, from (e).

Now eliminate l, and
$$\frac{2ax + 2by - 1}{ab} = \frac{2a'x + 2b'y - 1}{a'b'},$$

or
$$2\left(\frac{1}{b} - \frac{1}{b'}\right)x + 2\left(\frac{1}{a} - \frac{1}{a'}\right)y = \frac{1}{ab} - \frac{1}{a'b'},$$

the required equation of the locus of the centres is a straight line.

Since a, a', b, b' are the reciprocals of the intercepts which the lines (c), (d) (Art. 209), make on the coordinate axes, it is obvious that this centre-locus passes through the middle points of the three diagonals of the quadrilateral which circumscribes the system of conics.

211. *A system of conics passes through the four given points P, P', Q, Q'; if the sides PP', QQ', of this quadrilateral meet in the point O, the sides PQ, $P'Q'$, in the point O', and the diagonals PQ', $P'Q$, in the point O''; then either point O is the pole with respect to the conic of the line passing through the other two.*

The polar of any point $(x'y')$ with respect to the system of conics
$$(lx + my - 1)(l'x + m'y - 1) - kxy = 0 \qquad (a)$$

which passes through the four points P, P', Q, Q', is

$$(lx+my-1)(l'x'+m'y'-1)+(l'x+m'y-1)(lx'+my'-1) - k(x'y+xy')=0.$$

For the point O, $(x'y')$ is $(0, 0)$, and its polar reduces to

$$(l+l')x+(m+m')y-2=0,$$

or to the two equivalent equations

$$lx+my-1+l'x+m'y-1=0, \qquad (b)$$

$$lx+m'y-1+l'x+my-1=0. \qquad (c)$$

But (b) is O', the point of intersection of PQ and $P'Q'$; and (c) is O'', the point of intersection of PQ' and $P'Q$. Therefore the polar of O passes through O' and O''.

In the same way it can be shown that O' and O'' are respectively the poles of the lines passing through the other two points.

The triangle $OO'O''$, each of whose vertices is the pole of the opposite side with respect to a conic, is called a *self-conjugate* triangle (Art. 86, XII.).

212. *The intersections of the three diagonals of a quadrilateral are the vertices of a self-conjugate triangle with respect to the conic inscribed in the quadrilateral.*

Let $ABCD$ be the quadrilateral; let the sides AB, DC meet in N, the sides AD, BC meet in N'; then NN' is the external diagonal. Also let the diagonals AC, NN' meet in M, the diagonals BD, NN' meet in M', and the diagonals AC, BD meet in M''.

Further, let the inscribed conic touch AB in P, BC in P', CD in Q', and DA in Q. Then the polar of A is PQ, the polar of C is $P'Q'$, and AC is the polar of O', the intersection of PQ, $P'Q'$. The polar of B is PP', the polar of D is QQ', and BD is the polar of O, the intersection of PP', QQ'; therefore OO'

is the polar of M'', the intersection of AC, BD; but by Art. 211, the polar of OO' is O'', the intersection of PQ', $P'Q$, and therefore the diagonals AC, BD, and PQ', $P'Q$, meet in the same point. So also M' is the intersection of PP', QQ', and M of PQ, $P'Q'$. Therefore $MM'M''$ is a self-conjugate triangle with respect to the inscribed conic.

EXERCISES ON SYSTEMS OF CONICS.

1. Find the equation of the conic which passes through the five points $(2, 1)$, $(0, 3)$, $(-1, -3)$, $(1, 0)$, and $(3, -3)$.

Solution. The equations of the opposite sides $(2, 1)$, $(0, 3)$ and $(-1, -3)$, $(1, 0)$ are $y + x - 3 = 0$, $2y - 3x + 3 = 0$; and of the opposite sides $(0, 3)$, $(-1, -3)$ and $(1, 0)$, $(2, 1)$ are $y - 6x - 3 = 0$, $y - x + 1 = 0$. The system of conics passing through these four points is

$$(y + x - 3)(2y - 3x + 3) - k(y - 6x - 3)(y - x + 1) = 0. \quad (a)$$

But the required conic must pass through the fifth point $(3, -3)$, for which $k = \tfrac{3}{16}$, and (a) becomes the hyperbola

$$17y^2 + 11xy - 48x^2 - 24y + 129x - 81 = 0. \quad (b)$$

The equations of the diagonals through $(0, 3)$, $(1, 0)$ and $(-1, -3)$, $(2, 1)$ are $y + 3x - 3 = 0$, $3y - 4x + 5 = 0$, and the equation of the system of conics passing through these four points is

$$(y + 3x - 3)(3y - 4x + 5) - k(y + x - 3)(2y - 3x + 3). \quad (a')$$

For the conic of the system which passes through the fifth point $(3, -3)$, $k = -\tfrac{4}{3}$, for which value (a') reduces to (b), as it should.

2. Find the equation of the conic which passes through the five points $(2, 1)$, $(1, 0)$, $(3, -1)$, $(-1, 0)$, $(3, -2)$.

Ans. $x^2 + 19xy + 4y^2 - 45y - 1 = 0$.

3. Find the equation of the conic which passes through the five points $(1, 2)$, $(3, 5)$, $(-1, 4)$, $(-3, -1)$, $(-4, 3)$.

Ans. $79x^2 - 320xy + 301y^2 + 1101x - 1665y + 1586 = 0$.

EXERCISES ON SYSTEMS OF CONICS.

4. If the axes of $S = 0$ be parallel to those of $S' = 0$, so will also be the axes of $S - kS' = 0$.

5. If $S' = 0$ be a circle, the axes of $S - kS' = 0$ are parallel to the axes of $S = 0$.

6. If $S - kS' = 0$ represent a pair of straight lines, its axes become the internal and external bisectors of the angles between them.

7. If any chord of the conic $S = 0$ be drawn through any fixed point O, it will be cut harmonically by the curve and the polar of O.

Solution. Let any chord through O cut the conic $S = 0$ in the points P and R, and the polar of O in Q. Take O as the origin, and the chord $OPQR$ as the axis of X; then for $y = 0$, the conic $S = 0$, and $Gx + Fy + C = 0$, the polar of O, become

$$Ax^2 + 2Gx + C = 0, \quad Gx + C = 0,$$

the roots of which are OP, OR, OQ.

By Art. 20, $\quad OP + OR = -\dfrac{2G}{A}, \quad OP \cdot OR = \dfrac{C}{A}, \quad$ and $OQ = -\dfrac{C}{G}$.

$$\therefore \; \frac{1}{OP} + \frac{1}{OR} = \frac{2}{OQ}.$$

8. All conics drawn through the common intersections of two rectangular hyperbolas, are rectangular hyperbolas.

Solution. If $S = 0$, $S' = 0$, denote rectangular hyperbolas, then $A + B = 0$, $A' + B' = 0$; and for $S - kS' = 0$ the system of conics passing through the common intersections of $S = 0$, $S' = 0$, we shall also have $A + B - k(A' + B') = 0$.

9. If two conics have parallel axes, one conic of the system passing through their intersections will be a circle.

Solution. If the axes of $S = 0$ and $S' = 0$ are parallel to the coordinate axes, then $H = 0$, $H' = 0$, and the conic $S - kS' = 0$ of the system passing through the intersections of $S = 0$, $S' = 0$ will be a circle when $A - kA' = B - kB'$, or for the single value of $k = \dfrac{A - B}{A' - B'}$.

10. If the polars of the points $T(x'y')$ and $T'(x''y'')$ with respect to a conic cut the curve in P, Q and P', Q', a conic will pass through the six points T, P, Q, T', P', Q'.

Solution. The polars of $T(x'y')$, $T'(x''y'')$ with respect to the conic $Ax^2 + By^2 = 1$ are $Axx' + Byy' - 1 = 0$, $Axx'' + Byy'' - 1 = 0$.

The conic of the system

$$Ax^2 + By^2 - 1 = k(Axx' + Byy' - 1)(Axx'' + Byy'' - 1),$$

which passes through the four points P, Q, P', Q', will also pass through $T(x'y')$ if

$$Ax'^2 + By'^2 - 1 = \lambda(Ax'^2 + By'^2 - 1)(Ax'x'' + By'y'' - 1),$$

or $\quad \dfrac{1}{k} = Ax'x'' + By'y'' - 1.$

The symmetry of this value of k shows that the conic will also pass through $T'(x''y'')$.

11. If two chords of a conic be drawn through two points on a diameter equidistant from the centre, any conic drawn through the four ends of these chords will be cut by this diameter in points equidistant from the centre.

Solution. If the diameter and its conjugate are coordinate axes, the equation of the conic will be of the form $Ax^2 + By^2 = 1$, and the equations of the chords passing through the points $(a, 0)$, $(-a, 0)$ are

$$y - m(x - a) = 0, \quad y - m'(x + a) = 0.$$

The system of conics through the four ends of the chords is

$$Ax^2 + By^2 - 1 = k[y - m(x + a)][y - m'(x + a)].$$

The axis of X cuts the conics in the points for which $y = 0$,

and $\quad Ax^2 - 1 - kmm'(x^2 - a^2) = 0,$

which are equidistant from the origin.

12. If PSQ and $P'S'Q'$ are two focal chords of a conic, the lines PP' and QQ' cut the axis in points equidistant from the centre.

13. To find the equation of a pair of tangents drawn from any point $P(x', y')$ to the conic $S = 0$.

Solution. The polar of the point $P(x', y')$ with respect to $S = 0$ is

$$T(x', y') = 0, \text{ and } S - k[T(x', y')]^2 = 0 \qquad (a)$$

is the equation of a system of conics all tangent to $S = 0$ at the points in which it is cut by the polar $T(x', y') = 0$. But the required equation of the tangents is the conic of this system which also passes through the point $P(x', y')$.

Therefore $S(x', y') - k[S(x', y')]^2 = 0$, or $1 - k \cdot S(x', y') = 0$. For this value of k equation (a) becomes

$$S \cdot S(x', y') = [T(x', y')]^2,$$

the required equation.

14. The polar of a fixed point, with respect to a system of conics $S - kS' = 0$ passing through four given points, will pass through a fixed point.

Solution. If the origin is at the fixed point, then the polar of this point with respect to $S - kS' = 0$ is

$$Gx + Fy + C - k(G'x + F'y + C') = 0$$

which passes through the intersection of

$$Gx + Fy + C = 0, \quad G'x + F'y + C' = 0$$

for all values of k.

15. The locus of the poles of a given straight line with respect to a system of conics which pass through four given points is a conic.

Solution. Take the given straight line as the axis of x. The polar of $(x'y')$ with respect to $S - kS' = 0$ is

$$T(x'y') - k[T'(x'y')] = 0;$$

and if this line is the same as $y = 0$, then the coefficient of x and the independent term must each equal zero.

Therefore

$$Ax' + Hy' + G - k(A'x' + H'y' + G') = 0,$$
$$Gx' + Fy' + C - k(G'x' + F'y' + C') = 0.$$

Eliminate k, and the required locus is the conic

$$\frac{Ax + Hy + G}{Gx + Fy + C} = \frac{A'x + H'y + G'}{G'x + F'y + C'}.$$

16. The equation of the conic $Ax^2 + 2Hxy + By^2 = C'$ transferred to its axes, when the coordinate axes are oblique, is

$$\frac{(A+B-2H\cos\omega+Q)}{2C'\sin^2\omega}x^2 + \frac{(A+B-2H\cos\omega-Q)}{2C'\sin^2\omega} = 1,$$

if $\quad Q^2 = [2H-(A+B)\cos\omega]^2 + (A-B)^2\sin^2\omega.$

17. If a and b be the lengths of two tangents to the parabola $y^2 = 4px$ which intersects at right angles, then

$$\frac{a^{\frac{2}{3}}}{b^{\frac{1}{3}}} + \frac{b^{\frac{2}{3}}}{a^{\frac{1}{3}}} = \frac{1}{p^{\frac{2}{3}}}.$$

18. The locus of the points of contact of tangents to a series of confocal ellipses drawn from a fixed point on the major axis is a circle.

19. The length of the chord of an ellipse which touches a confocal ellipse, the squares of whose semi-axes are

$$a^2 - \lambda^2, \quad b^2 - \lambda^2, \quad \text{is } \frac{2\lambda b'^2}{ab},$$

in which b' denotes the length of the semi-diameter parallel to the given chord.

20. If any line cut two similar and concentric conics, its parts intercepted between the conics will be equal. Any chord of the outer conic which touches the inner one will be bisected at the point of contact.

21. If a tangent drawn at any point P of the inner of two concentric and similar ellipses meet the outer one in the points T and T', then any chord of the inner one drawn through P is half the algebraic sum of the parallel chords of the outer one through T and T'.

22. If the axes are oblique, the condition for the similarity of two conics is

$$\frac{AB - H^2}{(A+B-2H\cos\omega)^2} = \frac{A'B' - H'^2}{(A'+B'-2H'\cos\omega)'}.$$

23. What is the equation of the system of conics passing through the points where a given conic $S = 0$ meets the coordinate axes?

24. Show that when the values of (h, k), the coordinates of the centre of the conic $S = 0$ become indeterminate, its equation represents the two straight lines

$$A\left(x + \frac{Hy}{A}\right)^2 + 2G\left(x + \frac{Hy}{A}\right) + C = 0.$$

25. To determine the form of the equation $S = 0$ when the coordinate axes are tangents of the conic.

For tangency $A = \dfrac{G^2}{C}$, $B = \dfrac{F^2}{C}$, and $S = 0$ becomes

$$(Gx + Fy + C)^2 - 2(FG - HC)xy = 0.$$

26. Find the equations of the tangents of the conic

$$(lx + my - 1)(lx + m'y - 1) - kxy = 0,$$

at the points where it meets the axis of Y; and the equation of the line joining their point of intersection with the point where the conic touches the axis of X.

27. Show that the tangents at the points where the conic

$$(lx + my - 1)(l'x + m'y - 1) - kxy = 0$$

is cut by the straight line $y = k^2 x$, intersect on the straight line where $y = -k^2 x$ cuts it.

28. If conics be described touching two straight lines OX, OY in two points A and B, such that $OA = \dfrac{1}{l}$, $OB = \dfrac{1}{m}$, then the centres of the system will lie on the straight line $y = x$.

29. Two ellipses have a common focus and equal eccentricities; show that their common chord is parallel to the line bisecting the angle between their major axes.

30. The angle subtended at the focus by a chord of a conic is $120°$; show that the locus of the intersection of tangents at its extremities is

$$l = r(1 + 2e \cos\theta),$$

l being the semi-latus-rectum.

31. Find the latus-rectum and the coordinates of the vertex of the parabola whose equation is

$$(3x - 4y)^2 - 50ax + 25a^2 = 0.$$

32. Show that the latus-rectum of the conic $S = 0$ is $\dfrac{G + F}{(2)^{\frac{3}{2}}}$.

33. The equation of the directrix of the parabola $S = 0$ is

$$2(Hy - Ax) = \frac{H}{HF - BG}[(A + B)C - G^2 - F^2].$$

34. Find the equation of the axis of the parabola $(y - x)^2 = a(x + y)$.

35. If two hyperbolas have their axes parallel, the quadrilateral joining their points of intersection has the sum of its opposite angles equal to two right angles.

36. Two ellipses, referred, one to its axes $2a$, $2b$, and the other to its conjugate diameters $2a'$, $2b'$, have a common polar of the points (h, k), (h', k') respectively; show that

$$\frac{a}{a'} = \sqrt{\frac{h}{h'}}, \quad \frac{b}{b'} = \sqrt{\frac{k}{k'}}.$$

37. S, S' are the contiguous foci of two ellipses having the same major axis, and the eccentricities e, e' are complementary, so that $e^2 + e'^2 = 1$; B, B' are the ends of their minor axes. CP, CP' are drawn through the centre parallel to SB, $S'B'$ and meet the ellipses in P, P' respectively. Show that the tangents at P, P' are parallel.

38. If CP, CD be the semi-conjugate diameters of an ellipse, and the lines joining P, D and the ends of the major axis meet in Q, show that the locus of Q is an ellipse.

39. If A, B, C, D are four points on a rectangular hyperbola, such that AB is perpendicular to CD, show that BC, AD, and BD, AC are also perpendicular.

40. Three hyperbolas have parallel asymptotes; show that the three straight lines joining their points of intersection, taken two and two, all meet in a point.

EXERCISES ON SYSTEMS OF CONICS.   343

41. Any two parabolas which have a common focus and their axes in opposite directions intersect at right angles.

42. Two parabolas have a common focus and their axes in the same straight line; show that, if TP, TQ be perpendicular tangents one to each of the parabolas, the locus of T is a straight line.

43. TP, TQ are perpendicular tangents one to each of two confocal conics; show that CT bisects PQ.

Solution. If $\dfrac{x^2}{a^2}+\dfrac{y^2}{b^2}=1$, $\dfrac{x^2}{a'^2}+\dfrac{y^2}{b'^2}=1$, are two confocals, their tangents are

$$\frac{xx'}{a^2}+\frac{yy'}{b^2}=1, \quad \frac{xx''}{a'^2}+\frac{yy''}{b'^2}=1.$$

The equation of CT is

$$x\left(\frac{x'}{a^2}-\frac{x''}{a'^2}\right)+y\left(\frac{y'}{b^2}-\frac{y''}{b'^2}\right)=0.$$

The condition that CT shall pass through $\left(\dfrac{x'+x''}{2},\dfrac{y'+y''}{2}\right)$, the middle point of PQ, is

$$x'x''\left(\frac{1}{a^2}-\frac{1}{a'^2}\right)+y'y''\left(\frac{1}{b^2}-\frac{1}{b'^2}\right)=0,$$

or $\qquad \dfrac{x'x''}{a^2 a'^2}+\dfrac{y'y''}{b^2 b'^2}=0,$

since the conics are confocal. But this is the condition that the tangents shall be perpendicular.

44. TP, TQ are tangents to each of two parabolas which have a common focus and their axes in the same straight line; show that, if a line through P parallel to the axis bisect PQ, the tangents will be perpendicular.

45. If P, Q be any two points on an ellipse, and p, q, the corresponding points on a confocal ellipse, that is, points having the same eccentric angles, then $Pq = Qp$.

46. From any point T the two tangents TP, TP' are drawn to one conic, and the two tangents TQ, TQ' to a confocal conic, then the straight lines PQ, PQ' will make equal angles with the tangent at P.

47. If l and l' are the latus-rectums of two similar and similarly situated conics, and if r, r' are two parallel focal distances, then

$$\frac{r}{r'} = \frac{l}{l'} = \frac{a}{a'} = \frac{b}{b'}.$$

48. TP, TQ are perpendicular tangents, one to each of two confocal conics; show that the line PQ always touches a third confocal conic.

49. The confocal hyperbola through the point on the ellipse whose eccentric angle is ϕ has for its equation

$$\frac{x^2}{\cos^2\phi} - \frac{y^2}{\sin^2\phi} = a^2 - b^2.$$

50. If λ, μ be the parameters of the confocals which pass through two points P, Q on a given ellipse; show (1) that if P, Q are the extremities of conjugate diameters, then $\lambda + \mu$ is constant; and (2) that if the tangents at P and Q are perpendicular, then $\frac{1}{\lambda} + \frac{1}{\mu}$ is constant.

51. Show that the ends of the equi-conjugate diameters of a series of confocal ellipses are on a confocal rectangular hyperbola.

www.ingramcontent.com/pod-product-compliance
Lightning Source LLC
Chambersburg PA
CBHW030303240426
43673CB00040B/1050